# Geomaterials: Latest Advances in Materials for Construction and Engineering Applications

# Geomaterials: Latest Advances in Materials for Construction and Engineering Applications

Editor

**José Manuel Moreno-Maroto**

MDPI

Basel • Beijing • Wuhan • Barcelona • Belgrade • Novi Sad • Cluj • Manchester

*Editor*
José Manuel Moreno-Maroto
Autonomous University of Madrid
Madrid
Spain

*Editorial Office*
MDPI
St. Alban-Anlage 66
4052 Basel, Switzerland

This is a reprint of articles from the Special Issue published online in the open access journal *Applied Sciences* (ISSN 2076-3417) (available at: https://www.mdpi.com/journal/applsci/special_issues/Geomaterials_Construction).

For citation purposes, cite each article independently as indicated on the article page online and as indicated below:

| Lastname, A.A.; Lastname, B.B. Article Title. *Journal Name* **Year**, *Volume Number*, Page Range. |
| --- |

**ISBN 978-3-7258-0861-8 (Hbk)**
**ISBN 978-3-7258-0862-5 (PDF)**
**doi.org/10.3390/books978-3-7258-0862-5**

Cover image courtesy of José Manuel Moreno-Maroto

# Contents

# About the Editor

**José Manuel Moreno-Maroto**

José Manuel Moreno-Maroto, who earned his PhD in Materials Sci. and Environmental Eng. in 2019, currently holds the position of As. Professor at the Autonomous University of Madrid (Spain), where, in addition to teaching, he conducts research and supervises PhD, MSc, and BSc theses. Prior to this, between 2014 and 2022, he developed his research career in other institutions, namely, at the University of Castilla-La Mancha (Spain), the University of Cergy-Pontoise (France), the Innovarcilla Foundation (Spain), and the University of Jaén (Spain). Additionally, he worked in the private sector between 2006 and 2014, with roles including head of materials and geotechnical laboratory. His primary research focus is the development of high-performance geomaterials for sustainable construction and engineering applications. Particularly noteworthy is his research on lightweight aggregates, ceramic materials, geopolymers, artificial zeolites, and composites of the latter two, among others. In all cases, priority is given to the use of waste as a raw material. He has also advanced the understanding of the Atterberg limits of clays and soils, developing new testing methods and alternative classification systems. To date, he has developed four patents related to geomaterials and co-authored 34 articles in JCR-indexed journals, mainly Q1, leading most of them as first author and/or corresponding author. He has also presented 26 papers at national and international congresses and has been actively involved in various competitive research projects and fellowships. He is a regular reviewer for JCR journals and a member of several scientific societies and international committees. He has received three awards for his doctoral thesis, two from the University of Castilla-La Mancha, in 2020 and 2022, and one from the Spanish Ceramic and Glass Society in 2020, as well as an award for a conference presentation from the Spanish Clay Society in 2016.

# Preface

Welcome to this Special Issue on geomaterials, for which I have had the privilege to serve as a Guest Editor. Geomaterials, often overlooked but crucial in our daily lives, play a pivotal role in various fields, particularly in construction and civil engineering. They are the foundation upon which our structures stand, the earthworks that shape our landscapes, and the resources that fuel innovation in numerous industries.

This Special Issue delves into the world of geomaterials, exploring their diverse applications and properties, and the innovative research being conducted in this field. Geomaterials are not just abundant; they are strategically vital for the development of sustainable and resilient infrastructure, addressing environmental challenges, and supporting the ever-growing demands of urbanization.

I would like to extend my heartfelt gratitude to all the authors who have contributed their valuable research to this Special Issue. Your dedication and insights have enriched this collection and provided readers with a comprehensive view of the latest advancements in geomaterial science. I especially want to acknowledge the outstanding collaboration with Prof. Mark Tyrer, whose expertise and dedication have been instrumental in crafting the Editorial paper that serves as the gateway to this Special Issue.

In closing, I hope that this Special Issue inspires further exploration and innovation in the realm of geomaterials. May it serve as a valuable resource for researchers, practitioners, and anyone interested in the profound impact of these materials on our world. Together, we aim to shed light on the significance of geomaterials and their transformative potential in various sectors.

Thank you for joining us on this journey through the world of geomaterials.

**José Manuel Moreno-Maroto**
*Editor*

applied sciences

MDPI

*Editorial*

# Geomaterials: Latest Advances in Materials for Construction and Engineering Applications

**Mark Tyrer [1] and José Manuel Moreno-Maroto [2,*]**

[1] Collegium Basilea (Institute of Advanced Study) Hochstrasse, 51, 4053 Basel, Switzerland; m.tyrer@mtyrer.net
[2] Department of Geology and Geochemistry, Faculty of Sciences, Autonomous University of Madrid, Cantoblanco, 28049 Madrid, Spain
* Correspondence: josemanuel.moreno@uam.es

## 1. Introduction

The use of geomaterials spans long back into human history and relicts of man's endeavours remain as evidence of practical use of rocks and minerals for the benefit of evolving societies. As a discipline, the formal study and practice of geomaterials has evolved to be a distinct topic since the mid twentieth century. The practice of geotechnical and civil engineering gave rise to a sub-discipline, focused on the engineering properties of materials won from the earth (but largely excluding ores). Ore geology and production metallurgy have become subjects in their own right, focused on the recovery and processing of metals. Geomaterials, however, owe their early evolution to geotechnical activities which were concerned largely with the physical, rather than chemical, properties of consolidated and unconsolidated materials from the earth. Fookes [1] gives a substantial review of the topic as he saw it in 1991. Since then, this field of activity has expanded considerably, to include many new materials such as tyre crumb, ceramic wastes, combustion products and many other anthropogenic materials, which are not primary products of the extractive industries, but represent resource-efficient use of industrial by-products in much the same way as, for example, natural aggregates. The field currently occupies a space between four older disciplines and has become firmly established in its own right (Figure 1).

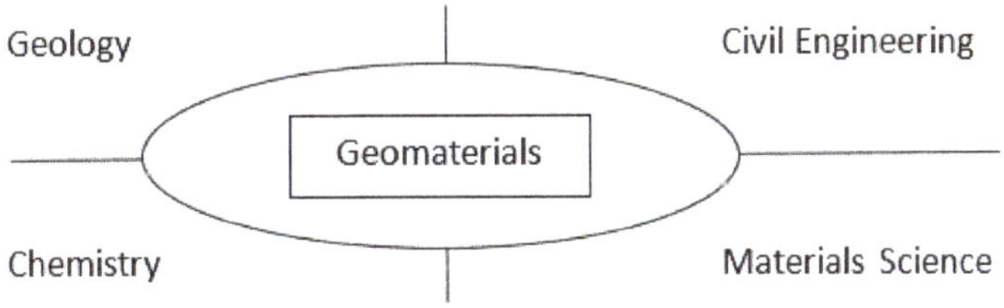


**Figure 1.** The discipline of geomaterials relates closely to other subjects.

**Citation:** Tyrer, M.; Moreno-Maroto, J.M. Geomaterials: Latest Advances in Materials for Construction and Engineering Applications. *Appl. Sci.* **2023**, *13*, 9129. https://doi.org/10.3390/app13169129

Received: 3 August 2023
Accepted: 6 August 2023
Published: 10 August 2023



In recent decades, our understanding of the importance of chemistry and microstructure in governing many engineering properties has grown, such that these aspects are firmly embedded in the study and practice of geomaterials. Before looking at geomaterials in any detail, it is appropriate to examine the use of the word over recent decades. Although the etymology of the word *'Geomaterials'* is obvious, its first written use remains elusive; however, it seems to have appeared regularly in the literature since the 1960s and 1970s. One of the earliest uses of the term (1964) appears in a report [2] by the Highway Research Board of the Division of Engineering and Industrial Research, National Academy of Sciences, National Research Council (USA) which discusses the mechanical behaviour and properties of various materials found in geotechnical engineering, referring to them collectively as "geomaterials." Analysis of word use in publications over the twentieth century allows an n-gram plot to be produced (Figure 2). This demonstrates the wide adoption of the term 'Geomaterials' in the 1980s and establishment of a burgeoning field of activity by the 21st century.

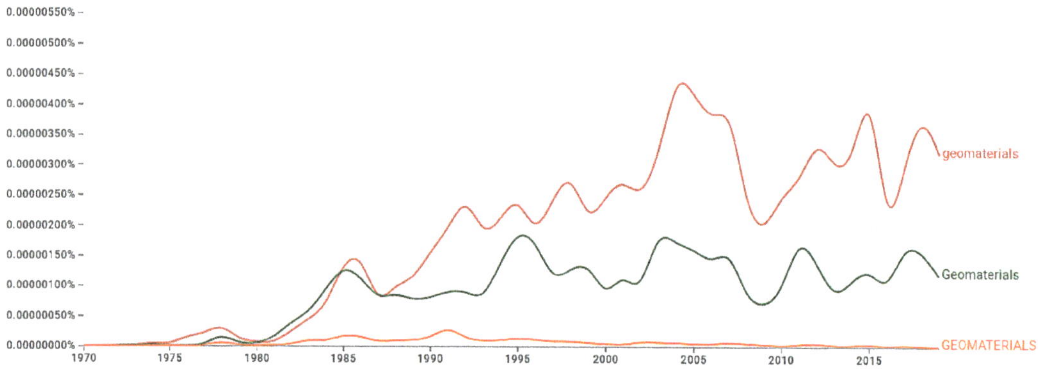


**Figure 2.** Use of the word 'Geomaterials' in English language books. Vertical axis is the fraction of the books examined by the Google$^{TM}$ n-gram generator (21 July 2023. https://books.google.com/ngrams/) per month, plotted against time.

Whilst the field of study is relatively recent, the practice dates back to the dawn of civilization. Both brick and stone have been in widespread use for at least 12,000 years, when bricks were initially hand moulded, often reinforced with plant fibres (commonly straw) and baked hard in the sun (Figure 3). Early fired bricks have been found in Banpo village [3], near Xi'an, China and these are 5500 years old. Brick making (both fired and unfired) was widespread by the Bronze Age and in the great empires of antiquity (Babylon, Assyria, Egypt, Persia, Phoenicia, etc.) whose people were skilled in both brick and stone masonry. Of similar antiquity is the calcining of limestone to make quick lime (CaO) and at a lower temperature, gypsum to make plaster. Both technologies were established in neolithic times and have a history of use spanning 10,000 years. By classical times (Greece, Rome), stone masonry was a major activity and many excellent examples of architectural and decorative stone work survive today.

The history of cement spans two periods of development. Lime mortar in construction was used since the Bronze Age and still survives to this day. Lime hardens by aerial carbonation to form a calcium carbonate–hydroxide binder which although robust, is moderately soluble. This left structures more susceptible to rain weathering than are their modern counterparts, hydraulic cements. A step change was seen in the Greek and Roman use of natural pozzolans (especially volcanic ash from Pozzuoli, Italy) that reacted with the alkaline pore solution to precipitate calcium silicate hydrate (CSH) gel which is much less soluble than lime mortar. The history and development of cement and concrete has been reviewed regularly and two contributions provide comprehensive summaries of this topic. Blezard [4] discusses the major developments in cement technology referring to the Lepinski Vir settlement in Serbia (7600 years ago) which he felt was the domain of the archaeologist, rather than scientist, but nonetheless contains the remnants of a 'concrete' floor. He also includes a photograph of an Egyptian mural from a site in Thebes, which shows workmen filling earthenware jars with water that is then mixed with lime and used as a mortar for stone masonry. Trout [5] gives an authoritative account of the development of cements from antiquity to the present and both publications report important milestones in the evolution of cement technology.

**Figure 3.** Çatalhöyük, Anatolia, Turkey. This neolithic settlement shows extensive use of mud brick construction from around 9000 years ago and is one of the best preserved neolithic/chalcolithic sites in the world. Image by Murat Özsoy (1958) Wikimedia Commons.

In the 1700s, John Smeaton, Bridley Higgins, Louis Viscat and Joseph Parker each made significant advances in producing hydraulic cements and in 1824 Joseph Aspdin was granted a patent on 'Patent Portland cement'. This was made by calcining limestone with clay to produce cement clinker that could be ground and mixed with water to produce a waterproof ('hydraulic') cement. Although Aspdin did not reach high enough temperatures to advance the clinkering reactions to near completion, this paved the way to modern cement production. Work by Le Chatelier in the 1880s into the chemistry of cements firmly established the foundations of our understanding of cement production and use. Developments throughout the late 19th and early 20th centuries saw industrialisation of cement production on a global scale, such that by the mid-twentieth century, cement and concrete were the preferred materials for construction in many projects.

Developments in processing technology greatly influenced the production and use of geomaterials. The mechanical crushing, screening and washing of quarried stone drove the expansion of aggregate production far beyond the use of naturally produced sand and gravel, although mechanised recovery of natural aggregates also resulted in considerable expansion of these industries. Similarly, mechanised cutting and dressing of stone for construction and of slates for roofing developed throughout the industrial revolution, driven by demand from new building in the late 18th and 19th centuries. At this time, demand for roadstone expanded alongside the need for building, requiring both dressed stone for road cobbles and compacted stone for their base layers. The introduction of tarmacadam (late 19th century; J.L. McAdam, improved and patented by E.P. Hooley; 1902) led to a rapid increase in road construction and asphalt-bound roadstone provided the wearing surface layer of roads from the mid-twentieth century.

The current practice of geomaterials concerns the influence of material properties on their engineering properties in service. This spans the geotechnical concerns of slope stability and earthen construction to the processing and durability of rocks and minerals in the myriad of applications for which they are used. Mechanical and thermal property studies, on which the discipline was built, have been supplemented by a much increased understanding of how chemistry and microstructure control durability and service life. Poole [6] and Ingham [7] published comprehensive works considering the chemical, mineralogical and microstructural properties of geomaterials and show how a detailed understanding of these aspects has led to greater confidence in explaining durability of the materials in service.

Figure 4 considers the major groups of geomaterials in terms of both their degree of consolidation and degree of processing. At the simplest level, unprocessed materials constitute the greatest proportion of geomaterials in use. Earth works, highway construction and excavation consume vast quantities of unconsolidated soil, aggregates and rock-fill, emplaced and compacted with long service lives in mind. An understanding of the mineralogy of such massive engineering projects ensures the risk of, for example, ground heave as a result of oxidation of pyrite in rocks may be minimised. Mineralogical instability is turned to advantage, however, in highway construction, where minerals in the igneous rocks used as road stone weather at different rates, ensuring the wear from vehicle tyres exposes fresh surfaces and maintain a high coefficient of friction between the tyre and the road. This gives rise to the concept of 'Polished Stone Value' (PSV) addressed in current standards for aggregates (e.g., BS EN 932 1997) [8].

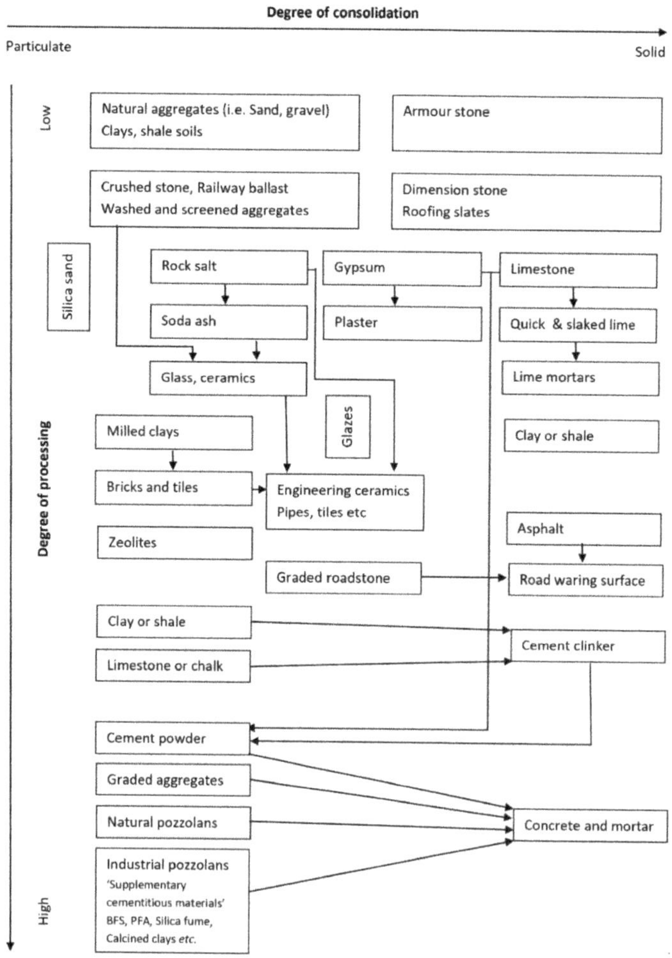


**Figure 4.** Major groups of geomaterials in terms of both their degree of consolidation and degree of processing.

Geomaterials are processed to varying degrees. Rocksalt (NaCl) is used directly in highway de-icing but is largely processed to form soda ash ($Na_2CO_3$), a major precursor for the chemical industries. In terms of geomaterials it is used in glass manufacture (it lowers the temperature of the silicate melt) which is used with other geomaterials as a glaze on many fired clay products such as pipes and tiles. These fired clay products, along with modern brick production, globally account for over 3,000,000,000 tonnes of clay per year. Gypsum is calcined to produce plaster products and is an essential component in cement manufacture. Blended with ground cement powder, it controls the hydration kinetics of the aluminates in the clinker, preventing 'flash set'. Limestone is an important geomaterial, both as a building stone and as a source of calcium in cement production and through simple calcining is the precursor for lime products (CaO and $Ca(OH)_2$). Zeolites should be considered amongst modern geomaterials, reflecting their use as ion exchange media in permeable reactive barriers (PRBs). This rather niche application is part of a much wider field of subsurface barrier construction to control the flow and quality of groundwater and, particularly, to constrain pollutant migration. Permeable reactive barriers allow the passage of ground water through chemically active media. Zeolites can sorb specific ions from a solution, whereas zero valent iron controls the redox potential of mobile solutions. Crushed limestone in PRB applications is used to raise the pH of moderately acid ground waters which may then precipitate metal ions from solution. An example of a non-permeable sub surface barrier is curtain wall construction. A trench is excavated into which a mineral mixture is poured. One example is a Portland cement—blast furnace slag and clay mixture—which never truly hardens. It remains compliant, setting to form a very low permeability semi-solid, which restricts the flow of groundwater through it.

Recent decades have seen the concept of resource efficiency become important to society and the economy. The re-use of industrial wastes and by-products is of mounting importance, driven in part by corporate social responsibility and in part by the rising taxation on waste disposal. This has led to the inclusion of many new materials in the geomaterials field. To this can be added spent foundry sands, ceramic wastes from many sources, waste sea shells, crushed brick and concrete wastes, highway 'blacktop' planings and a wide range of solid combustion products, all of which are reused in construction. Demolition waste has been used as bulk fill and highway sub-base layers for many years. The inclusion of supplementary cementitious materials has been driven by the desire of the cement industry to reduce the embedded carbon of cement, following a report by the World Business Council for Sustainable Development in 2002. This highlighted the $CO_2$ emissions released by the cement industry and the need for change. The solutions were considered by Gartner [9] who described low energy cement clinkers:

(A) Pozzolan-based cements.
(B) Calcium (sulfo)aluminate-based cements (made from low-grade alumina sources).
(C) Calcium sulfate-based cements.

Gartner compares the pros and cons of each approach and concludes that "*Based on the present analysis, the most promising low-$CO_2$ alternative cementing systems appear to be those that make use of large amounts of either natural or artificial pozzolans or those that effectively stabilise hydrated calcium sulfates (e.g., as ettringite)*". The United Nations Environment Programme commissioned a study (Scrivener et al. 2016) [10] which also considers a range of options. Most promising of these is the use of calcined, kaolinitic clays blended with (and importantly, replacing) much of the Portland cement clinker. The technology, known as $LC^3$, includes limestone flour in the mixture and is proving to be a step-change in $CO_2$ reduction from cements, as the raw materials are widely available and need simple processing. Lehna and Preston [11] similarly considered options for $CO_2$ reduction focusing on "*the potential to blend clinker with alternative materials, and on the use of 'novel cements'—two levers that can reduce the need for clinker itself by lowering the proportion of clinker required in particular cement mixtures*". These reports all agree that reduction in the amount of cement clinker used (the '*clinker factor*') is the key to reducing embedding carbon dioxide in cement production.

A wide range of industrial pozzolans have been considered but relatively few have been adopted commercially. Some (for example ground granulated blast furnace slag from the iron industry and pulverised fuel ash from coal fired power generation) have been in wide use for a century or so. They work by providing a source of silica (commonly aluminosilicate) which dissolves in the highly alkaline cement pore solution. The consequence is that more hydration products are formed but by using less cement clinker. Many other materials have a more modest place in the market. Silica fume, wood ash, rice husk ash and 'metakaolin' (calcined kaolin clays) are all in widespread use but limited, often local availability, or high price prevent their wider adoption. The specialist in geomaterials needs an ever-growing knowledge of the materials used to make blended cements.

To conclude, it is appropriate to consider new materials entering the spectrum of geomaterials. Waste tyres, for example, are both a waste and a resource. When disaggregated into tyre crumb, they can be used as a permeable (but not high load-bearing) medium and is commonly used in playground and footpath construction. Glass cullet is used to some extent as an aggregate but rarely in structural concrete, owing to the risk of alkali–silica reaction. However, when finely ground [12] (<40 μm), this material has considerable potential as a constituent in blended cements.

Ingenuity and the constant need to recycle materials at the end of their service lives will ensure new geomaterials enter the field continuously.

## 2. About This Special Issue

Geomaterials account for the majority of all goods consumed by mankind due to their massive application in the construction of buildings, the execution of civil and geotechnical works, as well as in environmental engineering projects. These kinds of materials may retain their primary characteristics (ornamental rocks, stone aggregates, clay barriers, compacted soils, etc.) or can be artificially manufactured either from natural raw materials or from wastes (cement, concrete, ceramics, expanded lightweight aggregates, geopolymers, mineral wool, etc.).

The growing increase in population and demand for natural resources means that these geomaterials, so widely used, need to be studied in depth in order to adapt to current needs, while contributing certain technological, environmental and economic benefits. Therefore, geomaterials *have been*, *are* and *will be* strategic assets with an enormous socio-economic and environmental impact.

In this regard, this Special Issue includes a series of articles focused on recent advances in geomaterials' research, with special emphasis on their potential application in the aforementioned sectors. The findings presented not only shed light on the knowledge we have had about geomaterials to date, but also pave the way for future research in order to delve deeper into their improvement, development and sustainability for future generations.

**Author Contributions:** Conceptualization: M.T. and J.M.M.-M.; writing—original draft preparation: M.T.; writing—review and editing: J.M.M.-M. All authors have read and agreed to the published version of the manuscript.

**Conflicts of Interest:** The authors declare no conflict of interest.

## References

1. Fookes, P.G. Geomaterials. *Q. J. Eng. Geol. Hydrogeol.* **1991**, *24*, 3–15. [CrossRef]
2. Highway Research Board of the Division of Engineering and Industrial Research, National Academy of Sciences. *Report 495*; National Research Council. National Cooperative Highway Research Program; Highway Research Board of the Division of Engineering and Industrial Research; National Academy of Sciences: Washington, DC, USA, 1964.
3. UNESCO. *Museum Vol XXXII, n° 4. Museums in China*; UNESCO: Paris, France, 1980.
4. Blezard, R.G. *Society of Chemical Industry Lecture Paper Series. Reflections on the History of the Chemistry of Cement*; Society of Chemical Industry: Cottingham, UK, 1998; ISSN 1353-114X.
5. Trout, E. The History of Calcareous Cements. In *Lea's Chemistry of Cement and Concrete Chemistry*, 5th ed.; Hewlett, P., Liska, M., Eds.; Elsevier: Amsterdam, The Netherlands, 2019; pp. 1–29.
6. Poole, A.B.; Sims, I. *Concrete Petrography A Handbook of Investigative Techniques*, 2nd ed.; CRC Press: London, UK, 2016.
7. Ingham, J. *Geomaterials under the Microscope*; Manson Publishing: London, UK, 2011.
8. *BS EN 932-1:1997*; Tests for General Properties of Aggregates Methods for Sampling. British Standards Institution: London, UK, 1997.

9. Gartner, E. Industrially interesting approaches to "low-$CO_2$" cements. *Cem. Concr. Res.* **2004**, *34*, 1489–1498. [CrossRef]
10. Scrivener, K.L.; John, V.M.; Gartner, E.M. *Eco-Efficient Cements: Potential, Economically Viable Solutions for a Low-$CO_2$, Cement Based Materials Industry*; United Nations Environment Programme: Paris, France, 2016.
11. Lehne, J.; Preston, F. *Making Concrete Change, Chatham House Report*; The Royal Institute of International Affairs: London, UK, 2018.
12. Idir, R.; Cyr, M.; Tagnit-Hamou, A. Pozzolanic properties of fine and coarse colour-mixed glass cullet. *Cem. Concr. Compos.* **2011**, *33*, 19–29. [CrossRef]

applied sciences

MDPI

Article

# Geotechnical Behaviour of Fly Ash–Bentonite Used in Layers

**Murtaza Hasan [1], Mehboob Anwer Khan [2], Abdullah H. Alsabhan [3], Abdullah A. Almajid [3], Shamshad Alam [3], Mohammad Amir Khan [4,*], Tinku Biswas [4] and Jaan Pu [5]**

[1] Department of Civil Engineering, Chandigarh University, Mohali 140143, India; murtazadce@gmail.com
[2] Department of Civil Engineering, Faculty of Engineering, Aligarh Muslim University, Aligarh 202002, India; mehboobcivil@yahoo.co.in
[3] Department of Civil Engineering, College of Engineering, King Saud University, P.O. Box 800, Riyadh 11421, Saudi Arabia; aalsabhan@ksu.edu.sa (A.H.A.); alabduallah@ksu.edu.sa (A.A.A.); salam@ksu.edu.sa (S.A.)
[4] Department of Civil Engineering, Galgotias College of Engineering and Technology, Greater Noida 201310, India; tinku.biswas@galgotiacollege.edu
[5] Faculty of Engineering and Informatics, University of Bradford, Bradford BD7 1DP, UK; j.h.pu1@bradford.ac.uk
* Correspondence: amirmdamu@gmail.com

**Abstract:** Increasing infrastructure growth has forced the construction industry to look for wasteful, cheap, and suitable materials for construction. An investigation into the geotechnical utilization of fly ash was carried out in the present study. Practical applications normally involve the use of large quantities of fly ash, so proper mixing of the fly ash with other materials may not be significantly achieved. Therefore, the present paper investigates the behaviour of a fly ash–bentonite layered system with different ratios. The physical properties and chemical composition of fly ash and bentonite were determined. SEM and energy dispersive X-ray experiments were also used to investigate the morphology and phase compositions of fly ash and bentonite. A series of consolidated undrained (CU) triaxial tests on fly ash–bentonite were carried out to investigate shear strength characteristics. Fly ash (F) and bentonite (B) were used in the following ratios: 1:1 (50% F:50% B), 2:1 (67% F:33% B), 3:1 (75% F:25% B), and 4:1 (80% F:20% B), with different numbers of interfaces (N), i.e., 1, 2, and 3 for each ratio. The deviator stress and cohesion value were found to increase with the number of interfaces for each ratio. The angle of shear resistance changed marginally with the increase in the fly ash–bentonite ratios and varying interfaces.

**Keywords:** fly ash; bentonite; shear strength; waste; triaxial tests; chemical; landfill

**Citation:** Hasan, M.; Khan, M.A.; Alsabhan, A.H.; Almajid, A.A.; Alam, S.; Khan, M.A.; Biswas, T.; Pu, J. Geotechnical Behaviour of Fly Ash–Bentonite Used in Layers. *Appl. Sci.* **2022**, *12*, 1421. https://doi.org/10.3390/app12031421

Academic Editor: José Manuel Moreno-Maroto

Received: 31 December 2021
Accepted: 26 January 2022
Published: 28 January 2022



## 1. Introduction

Around 70–75% of the electricity is generated by coal-based thermal power plants in India. Indian coal is of poor quality and contains almost 30–45% of ash [1,2]. Fly ash is also considered a hazardous water pollutant due to the presence of toxic heavy metals [3]. According to estimates from the Fly Ash Utilization Program (FAUP), the coal supplied to thermal power plants in the amount of 686.34 million tonnes in 2020–2021, which contains a significant amount of ash, had a fly ash generation of 232.56 million tonnes in 2020–2021 [4]. The Ministry of Power, Govt. of India estimates 1800 million tonnes of coal use every year and 600 million tonnes of fly ash generated by 2031–2032. Due to persistent efforts by the Fly Ash Mission in India and some other agencies, the utilization of fly ash has improved from a meagre 3% in 1994 to 92% in 2020, which is nowhere near its target of 100% fly ash utilization [5]. Various authors have successfully stabilized expensive soil by blending it with fly ash [6–8]. Fly ash has also been modified with lime, gypsum, and fibre [9,10].

Bentonite is an efficient clay substance for creating low-permeability barriers. Bentonite is a valuable material because of its great plasticity, strong dry-bonding strength, high shear and compressive strength, and low permeability [11–13]. Many researchers have carried out experimental investigations on fly ash mixed with bentonite [14–16]. The suitability of fly ash as a hydraulic barrier as well as the use of bentonite were investigated to improve the geotechnical qualities of fly ash [14]. It was established that 70% by weight of fly ash can be blended with bentonite to meet the requirements of compacted landfill liners [15,17].

The permeability of the compacted fly ash deposits is moderate. Fly ash is not suitable for use in seepage barriers such as liners. To lower the permeability of fly ash, clay-like admixtures should be used [2,18]. The utilization of a large quantity of fly ash means proper mixing of the fly ash with other materials may not be significantly achieved. To overcome this problem, the layered system has been adopted in the present study. When two layers of different materials interact at their interfaces, the deformation behaviour of the composite is influenced by the inherent properties of the materials. The effect of fly ash on bentonite ratio and the number of layers was studied.

## 2. Materials Used

### 2.1. Fly Ash

The fly ash came from a national thermal power plant in Dadri, India, with an installed capacity of 800 megawatts. The plant generates about 10 million tonnes of bottom ash and 39 million tonnes of fly ash each day. The fly ash is collected in a total of 24 electrostatic precipitator hoppers. To avoid deterioration over time, fly ash was collected in dry form and stored in moisture-proof containers. The results of the research into the chemical composition and physical properties of fly ash are shown in Tables 1 and 2, respectively. It was designated as class F, as defined by ASTM C 618 [19], indicating that it is typically obtained from bituminous and anthracite coals. The intrinsic diversity of fly ashes from different sources as well as from the same source is greater in India than in other nations such as Italy and the United States [20]. The chemical constituents of the Dadri fly ash were found to be within the range of other Indian fly ash chemical compounds. Scanning electron microscope (SEM) and Energy Dispersive X-ray (EDX) were carried out through JEOL, instrument JSM-6510 (Figure 1). The following settings were used to photograph the morphological characteristics of the fly: accelerating voltage = 15 kV, magnification = 3000, spot size = 40, μm marker = 5, and tilt = 0. The image clearly shows that fly ash is mostly made up of spherical particles of various sizes (Figure 1). Figure 2a shows the EDX of fly ash.

**Table 1.** Chemical composition of fly ash.

| Composition | Percentage |
|---|---|
| Silicon dioxide ($SiO_2$) | 59.00 |
| Alumina ($Al_2O_3$) | 29.00 |
| Iron oxide ($Fe_2O_3$) | 6.50 |
| Calcium oxide (CaO) | 1.80 |
| Magnesium oxide (MgO) | 1.44 |
| Sodium oxide ($Na_2O$) | 0.80 |
| Sulphur trioxide ($SO_3$) | 0.28 |

**Table 2.** Physical properties of fly ash and bentonite.

| Properties | Fly Ash | Bentonite |
|---|---|---|
| Specific gravity | 2.19 | 2.73 |
| Optimum moisture content (OMC), % | 22 | 28 |
| Unit weight, $kN/m^3$ | 14.39 | 21.5 |
| Maximum dry density (MDD), $kN/m^3$ | 11.8 | 16.8 |
| Liquid limit ($W_L$), % | 24 | 261 |
| Plastic limit ($W_P$), % | Non plastic | 38 |
| Plasticity index (P.I), % | 24 | 223 |
| Coefficient of permeability, ($10^{-10}$ m/sec) | 551 | 8.52 |
| Unconfined compression strength, $kN/m^2$ | 34 | 176 |
| Classification | Class F | CH |

**Figure 1.** Scanning electron micrographs of fly ash with different magnification factors: (**a**) 700; (**b**) 3000.

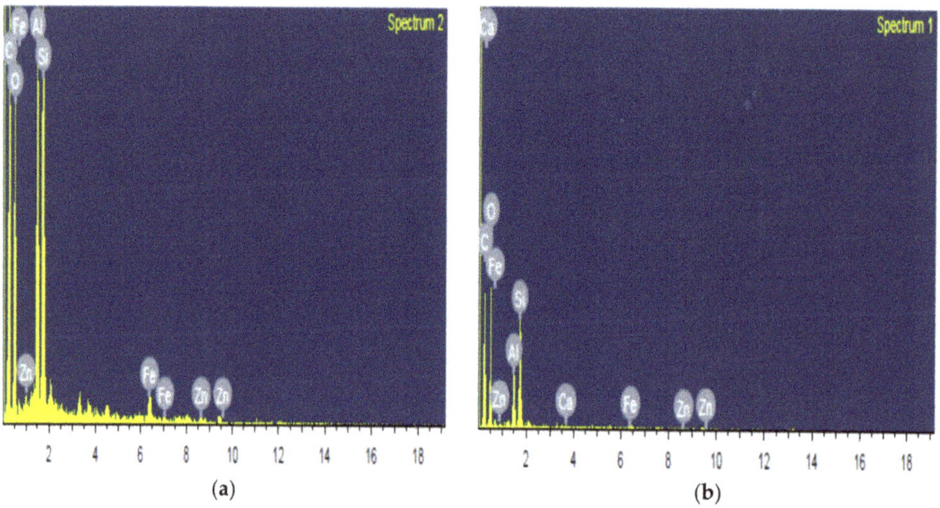


**Figure 2.** Energy-dispersive X-ray spectroscopy (EDS): (**a**) fly ash and (**b**) bentonite.

The density bottle method was used to determine the specific gravity of fly ash, which was found to be 2.19 according to IS: 2720-Part 3 [21]. To assess the grain size distribution of the Dadri fly ash, a hydrometer test was performed according to Indian Standard IS 2720-Part 4 [22]. The Atterberg limits were measured according to Indian Standard IS 2720-Part 4 [23]. The liquid limit ($W_L$) of fly ash was found to be 24% and non-plastic in nature. For typical soils, a light compaction standard Proctor test was performed according to Indian Standard IS 2720-Part 10 [24]. The optimal moisture content (OMC), unit weight, and maximum dry density (MDD) were found to be 22%, 14.39 kN/m$^3$, and 11.8 kN/m$^3$, respectively. Fly ashes have a lower MDD and a higher OMC than normal soils. This may be attributed to lower specific gravity and the porous nature of the particle. Unconfined compression tests were carried out in accordance with Indian Standard IS 2720-Part 10 [24]. Two specimens were tested, and at 2% axial strain, the failure axial stress was attained, causing the specimens to collapse. The Dadri fly ash specimens had an average unconfined compressive strength of 110 kN/m$^2$. At the MDD–OMC state, specimens were prepared and a falling head permeability test was performed. The coefficient of permeability of fly ash was determined to be around $5.51 \times 10^{-6}$ cm/s, which is normally in the range of non-plastic silts. Table 2 shows the physical properties of fly ash and bentonite.

### *2.2. Bentonite*

The bentonite utilized was dry, sieved through an IS sieve of 425 microns, and stored in the laboratory in sealed containers. The bentonite's morphological properties presented in Figure 3 were collected under the following conditions: accelerating voltage of 15 kV, 2700 magnification, 40 spot size, 5 µm marker, and tilt = 0. Energy dispersive X-ray analysis (EDS or EDX), as shown in Figure 2b, is based on the diffraction of very short-wave electromagnetic radiation in the regular, continuous mineral lattice. The specific gravity of bentonite was found to be 2.73. As illustrated in Figure 3, the particle size distribution curves of fly ash and bentonite were plotted as shown in Figure 4. The liquid limit ($W_L$) and plastic limit ($W_P$) of bentonite were found to be 261% and 38%, respectively. CH has been assigned to it. The compaction curves for bentonite and fly ash are shown in Figure 5. The optimum moisture content (OMC), unit weight, and maximum dry density (MDD) of bentonite were found to 28%, 21.52 kN/m$^3$, and 16.81 kN/m$^3$, respectively. The coefficient of permeability was found to be close to $8.52 \times 10^{-8}$ cm/s.

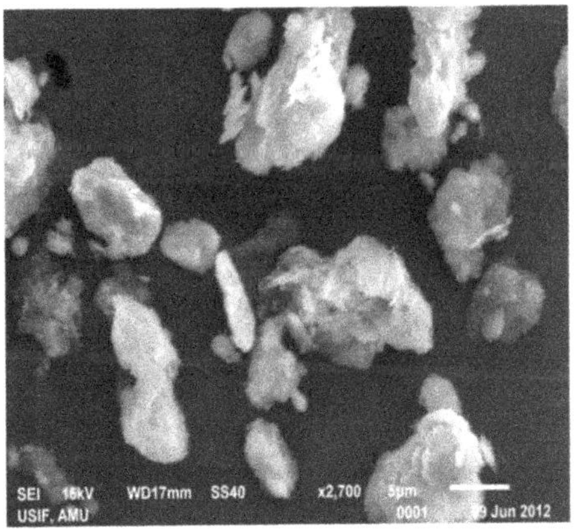


**Figure 3.** Scanning electron micrographs of bentonite at 2700 magnification factor.

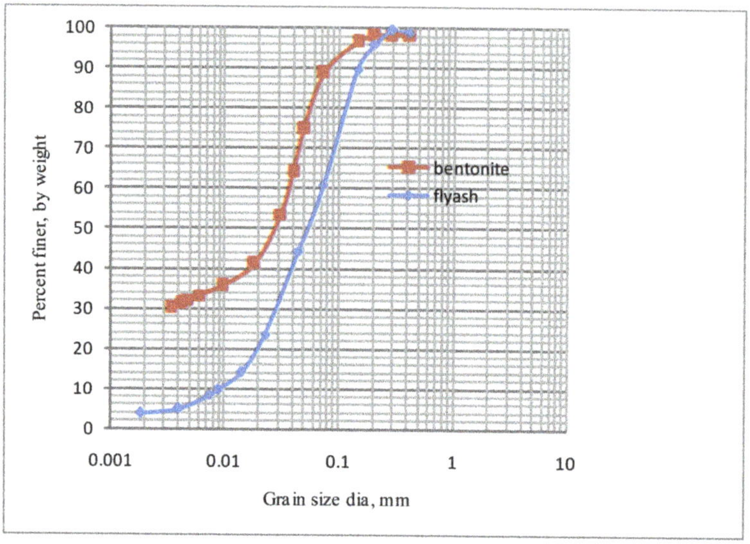


**Figure 4.** Particle size distribution curves for fly ash and bentonite.

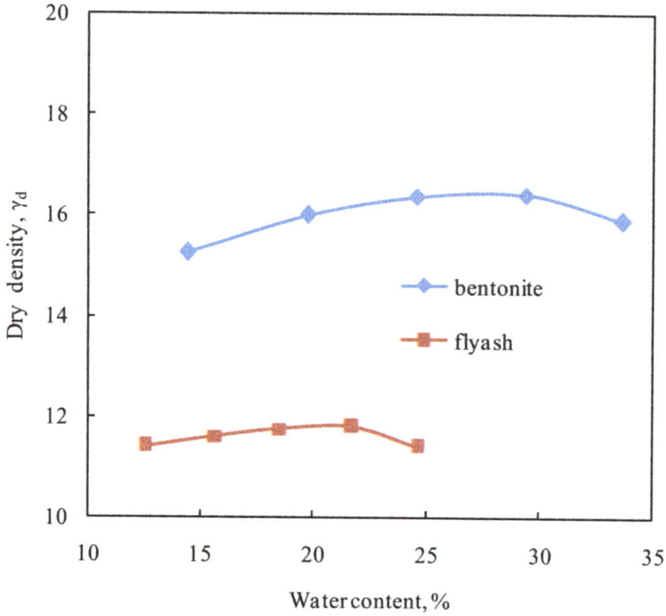


**Figure 5.** Proctor compaction curve of fly ash and bentonite.

## 3. Experimental Programme

Laboratory tests were conducted on the cylindrical specimen having a size of 39 mm in diameter and 84 mm in length. Experimental details are described in Table 3. Fly ash and bentonite were used in the following ratios (by volume): 1:1 (50% F:50% B), 2:1 (67% F:33% B), 3:1 (75% F:25% B), and 4:1 (80% F:20% B) in two layers, three layers, and four layers (N = 1, 2, and 3). Figure 6 explains the diagram presentation for F:B = 2:1 in different layer combinations. For F:B = 2:1 in two layers, the length of the fly ash and bentonite samples was kept at 56 mm and 28 mm, respectively. For F:B = 2:1 in three layers, the length of fly ash and bentonite in each layer was taken as 28 mm. However, for F:B = 2:1

in a four-layer combination, the length of each layer of fly ash and bentonite was kept at 28 mm and 14 mm, respectively. Each test was repeated once for better accuracy, so a total of 84 samples were tested under three different confining pressures of 100 kPa, 200 kPa, and 300 kPa.

**Table 3.** Experimental programme.

| Test Description | No. of Interfaces (N) | Confining Pressure (kPa) | | | Total No. of Tests |
|---|---|---|---|---|---|
| | | 100 | 200 | 300 | |
| Fly Ash Sample | - | ✔ | ✔ | ✔ | 6 |
| Bentonite Sample | - | ✔ | ✔ | ✔ | 6 |
| F:B = 1:1 | 1 | ✔ | ✔ | ✔ | 6 |
| | 2 | ✔ | ✔ | ✔ | 6 |
| | 3 | ✔ | ✔ | ✔ | 6 |
| F:B = 2:1 | 1 | ✔ | ✔ | ✔ | 6 |
| | 2 | ✔ | ✔ | ✔ | 6 |
| | 3 | ✔ | ✔ | ✔ | 6 |
| F:B = 3:1 | 1 | ✔ | ✔ | ✔ | 6 |
| | 2 | ✔ | ✔ | ✔ | 6 |
| | 3 | ✔ | ✔ | ✔ | 6 |
| F:B = 4:1 | 1 | ✔ | ✔ | ✔ | 6 |
| | 2 | ✔ | ✔ | ✔ | 6 |
| | 3 | ✔ | ✔ | ✔ | 6 |

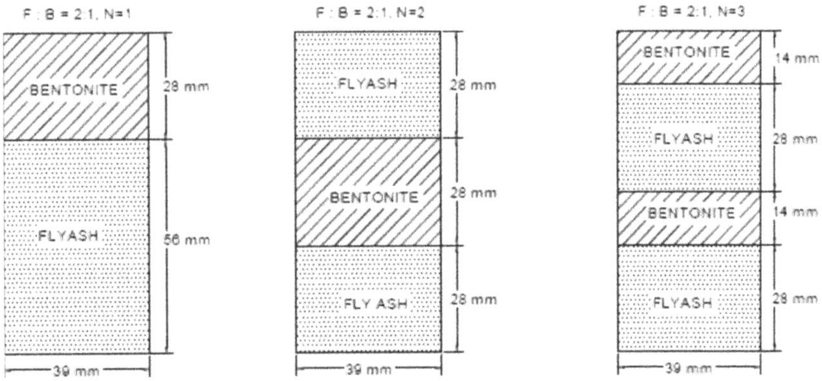


**Figure 6.** Diagrams representing the layered arrangement in samples for F:B = 2:1.

### *3.1. Sample Preparation*

Each sample was prepared on optimum moisture content. Firstly, Proctor compaction tests were carried out to find optimum moisture content and maximum dry density. The weights of the fly ash and bentonite were calculated for each layer by using max dry density of a sample 39 mm in diameter and 84 mm in length. A pre-determined amount of dry material was mixed properly by adding 28% (for bentonite) and 22% (for fly ash) water. Material was filled in a sampler at 5 mm thickness for each layer and compacted by an iron rod of diameter 20 mm of 250 gm weight to ensure effectiveness. Figure 7a–e represents a few specimens used in the present laboratory tests. The fly ash and bentonite samples are shown in Figure 7a,b. However, for F:B = 1:1 in the layer combination as N = 1, the length of each layer was taken as 42 mm, as shown in Figure 7c. For F:B = 1:1 with N = 2 and 3, the samples are shown in Figure 7d,e. To achieve homogeneity and thixotropy, the samples were left for 24 h. After that, the sample was placed in a pressure cell and exposed to a vertical compression force as well as all-around hydrostatic pressure. With end caps and rubber rings in place, the sample was correctly sealed, and the sealing rings of the cell were also properly positioned. Before testing, samples were saturated for 24 h.

**Figure 7.** Specimen before saturation of: (**a**) fly ash, (**b**) bentonite, (**c**) F:B = 1:1 in N = 1, (**d**) F:B = 1:1 in N = 2, and (**e**) F:B = 1:1 in N = 3.

*3.2. Test Procedure*

The consolidated undrained (CU) test was performed on samples (excluding simple fly ash) in which complete consolidation of the test specimen was allowed under confining pressure, but no drainage was allowed during shear. Back pressure was used to obtain full saturation. During the CU test, the effective stress parameters c and $\phi$ were determined by measuring the pressure in the pores.

The deviator stress and axial strain were calculated when the samples were exposed to confining pressures ($\sigma_3$) of 100 kPa, 200 kPa, and 300 kPa. The data collection was totally computer controlled, and the tests were run at a constant strain rate of 2% per minute. A regulated speed driven motor with a compression force of up to 100 kN was used to impose the axial displacement. As shown in Figure 8, the linear variable differential transformer (LVDT) and pore pressure transducers (PPT) were used to measure settlement and pore water pressure of the sample, respectively. All instruments were duly calibrated before testing.

**Figure 8.** Triaxial shear test apparatus with LVDT, PPT, and load cell.

## 4. Results and Discussion

### *4.1. Stress-Strain Behaviour*

Figure 9 shows the response of deviator stress to axial strain at confining pressures of 100, 200, and 300 kPa. Different layered systems with varying numbers of interfaces have been studied. Apart from the different layers used, the ratio of fly ash to bentonite has been subjected to variation from 1:1 to 4:1. Deviator stress at failure of fly ash was found to be 236, 380, and 495 kPa on confining pressures of 100, 200, and 300 kPa, respectively. As indicated in the graphs, samples confined to 300 kPa sustained greater axial strains and deviator stresses than samples confined to lower confining pressures of 100 kPa and 200 kPa. In addition, it can be seen that the stress endured by a plain fly ash sample is significantly lower than a plain bentonite specimen. Again, as the ratio of fly ash to bentonite increases, the deviator stress carried by the sample decreases.

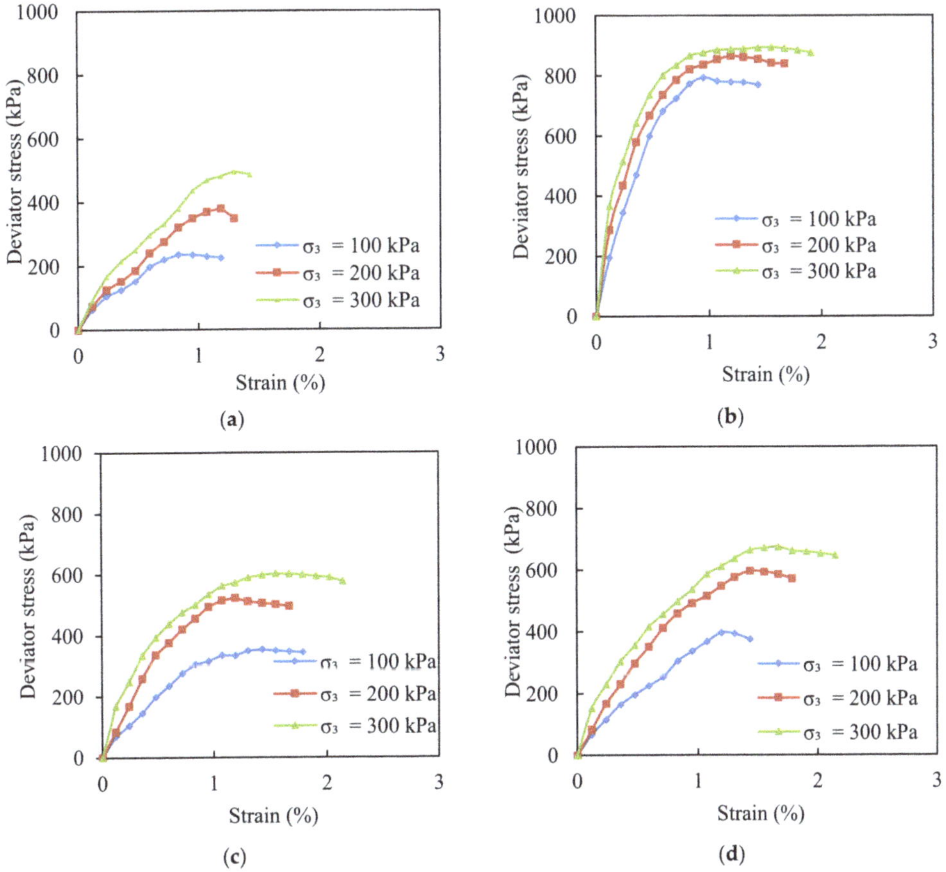


**Figure 9.** *Cont.*

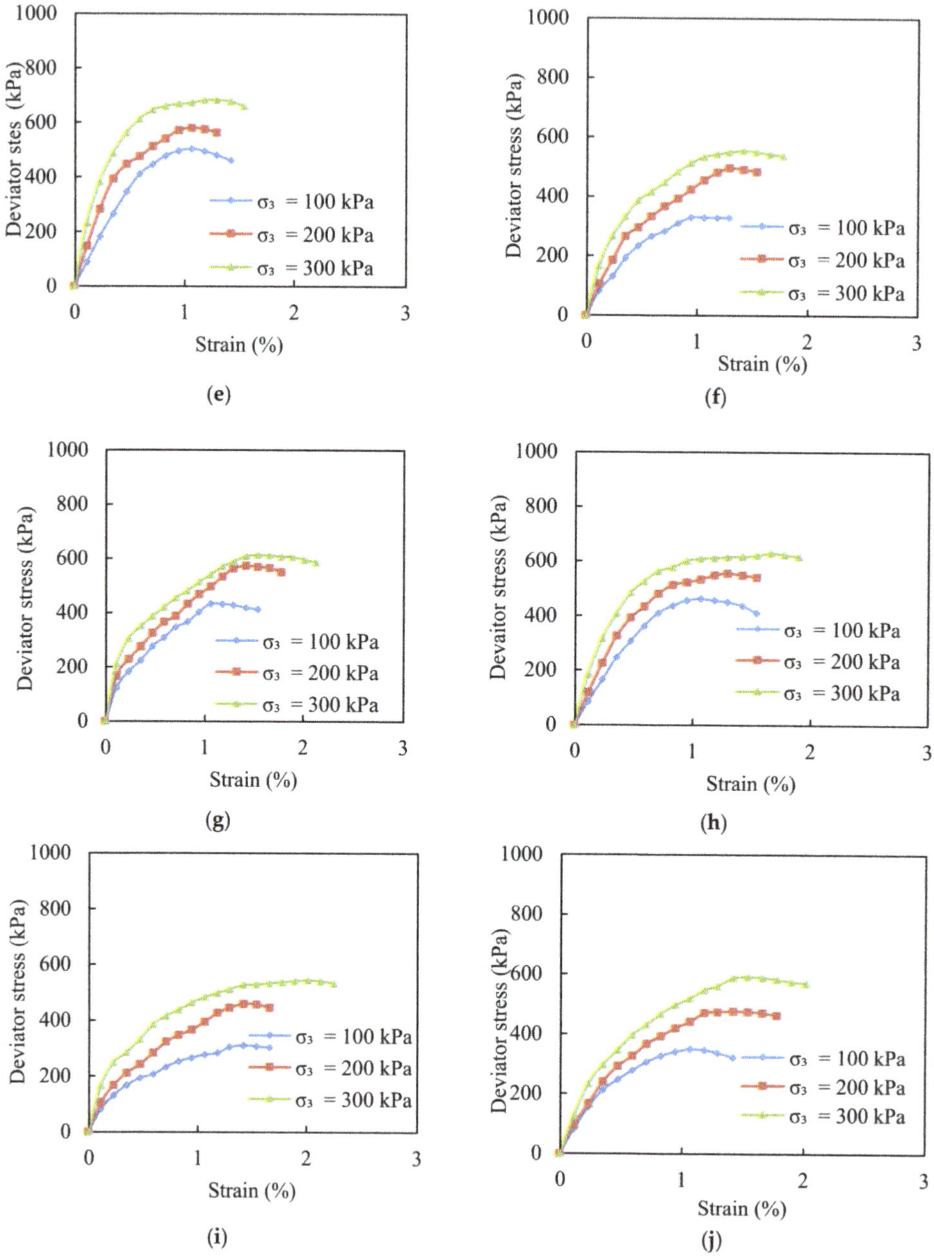


**Figure 9.** *Cont.*

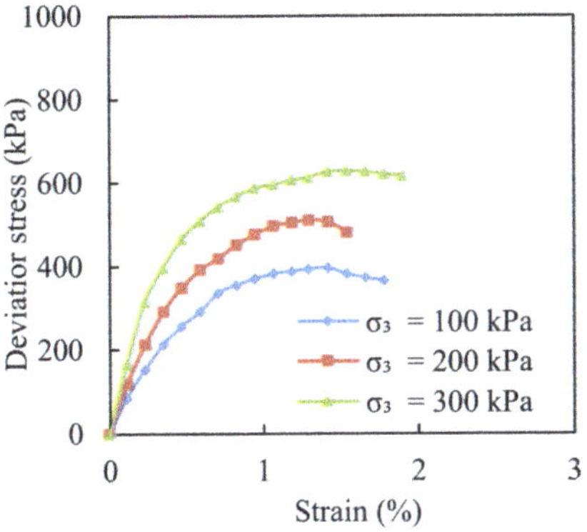


(**k**)

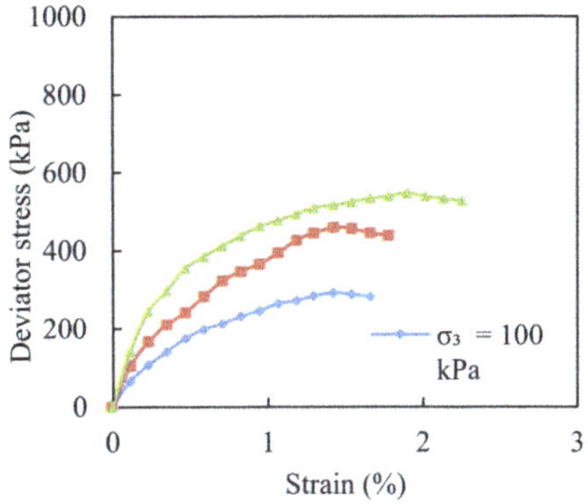


(**l**)

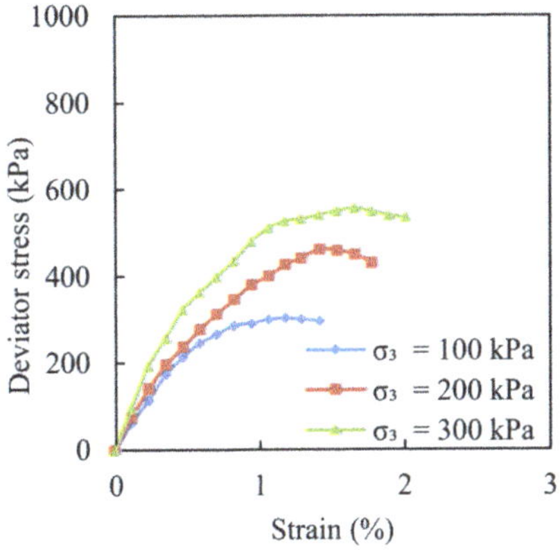


(**m**)

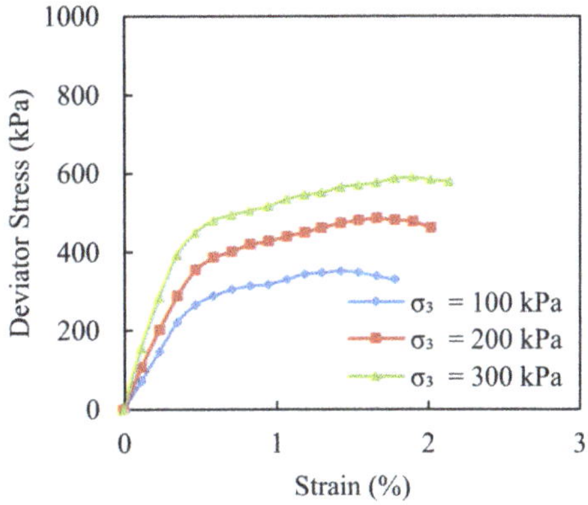


(**n**)

**Figure 9.** Deviator stress versus axial strain curves of fly ash, bentonite, and fly ash–bentonite in different ratios with layers. (**a**) Fly ash. (**b**) Bentonite. (**c**) F:B = 1:1, N = 1, F = 2L, B = 1L. (**d**) F:B = 1:1, N = 2, F = 2L, B = 1L. (**e**) F:B = 1:1, N = 3, F = 2L, B = 2L. (**f**) F:B = 2:1, N = 1, F = 1L, B = 1L. (**g**) F:B = 2:1, N = 2, F = 2L, B = 1L. (**h**) F:B = 2:1, N = 3, F = 2L, B = 2L. (**i**) F:B = 3:1, N = 1, F = 1L, B = 1L. (**j**) F:B = 3:1, N = 2, F = 2L, B = 1L. (**k**) F:B = 3:1, N = 3, F = 2L, B = 2L. (**l**) F:B = 4:1, N = 1, F = 1L, B = 1L. (**m**) F:B = 4:1, N = 2, F = 2L, B = 1L. (**n**) F:B = 4:1, N = 3, F = 2L, B = 2L.

For F:B = 1:1, the increments in deviator stress for N = 1, 2, and 3 with respect to fly ash are 50, 68.22, and 113.12%, respectively, on confining pressure 100 kPa; 38.42, 57.11, and 52.89% on confining pressure 200 kPa; and 21.62, 36.36, and 38.38% on confining pressure 300 kPa. For F:B = 2:1, the increments in deviator stress for N = 1, 2, and 3 with respect to fly ash are 40.25, 84.32, and 96.61%, respectively, on confining pressure 100 kPa; 30.79, 53.68, and 72.89% on confining pressure 200 kPa; and 12.12, 24.04, and 27.27% on confining pressure 300 kPa. For F:B = 3:1, the increments in deviator stress for N = 1, 2, and 3 with respect to fly ash are 32.20, 48.73, and 68.64%, respectively, on confining pressure 100 kPa; 17.37, 25.79, and 34.74% on confining pressure 200 kPa; and 10.10, 13.13, and 27.27% on confining pressure 300 kPa. For F:B = 4:1, the increments in deviator stress for N = 1, 2, and 3 with respect to fly ash are 24.58, 29.24, and 49.15%, respectively, on confining pressure 100 kPa; 11.32, 21.84, and 28.16% on confining pressure 200 kPa; and 10.71, 12.73, and 19.60% on confining pressure 300 kPa.

### 4.2. Shear Strength Parameters

The total axial stress results of the series of consolidated undrained triaxial tests at increasing confining pressure were plotted on a Mohr's circle diagram to evaluate effective shear strength parameters (c and ϕ). A straight line drawn as a tangent to the circles, with the equation $\tau = c + \sigma \tan \phi$, is used to simulate the condition of the sample's failure. The angle of shearing resistance (ϕ) is the angle between the tangent and a line parallel to the shear stress, and the value of cohesion (c) is read from the shear stress axis, where it is cut by the tangent to the Mohr's circles. The values of c and ϕ are summarized in Table 4.

**Table 4.** Shear strength parameters of fly ash bentonite in different ratios with layers.

| F:B | Cohesion (c), kPa | | | Angle of Internal Friction (ϕ), ° | | |
|---|---|---|---|---|---|---|
| | N = 1 | N = 2 | N = 3 | N = 1 | N = 2 | N = 3 |
| 1:1 | 76.8 | 83.4 | 107 | 22.5 | 22.2 | 20 |
| 2:1 | 75 | 76.1 | 110 | 21 | 20 | 17 |
| 3:1 | 66.6 | 86 | 105 | 21.5 | 20 | 21.5 |
| 4:1 | 55 | 59 | 78.2 | 22.8 | 22.8 | 22 |

As shown in Figure 10, for a particular ratio of fly ash to bentonite, the value of cohesion increased with the increase in the number of interfaces. When the number of interfaces is equal to one, it was noticed that the value of cohesion continuously decreased as the ratio of fly ash to bentonite was increased. At number of interfaces equal to 2, the value of cohesion decreased when the ratio F:B changed from 1:1 to 2:1. However, when the ratio of F:B became 3:1, the value of cohesion increased before reducing again at F:B = 4:1. At N = 3, it was found that, with increasing ratio of F:B, an increase in value of c was recorded until the ratio was 2:1. At a 3:1 ratio of F:B, the value of cohesion decreased and then fell drastically at F:B ratio = 4:1.

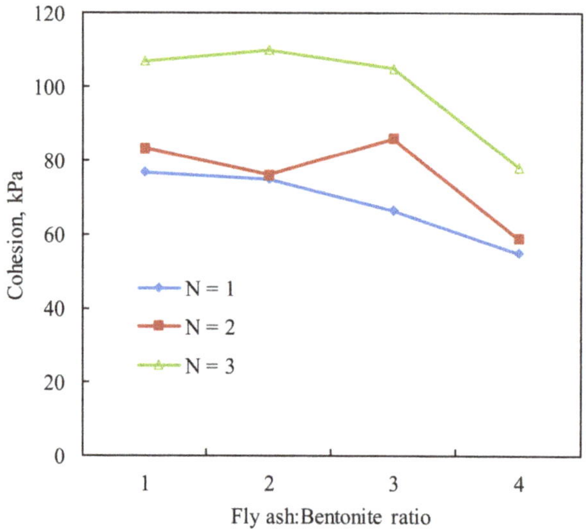


**Figure 10.** Effective cohesion values against fly ash–bentonite ratios.

As shown in Figure 11, it was also found that the value of the angle of shear resistance is largest for fly ash. For the same ratio of fly ash to bentonite, it decreases gradually as the number of interfaces was increased. However, the values of ϕ are quite close to each other, i.e., not much drastic change is observed in their response to changing ratios and varying interfaces.

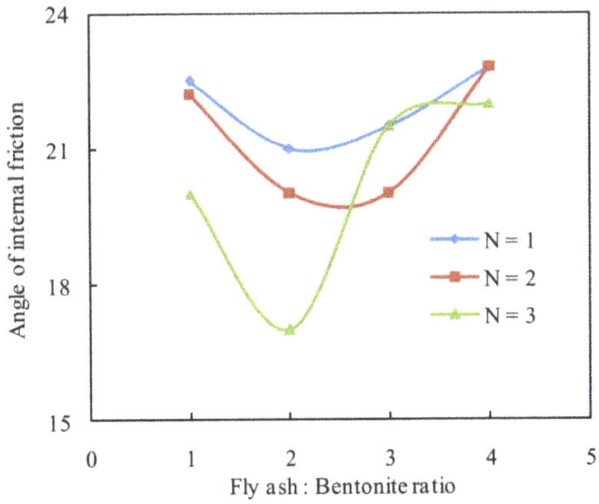


**Figure 11.** Effective angle of shear resistance values against fly ash–bentonite ratios.

## 5. Conclusions

The following conclusions are drawn based on the results of the experimental investigation:

1. Fly ash is a nearly cohesionless substance that has a binding effect at the fly ash–bentonite layered system interfaces.
2. For F:B = 1:1, 1:2, 1:3, and 1:4, the value of cohesion has been varied in the ranges of 76.8–107 kPa, 75–110 kPa, 66.6–105 kPa, and 55–78.2 kPa, respectively, for different layer arrangements. As the ratio of fly ash to bentonite is increased, the value of cohesion is decreased, while at the same ratio, the value of cohesion is increased by increasing the number of interfaces.
3. For F:B = 1:1, 1:2, 1:3, and 1:4, the value of angle of shearing resistance has been varied in the range of 17–22.8° for different layer arrangements. The value of the angle of shearing resistance has been found to decreases with the increase in the number of interfaces for a specified ratio of fly ash to bentonite.
4. The stress–strain behaviour of the fly ash–bentonite layered system is initially elastic at low stress levels and becomes non-linear at higher stress levels.
5. As the number of interfaces and confining pressures grow, the shear strength and net safe bearing capacity of the fly ash–bentonite layered system increases.
6. Based on laboratory model tests, it has been suggested that the embankment may provide greater stability if it is constructed for a fly ash to bentonite ratio of 3:1 (75% of fly ash and 25% of bentonite in layers) by keeping the number of interfaces as N = 3.
7. It has been found that a 3:1 ratio of fly ash to bentonite with three interfaces can be successfully used for filling low-lying areas and in many other engineering constructions, while also providing a means of using the material without negatively impacting the environment. However, the authors recommend that large-scale studies be conducted in the future.

**Author Contributions:** M.H.: drafting—data collection and preparation of the manuscript, writing-review and editing, M.A.K. (Mehboob Anwer Khan): drafting—preparation of the manuscript, revision, and correction; A.H.A.: composing—reviewing and modifying; A.A.A.: composing—reviewing and modifying; S.A.: composing—reviewing and modifying and M.A.K. (Mohammad Amir Khan): composing—reviewing and modifying., T.B. reviewing and modifying, J.P.: reviewing and modifying. All authors have read and agreed to the published version of the manuscript.

**Funding:** The authors would like to acknowledge the support provided by Researchers Supporting Project Number (RSP2022R473), King Saud University, Riyadh, Saudi Arabia.

**Institutional Review Board Statement:** Not applicable.

**Informed Consent Statement:** Not applicable.

**Data Availability Statement:** Not applicable.

**Conflicts of Interest:** The authors declare no conflict of interest.

## References


1. Bhatt, A.; Priyadarshinia, S.; Mohanakrishnana, A.C.; Abria, A.; Sattlera, M.; Techapaphawitc, S. Physical, Chemical, and Geotechnical Properties of Coal Fly Ash: A Global Review. *Case Stud. Constr. Mater.* **2019**, *11*, e00486. [CrossRef]
2. Rajak, T.K.; Yadu, L.; Chouksey, S.K. Effect of fly ash on geotechnical properties and stability of coal mine overburden dump: An overview. *Appl. Sci.* **2020**, *2*, 973. [CrossRef]
3. Lauzurique, Y.; Montalvo, S.; Salazar, R.; García, V.; Huiliñir, C. Fly ash from coal combustion as improver of anaerobic digestion: A review. *J. Environ. Chem. Eng.* **2021**, *9*, 106422. [CrossRef]
4. Mathur, N. *Report on Fly Ash Generation at Coal/Lignite Based Thermal Power Stations and its Utilization in the Country for the Year*; Ministry of Power, Government of India: New Delhi, India, 2020; p. 110016.
5. Vittal, G. Bulk utilization of fly ash in construction of road embankments in India. *Coal Ash India* **2001**, *2*, 6–7.
6. Indraratna, B.; Nutalaya, P.; Kuganenthira, N. Stabilization of a dispersive soil by blending with fly ash. *Q. J. Eng. Geol. Hydrogeol.* **1991**, *24*, 275–290. [CrossRef]
7. Sivapullaiah, P.V.; Prashant, J.P.; Sridharan, A. Effect of fly ash on the index properties of BC soil. *Soils Found.* **1996**, *36*, 97–103. [CrossRef]
8. Kaniraj, S.R.; Havanagi, V.G. Behavior of cement-stabilized fiber-reinforced fly ash–soil mixtures. *J. Geotech. Geoenviron. Eng. ASCE* **2001**, *127*, 574–584. [CrossRef]
9. Ghosh, A.; Subbarao, C. Strength Characteristics of Class F Fly ash Modified with Lime and Gypsum. *J. Geotech. Geoenvironmental. Eng.* **2007**, *133*, 7. [CrossRef]
10. Senol, A. Effect of fly ash and polypropylene fibres content on the soft soils. *Bull. Eng. Geol. Environ.* **2012**, *71*, 379–387. [CrossRef]
11. Sivapullaiah, P.V.; Sridharan, A.; Stalin, V.K. Hydraulic conductivity of bentonite-sand mixtures. *Can. Geotechn. J.* **2000**, *37*, 406–413. [CrossRef]
12. Pandian, N.S.; Nagaraj, T.S.; Raju, P.N. Permeability and compressibility behavior of bentonite-sand/soil mixes. *Geotechnical. Test. J.* **1995**, *18*, 86–93.
13. Kashir, M.; Yanful, E.K. Hydraulic conductivity of bentonite permeated with acid mine drainage. *Can. Geotech. J.* **2001**, *38*, 1034–1048. [CrossRef]
14. Sivapullaiah, P.V.; Lakshmikantha, H. Properties of Fly ash as Hydraulic Barrier. *Soil Sediment. Contam.* **2004**, *13*, 489–504. [CrossRef]
15. Younus, M.M.; Sreedeep, S. Evaluation of Bentonite-Fly ash Mix for Its Application in Landfill Liners. *J. Test. Eval.* **2012**, *40*, 40–49. [CrossRef]
16. Mollamahmutoglu, M.; Yilmaz, Y. Potential Use of Fly ash and Bentonite Mixture as Liner or Cover at Waste Disposal Areas. *Environ. Geol.* **2001**, *40*, 1316–1324. [CrossRef]
17. Garg, A.; Reddy, N.G.; Huang, H. Modelling contaminant transport in fly ash–bentonite composite landfill liner: Mechanism of different types of ions. *Sci. Rep.* **2020**, *10*, 11330. [CrossRef] [PubMed]
18. Wasil, M. Effect of Bentonite Addition on the Properties of Fly Ash as a Material for Landfill Sealing Layers. *Appl. Sci.* **2020**, *10*, 1488. [CrossRef]
19. *ASTM C 618*; Standard Specification for Coal Fly Ash and Raw or Calcined Natural Pozzolan for Use in Concrete. American Society for Testing and Materials: Philadelphia, PA, USA, 2019.
20. Zabielska-Adamska, K. Characteristics of Compacted Fly Ash as a Transitional Soil. *Materials* **2020**, *13*, 1387. [CrossRef] [PubMed]
21. *IS: 2720-Part 3*; Methods of Test for Soils: Determination of Specific Gravity, Section 1 Fine Grained Soils. Indian Standards Institution: New Delhi, India, 1980.
22. *IS: 2720-Part 4*; Methods of Test for Soils: Grain Size Analysis. Indian Standards Institution: New Delhi, India, 1985.
23. *IS: 2720-Part 5*; Code of Practice for Determination of Liquid Limit and Plastic Limit. Indian Standards Institution: New Delhi, India, 1985.
24. *IS: 2720-Part 10*; Methods of Test for Soils: Determination of Unconfined Compressive Strength. Indian Standards Institution: New Delhi, India, 1991.

*applied sciences*

MDPI

*Article*

# Low Compressibility at the Transition Zone of Railway Tracks Reinforced with Cement-Treated Gravel and a Geogrid under Construction

**Seongyong Park [1], Dae Sang Kim [2,*], Ungjin Kim [2] and Sangseom Jeong [3]**

[1] Hanwha Engineering and Construction, Seoul 04541, Korea
[2] Advanced Railroad Civil Engineering Division, Korea Railroad Research Institute, #176 Railroad Museum St., Uiwang 16105, Korea
[3] School of Civil and Environmental Engineering, Yonsei University, Seoul 03722, Korea
* Correspondence: kds@krri.re.kr; Tel.: +82-31-460-5305; Fax: +82-31-460-5408

**Abstract:** In the transition zone of railway tracks, track irregularities occur frequently due to differential settlement, which arises from the difference between the vertical supporting stiffness of the abutment and the backfill. This is disadvantageous because it increases the maintenance requirements and deteriorates the ride quality. To address this challenge, this study proposes a strategy involving the application of cement-treated gravel reinforced with geogrids and rigid facing walls. The reinforced subgrade for railways (RSR), which can reduce residual settlement through the initial construction of the backfill reinforced with geogrids and the subsequent development of the rigid facing wall, was constructed at the transition zone with cement-treated gravel as the backfill material. The long-term behaviors during and after construction on the RSR for a period of 16 months were evaluated by analyzing the surface and ground settlements, horizontal earth pressure, and geogrid strain. The minor net settlement of the reinforced backfill converges at the early stage of subgrade construction, and the horizontal earth pressure was approximately reduced to the level of 54–63% of the Rankine active earth pressure.

**Keywords:** transition zone; differential settlement; horizontal earth pressure; cement-treated gravel; geogrid

**Citation:** Park, S.; Kim, D.S.; Kim, U.; Jeong, S. Low Compressibility at the Transition Zone of Railway Tracks Reinforced with Cement-Treated Gravel and a Geogrid under Construction. *Appl. Sci.* **2022**, *12*, 8861. https://doi.org/10.3390/app12178861

Academic Editor: José Manuel Moreno-Maroto

Received: 28 July 2022
Accepted: 26 August 2022
Published: 3 September 2022



## 1. Introduction

In transition zones, which are the weakest zones in the railway roadbed, track irregularities frequently occur because of differential settlements, which are induced by the difference in the vertical stiffness of the concrete abutment and the earth transition zone [1]. To mitigate differential settlement, several design codes suggest various measures, such as the implementation of approach blocks, approach slabs, and reinforced materials nd structures.

The approach block is used to mitigate differential settlement; material with medium stiffness is placed between the abutment concrete structure and the subgrade on the back of the abutment, thereby smoothing the change in stiffness of the subgrade. Studies related to the approach block were conducted which focused on several materials [2–4] and ground-reinforcement strategies [5,6]. Cement-treated gravel is a material typically used for the construction of approach blocks [7–9]. When graded gravel is mixed with a certain amount of cement (3–5%) and compacted, it exhibits a stiffness greater than that of graded gravel (without cement) [10,11]; thus, the difference in stiffness between backfill and concrete structures can be mitigated.

The approach slab is a concrete structure in the shape of a slab and is supported at one end by the bridge abutment and at the other end by the embankment soil [12]. By connecting the abutment and embankment with a slab, this structure is used to reduce the difference in stiffness of the subgrade. A nonlinear finite element analysis [13], a field survey [14], and dynamic analysis using a three-dimensional finite element model [15] have been performed to investigate the approach slab. These studies reported on the effect of reducing differential settlement in transition zones.

In another approach to improve the performance of the transition zone, geosynthetics-based reinforcement methods have been explored. Watanabe et al. performed shaking table tests of an abutment reinforced with a geogrid and discovered that the structural stability may be secured by utilizing the tension of the geogrid [16]. Moreover, Feng et al. investigated the performance of a geogrid-reinforced and pile-supported bridge approach via field monitoring [17]. In addition, the performance and stability of geosynthetic-reinforced soil bridges have been evaluated via numerical analyses [18] and their life cycle assessment [19]. However, only a few studies have focused on the installation of geosynthetic-reinforced retaining walls at the transition zone of the railway abutment.

The above studies have something in common that they make the change of stiffness smooth in the transition zone. Because the transition zone composed of soil material has a lower stiffness than a bridge with concrete, structural and material methods have been employed to increase the stiffness of the transition zone.

In this study, the low compressibility and applicability of the geosynthetic-reinforced retaining wall using cement-treated gravel as a backfill material in the transition zone was evaluated. For this study, a reinforced subgrade for railway (RSR) was constructed in an abutment transition zone at the Osong railway test line in Korea, and long-term measurement was performed during and after the construction.

## 2. Design and Construction of RSR Abutment and Transition Zone

By initially constructing the backfill which is reinforced with a geogrid, followed by a rigid facing wall, the resultant RSR can minimize the residual settlement, as illustrated in Figure 1 [20,21]. Through the integration of the geogrid and the rigid facing wall, the bearing capacity of the subgrade can be increased, and horizontal earth pressure can be reduced [22,23]. Moreover, unlike segmental retaining walls reinforced with geosynthetics, the RSR eliminates any occurrences of local failure; thus, this subgrade is suitable for supporting large and frequent railway loads.

subsectionDesign of RSR Abutment and Transition Zone

By issuing a field change request, the conventional reinforced concrete (RC) retaining wall was changed to an RSR. In the previous design, because of the short distance between the retaining wall and the adjacent road, the foundation of the retaining wall intruded onto the adjacent road. In addition, owing to the proximity of the site to a high-speed railroad, there were concerns that large pieces of equipment such as a crane may fall toward the high-speed railroad; thus, the construction was conducted at night during a 3 h period, from 01:00 to 03:00 AM, to obey Korean regulations. Nevertheless, the RSR could be constructed in a narrow working space without the need for any large equipment.

The site investigation results are listed in Table 1. The landfill and sedimentary layers comprised clayey sand and sandy clay, respectively. Although the ground contains clayey soil, it was in a stiff state with a high $n$ value ($n$ = 8–13), but the layer thickness was not thick.

The RSR was constructed at a height of 2.6–7.4 m and, including abutment transition zones, the total length was 156.0 m in STA.1k662.90–1k820.00 of the Osong railway test line in Korea. In this study, the STA.1k665 cross section in the abutment transition zone was selected for stability analysis, and its behaviors were subsequently evaluated.

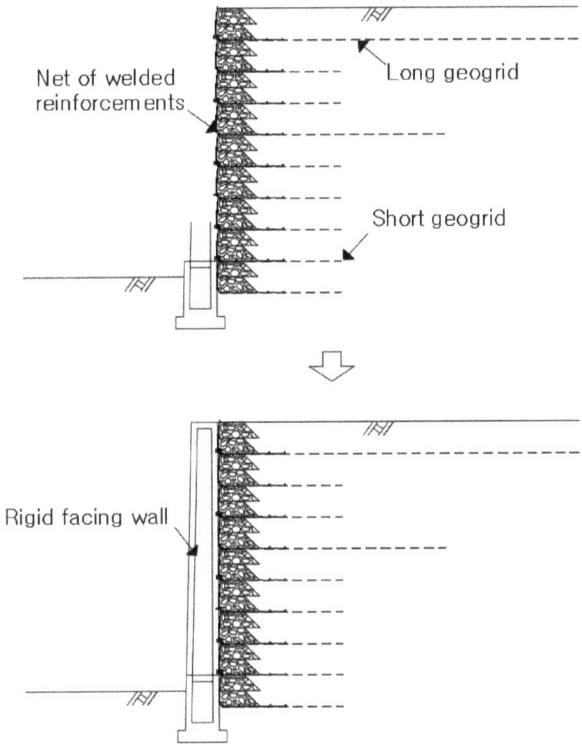


**Figure 1.** Construction of reinforced subgrade for railways (RSR) [23].

**Table 1.** Site investigation results.

| Depth (m) | Thickness (m) | Layer Descriptions | *n* Values [1] (Number of Blows/cm) |
|---|---|---|---|
| 0–1.4 | 1.4 | Landfill layer<br>Medium stiff clayey sand | 13/30 |
| 1.4–3.7 | 2.3 | Sedimentary layer<br>Stiff sandy clay | 8/30–12/30 |
| 3.7–12.0 | 8.3 | Weathered soil layer<br>Soft to very stiff silty sand | 16/30–50/9 |
| 12.0–15.0 | 3 | Weathered rock layer<br>Very stiff weathered<br>Granite | 50/10–50/7 |

[1] The number of hammer blows required to penetrate the sampler by 0.3 m in the standard penetration test.

As depicted in Figure 2, the bench cut size for the foundation is reduced, and instead of installing steel piles with a length of 11 m, the ground stability was improved by installing cast-in-place (CIP) concrete piles with lengths of 1.5–2.0 m. The length of the short geogrids is 2.60 m (35% of the wall height), and the length of the long geogrids is selected to satisfy the criteria for safety against circular failure, sliding, and overturning. These geogrids provide a tensile strength of 100 kN/m and a design tensile strength of 50 kN/m. Kim (2017) verified the stability of this RSR section during the dry and wet seasons, as well as during earthquakes using the RSR design program developed by the Korea Railroad Research Institute, as summarized in Table 2 [24].

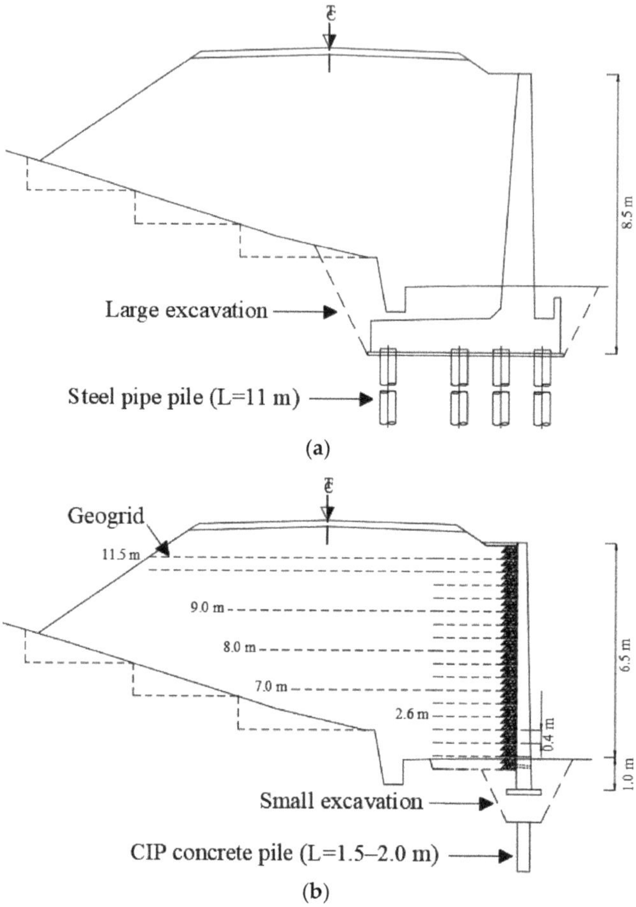


**Figure 2.** Design of RC retaining wall and RSR: (**a**) originally designed RC retaining wall; (**b**) RSR design.

**Table 2.** Results of stability analysis [24].

| Factors of Safety | Loading Conditions | | |
|---|---|---|---|
| | **Dry** | **Wet** | **Earthquake** |
| Circular Failure | 2.011 > 1.500 | 1.802 > 1.300 | 1.681 > 1.100 |
| Sliding | 3.546 > 2.000 | - | 3.793 > 1.500 |
| Overturning | 3.712 > 1.500 | - | 2.715 > 1.100 |

### *2.1. Construction of RSR Abutment Transition Zone*

The construction details of the RSR are as follows: (1) first, the cast-in-place concrete pile was installed. (2) Subsequently, the bench cut was applied, and the 3% cement-treated gravel was spread out and compacted. (3) Thereafter, a 1 m-high small-strip foundation was installed. (4) Geogrids and nets of welded rebars were installed, (5) the cement-treated gravel was laid, (6) followed by compaction, (7) and then the geogrid was rolled up. (8) Steps (4) to (7) were repeated to obtain the 17th layer. (9) Subsequently, the backfill was stabilized, (10) the rebars and connectors were installed, (11) and the forms and placement of the concrete for the rigid wall were carried out. Finally, (12) the RSR was completed.

The construction stages are illustrated in Figure 3, and the time schedules and major milestones in the RSRs construction are listed in Table 3. The backfill materials for the RSR that were applied to the sections in accordance with the Korean Railway Design Standard for Roadbed are depicted in Figure 4 [25]. The properties and compaction conditions for backfill materials are summarized in Table 4. The 3% cement-treated gravel was applied for the ⓑ section, and gravel with a maximum diameter of 63 mm was applied for the ⓒ section.

**Figure 3.** RSR construction process. (**1**) Installation of cast-in-place concrete pile; (**2**) bench cut, spreading, and compaction processes of 3% cement-treated gravel; (**3**) installation of small strip foundation; (**4**) installation of geogrids and nets of welded rebars; (**5**) laying of 3% cement-treated gravel; (**6**) compaction; (**7**) rolling up the geogrid; (**8**) repetition of steps (**4**) to (**7**) to obtain the 17th layer; (**9**) stabilization of the embankment; (**10**) installation of rebars and connectors; (**11**) installation of forms and placing of concrete for rigid wall; (**12**) curing and completion of RSR.

**Table 3.** Construction working schedules.

| No. | Measurement Date | Works |
|---|---|---|
| ① | | Installation of CIP concrete pile |
| ② | Day 0 | Work starts on the reinforced backfill |
| ③ | Day 23 | Completion of reinforced backfill |
| | | Stabilization period |
| ④ | Day 185 | Work starts on the rigid facing wall |
| ⑤ | Day 242 | Completion of the rigid facing wall |
| ⑥ | Day 308 | Completion of the reinforced roadbed |
| ⑦ | Day 428 | Completion of the track |

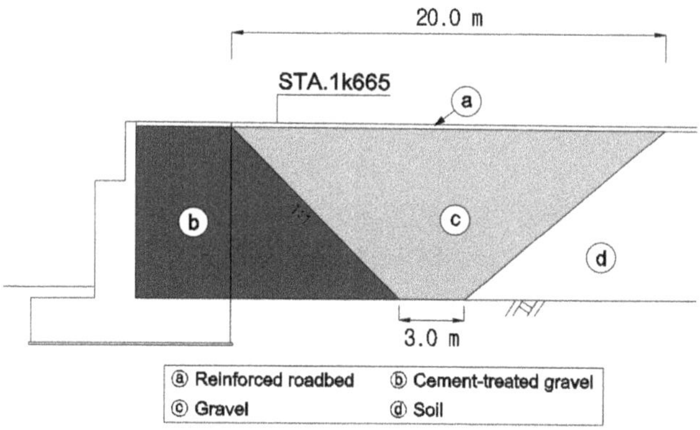


**Figure 4.** Backfill materials of transition zone.

**Table 4.** Properties and compaction conditions for backfill materials.

| Zones | Materials | Properties and Compaction Conditions |
|---|---|---|
| ⓐ | Reinforced roadbed (Crushed gravel) | $D_{max}$ [1] = 31.5 mm, over 100% OPM [2] |
| ⓑ | Cement-treated gravel | $D_{max}$ = 63 mm (Cement 3%), $E_{v2}$ [3] $\geqq$ 120 Mpa, $E_{v2}/E_{v1}$ [3] < 2.2 |
| ⓒ | Gravel | $D_{max}$ = 63 mm, $E_{v2} \geqq 80$ Mpa, $E_{v2}/E_{v1} < 2.2$ |
| ⓓ | Soil | (Upper roadbed) $E_{v2} \geqq 80$ Mpa, $E_{v2}/E_{v1} < 2.3$, over 95% $\rho_{d,max}$ [4] (Lower roadbed) $E_{v2} \geqq 60$ Mpa, $E_{v2}/E_{v1} < 2.7$, over 90% $\rho_{d,max}$ |

[1] $D_{max}$: maximum particle diameter; [2] OPM: optimum moisture content; [3] $E_{v1}$ and $E_{v2}$: deformation moduli; [4] $\rho_{d,max}$: maximum dry density.

## 3. Installation of Sensors and Measuring Devices

To analyze the behaviors of the RSR which was applied to the abutment transition zone, 19 geogrid strain gauges, 3 horizontal earth pressure gauges, and settlement rods for the ground and the backfill surface were installed and measured during the construction process. The long-term measurements were conducted over a period of 16 months, including the 2.6 month-long construction period, and the RSR behaviors were identified. A schematic view of the measuring sensors is illustrated in Figure 5.

To distinguish the settlement of the ground and embankment, the settlement rods were placed on the ground and backfill surface, respectively. The settlement rod for the ground comprised a 1 m × 1 m square plate and a rod with a 15 mm diameter that could be extended as the backfill height was increased by 1 m. The surface settlement rod, featuring a circular steel plate with a 300 mm diameter and an attached 400 mm high rod, was installed at a distance of 0.5 m from the edge of the backfill. Considering the field conditions, the settlement measurements were performed after the completion of the embankment construction (from day 23 to day 274). The horizontal earth pressure gauges were installed at the net of the welded rebars at the 2nd, 10th, and 16th layers to observe the variation in the horizontal earth pressure with respect to the height. The strain gauges were installed on the 2nd and 5th layers and the 10th, 13th, and 16th layers for the short and long geogrids, respectively. The strain gauges on the short geogrids were installed at a distance of 0.5, 1.5, and 2.5 m from the exterior of the reinforced backfill. The strain gauges on the long geogrids were installed at a distance of 0.5, 1.5, and 2.5 m for the 10th layer, 0.5, 1.5, 3.5, and 5.5 m for the 13th layer, and 0.5, 1.5, 3.5, 5.5, 7.5, and 9.5 m for the 16th layer. The installed sensors used for the measurements are shown in Figure 6.

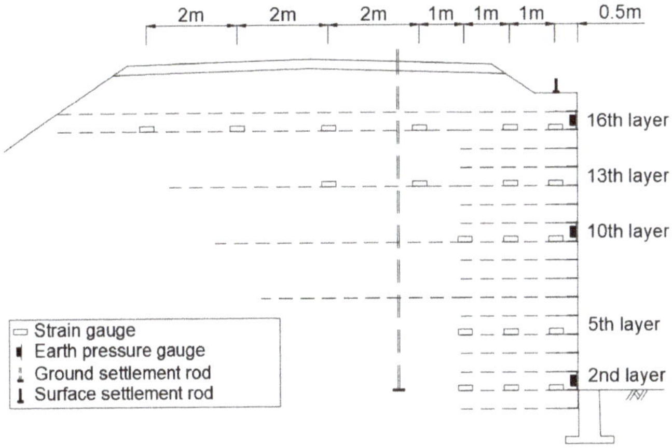


**Figure 5.** Location of installed sensors.

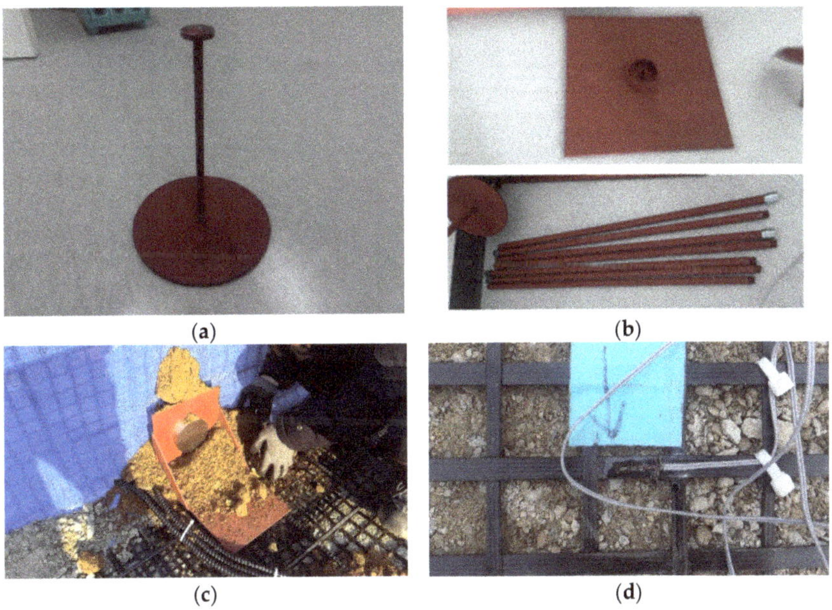


**Figure 6.** Installed sensors: (**a**) surface settlement rod; (**b**) ground settlement rod; (**c**) horizontal earth pressure meter; (**d**) strain gauge.

## 4. Results and Analysis

### *4.1. Settlement*

The settlements of the ground and surface of the embankment over time after the construction of the reinforced backfill are plotted in Figure 7, and the circled numbers indicate the work schedules stated in Table 3. Settlement measurements were performed for 341 days and for 252 days for the ground and surface, respectively, after the reinforced backfill was completed. The surface settlement changes considerably immediately after the completion of the reinforced backfill construction and converges over time. In particular, the maximum surface settlement is 23.1 mm less than the allowable residual settlement stated in the Korean Railway Design Standard for roadbed construction (30 mm).

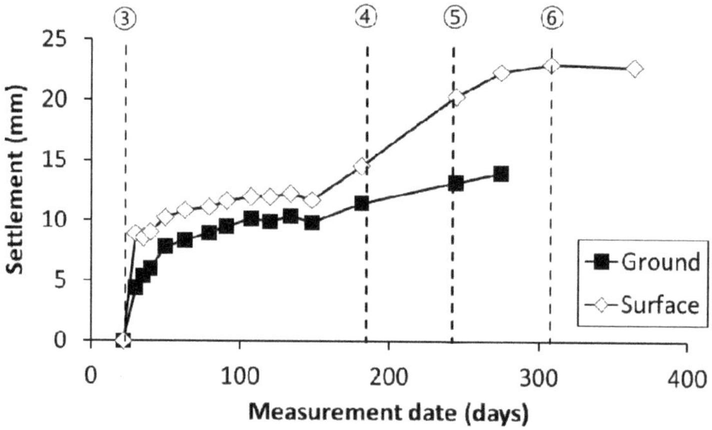


**Figure 7.** Settlements of RSR over time.

The difference in settlements of the ground and surface of the backfill was considered as the net settlement of the backfill. The variation of net embankment settlement and rainfall over time is plotted in Figure 8. Immediately after the completion of the reinforced backfill construction, a compression settlement of 4.5 mm occurred and exhibited a stable condition (1.82–3.09 mm) prior to the wall's construction. The rate of total net settlement of the reinforced backfill was 0.12%. Moreover, additional settlements of 4.06 mm and 1.19 mm occurred owing to the construction of the rigid facing wall and heavy rainfall, respectively. Because the reinforced backfill supports the wall in the RSR structure, additional settlement occurs under the weight of the wall. During heavy rainfall, additional settlement occurred as the soil particles were relocated by the water penetrating through the reinforced backfill [26,27]. Considering the reinforced backfill height of 6.8 m, the rates of each additional settlement were 0.06% and 0.02%, respectively. The low compressibility of the RSR when using cement-treated gravel as a backfill was confirmed.

### *4.2. Horizontal Earth Pressure*

The variation in the horizontal earth pressure as a function of the height over time is represented in Figure 9. During the construction of the reinforced backfill, the pressure increases with the weight of the soil, which is the largest at the second layer. Even after the completion of the reinforced backfill, the horizontal earth pressure increases moderately without any increase in the overburden load. The increase in earth pressure generated during the backfill stabilization process was similar to that of the RSR using sand soil as a backfill [23]; however, this value was considerably smaller for the backfill using cement-treated gravel.

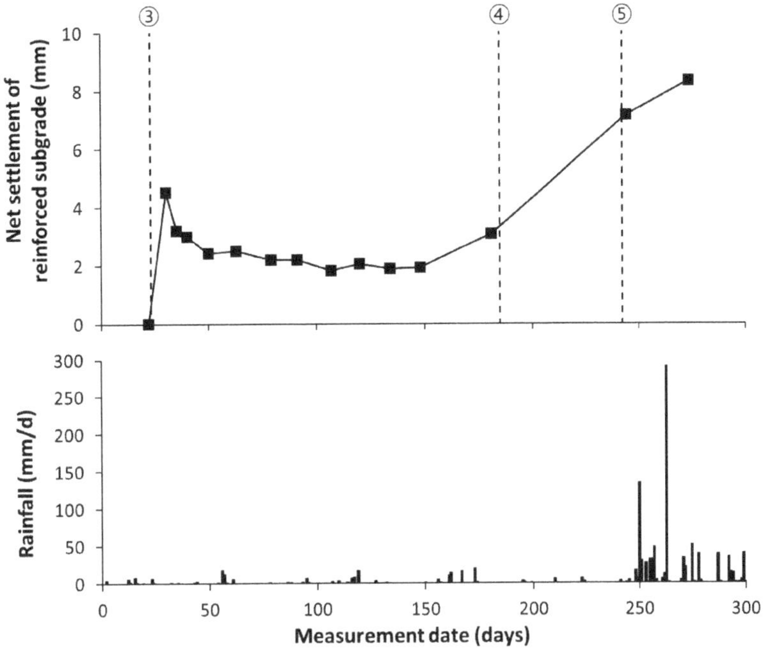


**Figure 8.** Net settlement of reinforced backfill and rainfall over time.

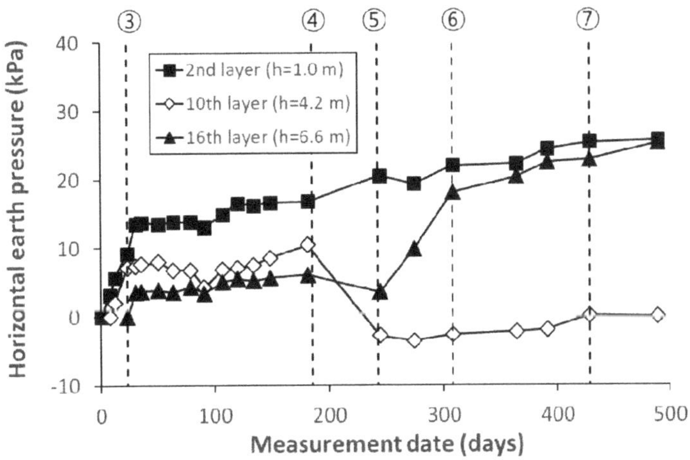


**Figure 9.** Variation of the horizontal earth pressure over time.

Upon the completion of the rigid facing wall, the horizontal earth pressure tends to increase at the 2nd layer, whereas the horizontal earth pressure tends to decrease at the 10th and 16th layers. Notably, the RSR supports the concrete pouring pressure by connecting the net of welded rebars to the formwork. Considering this fact, the decrease in the horizontal earth pressure presumably occurred because the concrete pouring pressure applied to the net of welded rebars resulted in the horizontal displacement of the reinforced backfill.

The horizontal earth pressure at the 16th layer increased immediately after the completion of the rigid facing wall. Presumably, these effects, owing to rainfall and the overburden pressure exerted by the reinforced roadbed and track on the 16th layer, were prominent because of this layer's proximity to the surface. In the interim between the construction of the rigid facing wall and the reinforced roadbed, the horizontal earth pressure increases to 5.1 kPa in the 2nd layer, 2.8 kPa in the 10th layer, and 21.7 kPa in the 16th layer, which was 4.2–7.8 times larger than the pressures of other layers.

For the period between the completion of the reinforced backfill and the initiation of the rigid face wall, the variation in the distribution of the horizontal earth pressure with respect to the height is plotted in Figure 10. The horizontal earth pressure in the second layer (1.0 m) increases from 41% to 54% of the Rankine's active earth pressure considering the friction angle of 35° used in an existing design. Moreover, it increases from 45% to 63% and from 87% to 148% of the Rankine's active earth pressure in the 10th layer (4.2 m) and 16th layer (6.6 m), respectively. Except for the 16th layer, which is heavily influenced by the external environment, such as construction works, the horizontal earth pressure in all layers is similar to that of granular soil with a friction angle of 45–50°. This entails an increase in the shear strength of the backfill in the RSR using cement-treated gravel. Notably, the higher the shear strength, the higher the stiffness encountered in the soil will be [28,29], and, accordingly, the low compressibility of the RSR can be confirmed indirectly.

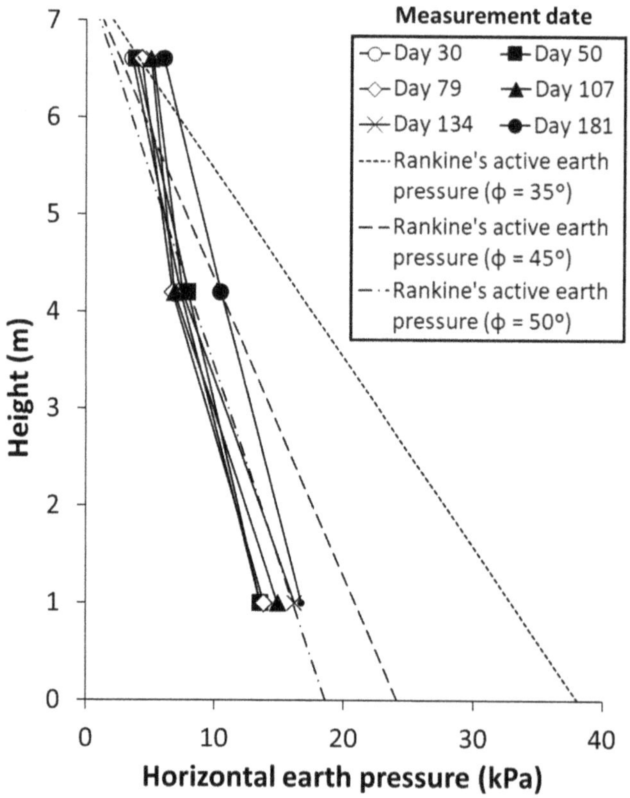


**Figure 10.** Variation in horizontal earth pressure over height.

## 4.3. Geogrid Strain

The variation in the geogrid strain measured at the layers over time is illustrated in Figure 11. The maximum increase in the geogrid strain of 0.811% during and after construction was measured at the fifth layer. This value is equivalent to 16% of the geogrids' design tensile strain (5%). In addition, the maximum geogrid strain of 0.784% was measured at the fifth layer (2.2 m) during the construction of the reinforced backfill.

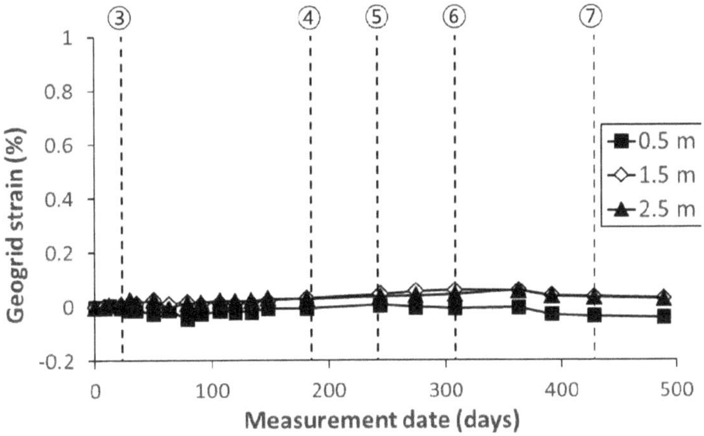


(a)

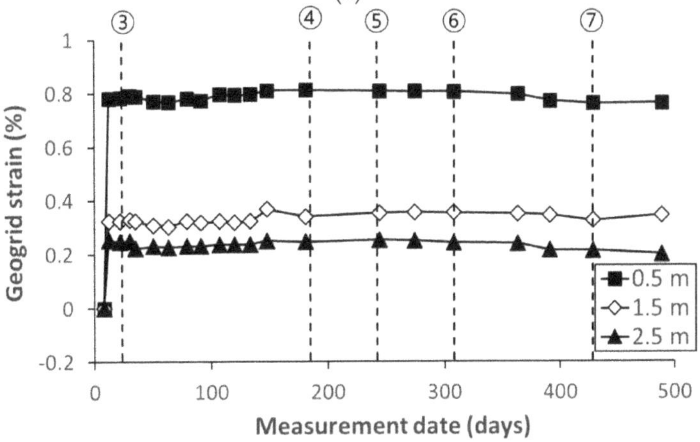


(b)

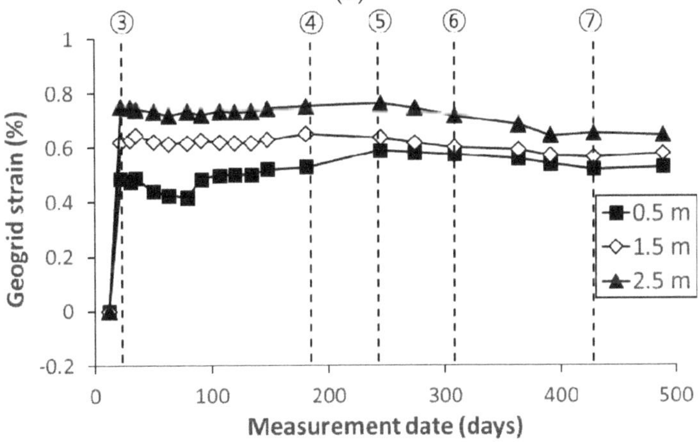


(c)

**Figure 11.** *Cont.*

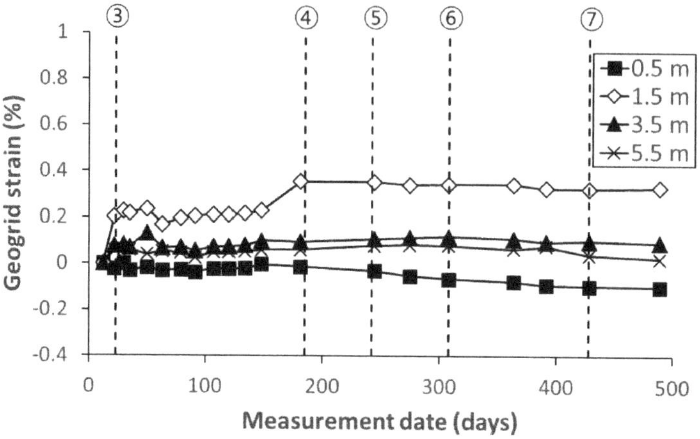


(d)

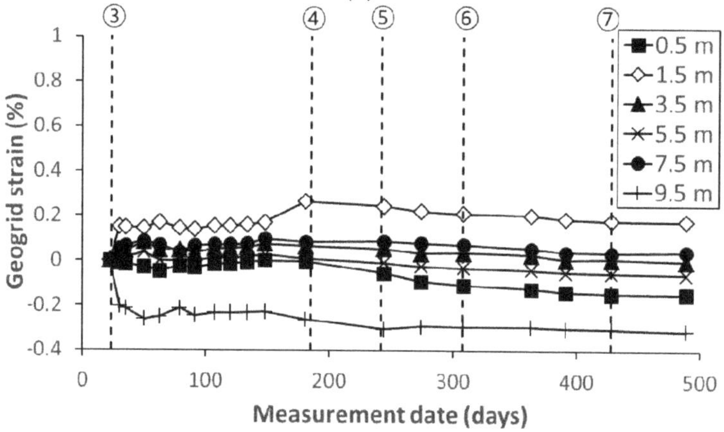


(e)

**Figure 11.** Variation of geogrid strain over time: (**a**) 2nd layer (short geogrid); (**b**) 5th layer (short geogrid); (**c**) 10th layer (long geogrid); (**d**) 13th layer (long geogrid); (**e**) 16th layer (long geogrid).

Except for the second layer, the maximum strain increases were measured for all layers during the construction of the reinforced backfill. The results demonstrate that the increase in the overburden load with the backfill height significantly influences the deformation of the geogrids. Considering the limited soil deformations in the second layer owing to the confining effects of the foundation, the geogrid deformation at the second layer was assumed to be minimal. Following the completion of the backfill construction process, the deformation of the geogrid resulting from the wall load increment, reinforced roadbed, and track construction is approximately negligible.

## 5. Conclusions

In this study, the RSRs performance at the abutment transition zone of the railway was evaluated. Notably, this subgrade was reinforced with cement-treated gravel and a geogrid. The conclusions from the analysis results of the long-term measurement of settlement and horizontal earth pressure are as follows:

(1) During the stabilization period, the net settlement of the reinforced backfill converged at the early stage of subgrade construction, and the subsequent increase in settlement was dominated by that occurring in the ground. The rate of total net settlement of the

reinforced backfill was 0.12%. In the RSR, because the reinforced backfill supported the rigid facing wall, additional settlement occurred as a result of the wall load and rearrangement of soil particles owing to rainfall. The additional settlement caused by the wall load and rainfall was 0.06% and 0.02%, respectively. The results confirm the low compressibility of the RSR reinforced with cement-treated gravel and geogrid.

(2) The horizontal earth pressure was reduced to a level of 54–63% of Rankine's active earth pressure. In view of the decrease in active earth pressure, the shear strength of the cement-treated gravel was considered similar to that of granular soil with a friction angle of 45–50° for practical designs. In particular, the higher the shear strength, the higher the stiffness encountered in the soil would be; thus, the low compressibility of the RSR could be confirmed indirectly.

(3) The measured maximum geogrid strain was equivalent to 0.784%, which indicated the stability of the RSR in the event of pullout and rupture of the geogrid. Upon the completion of the backfill, the geogrid deformation resulting from the increase in the overburden load was approximately negligible.

**Author Contributions:** Conceptualization and methodology, D.S.K.; validation, D.S.K. and S.J.; investigation, S.P. and U.K.; resources, S.P. and U.K.; data curation, S.P., D.S.K. and U.K.; writing—original draft preparation, S.P. and U.K.; writing—review and editing, D.S.K. and S.J.; supervision, D.S.K.; project administration and funding acquisition, D.S.K. All authors have read and agreed to the published version of the manuscript.

**Funding:** This research was supported by a grant from the R&D Program (Development of long-life technology for ballasted tracks using EPDM rubber pad, PK2204B1I) of the Korea Railroad Research Institute.

**Conflicts of Interest:** The authors declare no conflict of interest.

## References

1. Li, D.; Davis, D. Transition of railroad bridge approaches. *J. Geotech. Geoenvironmental Eng.* **2005**, *131*, 1392–1398. [CrossRef]
2. Jing, G.; Siahkouhi, M.; Wang, H.; Esmaeili, M. The improvement of the dynamic behavior of railway bridge transition zone using furnace slag reinforcement: A numerical and experimental study. *Proc. Inst. Mech. Eng. Part F J. Rail Rapid Transit* **2022**, *236*, 362–374. [CrossRef]
3. Jamnongpipatkul, P.; Dechasakulsom, M.; Sukolrat, J. Application of air foam stabilized soil for bridge-embankment transition zone in Thailand. In Proceedings of the Asphalt Material Characterization, Accelerated Testing, and Highway Management: Selected Papers from the 2009 GeoHunan International Conference, Changsha, China, 3–6 August 2009; pp. 181–193. [CrossRef]
4. Zhang, J.P.; Liu, T.; Pei, J.Z.; Li, R.; Zou, D.G.; Zhang, Y.Q. Settlement characteristics of bridge approach embankment based on scale model test. *J. Cent. South Univ.* **2020**, *27*, 1956–1964. [CrossRef]
5. Liu, K.; Su, Q.; Ni, P.; Zhou, C.; Zhao, W.; Yue, F. Evaluation on the dynamic performance of bridge approach backfilled with fibre reinforced lightweight concrete under high-speed train loading. *Comput. Geotech.* **2018**, *104*, 42–53. [CrossRef]
6. Xiao, C.; Gao, S.; Liu, H.; Du, Y. Case history on failure of geosynthetics-reinforced soil bridge approach retaining walls. *Geotext. Geomembr.* **2021**, *49*, 1585–1599. [CrossRef]
7. Hu, P.; Zhang, C.; Wen, S.; Wang, Y. Dynamic responses of high-speed railway transition zone with various subgrade fillings. *Comput. Geotech.* **2019**, *108*, 17–26. [CrossRef]
8. Xiao, D.; Jiang, G.L.; Liao, D.; Hu, Y.F.; Liu, X.F. Influence of cement-fly ash-gravel pile-supported approach embankment on abutment piles in soft ground. *J. Rock Mech. Geotech. Eng.* **2018**, *10*, 977–985. [CrossRef]
9. Tatsuoka, F.; Tateyama, M.; Aoki, H.; Watanabe, K. Bridge abutment made of cement-mixed gravel back-fill. In *Elsevier Geo-Engineering Book Series*; Elsevier: Amsterdam, The Netherlands, 2005; pp. 829–873. [CrossRef]
10. Kongsukprasert, L.; Tatsuoka, F.; Takahashi, H. Effects of curing period and stress conditions on the strength and deformation characteristics of cement-mixed soil. *Soils Found.* **2007**, *47*, 577–596. [CrossRef]
11. Watanabe, K.; Tateyama, M.; Yonezawa, T.; Aoki, H. Strength characteristics and construction management of cement-mixed gravel. In Proceedings of the 16th International Conference on Soil Mechanics and Geotechnical Engineering, Osaka, Japan, 12–16 September 2005; IOS Press: Amsterdam, The Netherlands, 2005; pp. 619–622. [CrossRef]
12. Hoppe, E.J. *Guidelines for the Use, Design, and Construction of Bridge Approach Slabs*; Virginia Transportation Research Council: Charlottesville, VA, USA, 1999.
13. Roy, S.; Thiagarajan, G. Nonlinear finite-element analysis of reinforced concrete bridge approach slab. *J. Bridge Eng.* **2007**, *12*, 801–806. [CrossRef]

14. Heydari-Noghabi, H.; Zakeri, J.A.; Esmaeili, M.; Varandas, J.N. Field study using additional rails and an approach slab as a transition zone from slab track to the ballasted track. *Proc. Inst. Mech. Eng. Part F J. Rail Rapid Transit* **2018**, *232*, 970–978. [CrossRef]
15. Asghari, K.; Sotoudeh, S.; Zakeri, J.A. Numerical evaluation of approach slab influence on transition zone behavior in high-speed railway track. *Transp. Geotech.* **2021**, *28*, 100519. [CrossRef]
16. Watanabe, K.; Tateyama, M.; Yonezawa, T.; Aoki, H.; Tatsuoka, F.; Koseki, J. Shaking table tests on a new type bridge abutment with geogrid-reinforced cement treated backfill. In Proceedings of the 7th International Conference on Geosynthetics Society, Nice, France, 22–27 September 2002; pp. 119–122.
17. Feng, S.; Xu, R.; Yu, J.; Zhang, C.; Cheng, K. Field monitoring of geogrid-reinforced and pile-supported embankment at bridge approach. *Int. J. Geosynth. Ground Eng.* **2021**, *7*, 2. [CrossRef]
18. Zheng, Y.; Fox, P.J.; McCartney, J.S. Numerical simulation of deformation and failure behavior of geosynthetic reinforced soil bridge abutments. *J. Geotech. Geoenvironmental Eng.* **2018**, *144*, 04018037. [CrossRef]
19. Bizjak, K.F.; Lenart, S. Life cycle assessment of a geosynthetic-reinforced soil bridge system—A case study. *Geotext. Geomembr.* **2018**, *46*, 543–558. [CrossRef]
20. Tatsuoka, F.; Tateyama, M.; Uchimura, T.; Koseki, J. Geosynthetic-reinforced soil retaining walls as important permanent structures 1996-1997 mercer lecture. *Geosynth. Int.* **1997**, *4*, 81–136. [CrossRef]
21. Tatsuoka, F.; Tateyama, M.; Moltri, Y.; Matsushima, K. Remedial treatment of soil structure using geosynthetic-reinforcing technology. *Geotext. Geomembr.* **2007**, *25*, 204–220. [CrossRef]
22. Kim, U.; Kim, D.S. Evaluation of Deformation Characteristic of Railway Subgrade Using Reinforced Rigid Walls with Short Reinforcement under Repetitive and Static Loads. *Appl. Sci.* **2021**, *11*, 3615. [CrossRef]
23. Kim, U.J.; Kim, D.S. Load sharing characteristics of rigid facing walls with geogrid reinforced railway subgrade during and after construction. *Geotext. Geomembr.* **2020**, *48*, 940–949. [CrossRef]
24. Kim, D.S. Performance Evaluation on Deformation Control of Reinforced Subgrade for Railways under Construction. *J. Korean Soc. Hazard Mitig. Korean Soc. Hazard Mitig.* **2017**, *17*, 17–22. [CrossRef]
25. Korea Railway Network Authority. *Railway Design Standard for Roadbed*; Korea Railway Network Authority: Daejeon, Korea, 2013; ISBN 978-89-91723-95-5.
26. Tanaka, T.; Verruijt, A. Seepage Failure of Sand Behind Sheet Piles—The Mechanism and Practical Approach to Analyze—. *Soils Found.* **1999**, *39*, 27–35. [CrossRef]
27. Yin, G.; Zhang, Q.; Wang, W.; Chen, Y.; Geng, W.; Liu, H. Experimental study on the mechanism effect of seepage on microstructure of tailings. *Saf. Sci.* **2012**, *50*, 792–796. [CrossRef]
28. Hara, A.; Ohta, T.; Niwa, M.; Tanaka, S.; Banno, T. Shear modulus and shear strength of cohesive soils. *Soils Found.* **1974**, *14*, 1–12. [CrossRef]
29. Seng, S.; Tanaka, H. Properties of cement-treated soils during initial curing stages. *Soils Found.* **2011**, *51*, 775–784. [CrossRef]

applied sciences

MDPI

Article

# Zeolitization of Diatomite Residues by a Simple Method

José Manuel Moreno-Maroto [1,2,*], Jacinto Alonso-Azcárate [3], Carmen Martínez-García [2], Maximina Romero [4], Aurora López-Delgado [4] and Teresa Cotes-Palomino [2]

1 Department of Geology and Geochemistry, Faculty of Sciences, Autonomous University of Madrid, Cantoblanco, 28049 Madrid, Spain
2 Department of Chemical, Environmental and Materials Engineering, Higher Polytechnic School of Linares, Scientific and Technological Campus of Linares, University of Jaén, 23700 Linares (Jaén), Spain
3 Department of Physical Chemistry, Faculty of Environmental Sciences and Biochemistry, University of Castilla-La Mancha, Avenida Carlos III, s/n, 45071 Toledo, Spain
4 Eduardo Torroja Institute for Construction Sciences, IETcc-CSIC, 28033 Madrid, Spain
* Correspondence: josemanuel.moreno@uam.es

**Abstract:** The possibility of transforming a diatomite-rich waste from the brewing industry into synthetic zeolites has been investigated. After precalcination at 550 °C to eliminate the retained organic matter, the clean diatomite (Dt; with a Si/Al molar ratio of 17.4), was hydrothermally treated for 24 h with continuous stirring in a 3M NaOH solution at 80 °C. The results of mineralogical characterization by X-ray diffraction with Rietveld refinement have shown a crystallization of 55% of zeolite P, which was neoformed from the amorphous phase, opal-CT and quartz of the starting sample. The spectra obtained by Fourier Transform Infrared Spectrometry have corroborated such zeolitization. $N_2$ adsorption–desorption isotherms have shown that the zeolitized material (Dt-Z) is mesoporous, with almost 60% more specific surface area than Dt (62.6 $m^2/g$ vs. 39.4 $m^2/g$), greater microporosity and 40% smaller average pore size than Dt (71 Å vs. 118 Å). This study is a first approximation to know the potential of diatomite wastes as zeolite precursors, for which additional research including an aluminum source will be required.

**Keywords:** zeolite; diatomite; waste valorization; hydrothermal synthesis; NaOH

**Citation:** Moreno-Maroto, J.M.; Alonso-Azcárate, J.; Martínez-García, C.; Romero, M.; López-Delgado, A.; Cotes-Palomino, T. Zeolitization of Diatomite Residues by a Simple Method. *Appl. Sci.* **2022**, *12*, 10977. https://doi.org/10.3390/app122110977

Academic Editors: Theodore E. Matikas and Nikolaos Koukouzas

Received: 8 September 2022
Accepted: 28 October 2022
Published: 29 October 2022



## 1. Introduction

Diatomite, also called diatomaceous earth, is a type of sedimentary rock formed by the fossilization of unicellular algae called diatoms. Its composition is mainly rich in hydrated amorphous or opaline silica, being chemically stable and inert. The holed structural pattern of diatomite gives it very interesting physical properties, which has led this material to be commonly used in different industrial processes as an absorbent, filter medium, thermal insulator, filler, catalyst carrier, among others [1,2]. Accordingly, diatomite has been exploited for more than a century and is used in large quantities due to its low cost [3,4]. Among other applications, diatomite is used in the agri-food industry as a filter medium in the production of wine and beer, generating large quantities of waste rich in this material, in this case saturated in organic matter. From an environmental and economic perspective, different strategies are being considered to valorize these wastes, not only to obtain new products with added value, but also to reduce their potential hazardousness [3,4].

Among the potential options is the possibility of synthesizing zeolites from raw materials rich in diatomite, preferably in residual form. Zeolites are microporous aluminosilicate minerals, made up of $TO_4$ tetrahedrons (where T is Si or Al) assembled with oxygen bonds. Zeolites, whether natural or synthesized, present a wide variety of nanoporous crystalline structures with interconnected cages. Due to their nanostructure, zeolites can be considered advanced materials whose application is very wide as a catalyst in chemical processes, adsorbent of cations and organic molecules, desalination processes, gas separation, etc. [4,5]. There are several publications that cover the study of the synthesis of zeolites from (fundamentally natural) diatomites, some of which are summarized below.

Shan et al. [6] manufactured (Fe,Al)-ZSM-5 zeolite microspheres with meso- and macroporosity from natural diatomite, using different reagents (NaOH, tetrapropylammonium bromide (TPABr), NaCl and distilled water). The zeolite was hydrothermally synthesized in an autoclave for 6 days at 180 °C, with subsequent removal of TPABr by calcination at 550 °C. Paper sludge ash in combination with different proportions of diatomite as silica source were used by Wajima et al. [7] to obtain zeolite Na-P1 (zeolite P). The zeolite crystallization process was carried out by hydrothermal means in an Erlenmeyer flask for 24 h at 90 °C, using a 3M NaOH solution as alkaline medium. The zeolite obtained presented the capacity to eliminate $NH_4^+$ and $PO_4^{3-}$, having potential application in water filtration. Jia et al. [8] produced zeolites with hierarchical porosity from a commercial diatomite. Once several steps were completed, the final zeolitization process was carried out in an autoclave at 150 °C for 72 h. After applying a scrubbing treatment on a commercial diatomite in order to remove impurity phases, Du et al. [9] synthesized zeolite P using NaOH, $Al(OH)_3$ and deionized water as chemical reagents. The zeolitization was carried out by the authors using a water bath at 90 °C, studying different treatment times (4–24 h).

For their part, Yao et al. [10] used a commercial diatomite, NaOH and $Al(OH)_3$ to synthesize zeolite X. After a first stage of aging at different times and temperatures between 0–120 min and 30–60 °C, zeolites were obtained by hydrothermal treatment also studying different conditions, between 90–120 °C and 3–9 h. Servatan et al. [11] synthesized mesoporous zeolites using a diatomite exploited in an Iranian mine. The synthesis process covered different stages, including, precalcination; acid treatment; treatment with TPABr, sodium aluminate and NaOH; hydrothermal treatment at 170 °C for 60 h; treatment with $NH_4NO_3$ and final calcination. As a result, the authors obtained ZSM-5 zeolites of good quality. In a relatively recent study, Stafin et al. [4] used diatomaceous earth wastes from beer filtration. Apart from distilled water, the diatomite was mixed with NaOH and $Al(NO_3)_3 \cdot 9H_2O$ as aluminum source. Both a conventional hydrothermal treatment route (100 °C at different times between 3 and 24 h) and through microwaves for 170 s were investigated. Zeolite NaP1, hydrosodalite, sodalite and cancrinite were obtained as products of interest, concluding that microwave treatment can also be very effective.

The present study aims to deepen the knowledge of the synthesis of zeolites from diatomite. In this case, a primary route to know the potential of diatomite wastes as zeolite precursors will be looked for. If adequate, it will pave the way for additional research to obtain quality zeolites from diatomite wastes. It will be especially relevant the non-addition of any aluminum source, so that, unlike what has been presented so far in the literature, silicon-rich diatomaceous earths will be used as the only precursor in the zeolite synthesis. Furthermore, considering the current guidelines of sustainability and circular economy, another of the main objectives is not to use natural raw materials, so the feasibility of a residual diatomite from the brewing industry will be studied as a precursor in zeolitization.

## 2. Materials and Methods

The diatomite residue used in this investigation came from the filter earths used in the beer brewing process of the Heineken company at its plant located in Jaén (Spain). Once dried in an oven at 105 °C, this material was ready to be used without the need for grinding, since its particle size presented a high fineness (all the material passed through the 63-μm sieve). Thus, the energy and economic costs of milling would be eliminated, an aspect that is especially positive if the zeolitization protocol described were to be scaled up to an industrial level. The diatomite's chemical composition (obtained by X-ray fluorescence using a Thermo ARL ADVANT'XP Sequential XRF equipment) is shown in Table 1.

**Table 1.** Chemical composition of the diatomite residue used in this research as received and after calcination at 550 °C (data in percentage).

| Diatomite | $SiO_2$ | $Al_2O_3$ | $Fe_2O_3$ | $K_2O$ | MgO | CaO | $TiO_2$ | $Na_2O$ | $P_2O_5$ | $SO_3$ | LOI |
|---|---|---|---|---|---|---|---|---|---|---|---|
| As received | 76.8 | 3.8 | 1.5 | 1.0 | 0.2 | 0.8 | 0.4 | 0.7 | 0.3 | 0.03 | 14.5 |
| Calcined (Dt) | 89.8 | 4.4 | 1.8 | 1.1 | 0.2 | 0.9 | 0.4 | 0.8 | 0.4 | 0.03 | 0.0 |

The high percentage of loss on ignition (LOI) is attributable to the organic matter retained in the diatomite mineral matrix during the filtration process in beer production. Accordingly, the diatomite waste was calcined at 550 °C for 5 h in a muffle to remove the organic matter. On the other hand, as expected, the $SiO_2$ content is predominant, reaching a value of almost 90% after calcination (Table 1), since silicon is the main component of diatomaceous earths. Far below, the second element present is aluminum (~4%), followed by lower proportions of iron and alkali and alkaline earth metals, among others. The presence of these elements could be due to impurities in the diatomite itself, as well as to substances retained in its structure during filtration. Despite this, the chemical composition of Table 1 is in the same order as others found in the literature for diatomaceous earths from different sources and countries [12]. As a result, the Si/Al molar ratio of the raw material is high (Si/Al = 17.4).

After that, the clean diatomite (Dt) was subjected to hydrothermal treatment with a 3M NaOH solution. The liquid (NaOH solution)/solid (mass of Dt) ratio used was 10, specifically 10 g of Dt in 100 mL of NaOH solution. The mixture was kept under constant magnetic stirring for 24 h at 80 °C in a closed Erlenmeyer flask [7]. On completion of the 24 h of hydrothermal stirring, the suspension was filtered and washed with distilled water until a neutral pH was reached. The sediment finally obtained was dried at 105 °C in an oven and subsequently characterized. The sample obtained after this treatment was named Dt-Z.

The quantitative mineralogical composition of the original Dt and the phases formed from it was determined by X-ray diffraction (XRD) using a PANalytical® diffractometer X'Pert Pro model. For this analysis, polycrystalline powder diffractograms were obtained after adding ~25 wt.% alumina to the sample in order to determine both the crystalline and amorphous phases by Rietveld refinement [13,14]. The measurement conditions were as follows: 45 kV, 40 mA, Cu radiation, soller slit of 0.04 rad, antiscattering slit of 1/2°, divergence slit of 1/8°, step size of 0.0167, counting time (s) of 130, Bragg-Brentano HD module and X'celerator detector. The Rietveld analysis of the sample was carried out with the HSP program of Panalytical. It is important to note that the quantification of mineral/amorphous phases is not an aspect usually addressed in other studies shown in the literature, which are limited exclusively to phase identification, making this a point of special interest in this research. The $N_2$ adsorption–desorption isotherms at 77 K of both Dt and the resulting zeolites (Dt-Z) were obtained using Surface Area and Porosity Analyzer–Model ASAP 2420 manufactured by Micromeritics Instrument Corporation, deducing the Brunauer–Emmett–Teller (BET) surface area, the Barrett–Joyner–Halenda (BJH) mesopore characteristics, as well as the micropore volume based on the Density Functional Theory (DFT). The chemical bonds generated during the zeolitization was studied by Fourier Transform Infrared Spectrometry (FTIR) using a Bruker Vertex 70v equipment, applying the Attenuated Total Reflection (ATR) method. The spectra were obtained between 4000 and 400 $cm^{-1}$ and a resolution of 2 $cm^{-1}$.

## 3. Results and Discussion

### 3.1. XRD Analysis: Mineralogical Transformations

Figure 1 shows the diffractograms obtained from the XRD test applied to Dt and Dt-Z, while a summary of the quantification by Rietveld method of the different phases is shown in Table 2. The diatomite used in this research has a fundamentally amorphous nature (81.2%), which makes it a material with, a priori, good zeolitization potential. Considering the chemical composition shown in Table 1, such an amorphous phase would be mainly composed of $SiO_2$. In smaller proportions, Dt also presents opal-CT (11.1%) and quartz (7.7%) as the only crystalline phases present in the original sample. However, as can be seen in Figure 1 and Table 2, the alkaline hydrothermal treatment has produced very significant changes both in the proportions of these original phases and in the neoformation of new ones. Thus, Dt-Z contains only 39.2% and is amorphous phase; it does not contain opal-CT and its quartz proportion has also been reduced to 4.3 %. In addition, new minerals have crystallized. As a minor species, calcite (1.6%) has been neoformed, whose crystallization could have been due to the presence of $Ca^{2+}$ ions that have reacted with dissolved $CO_2$ during the hydrothermal agitation. However, calcite's contribution to the mineralogy of Dt-Z is not very relevant if compared to zeolite-P, whose crystallization fulfills the objective of this research, also being the predominant phase in Dt-Z, representing 55% of the total sample.

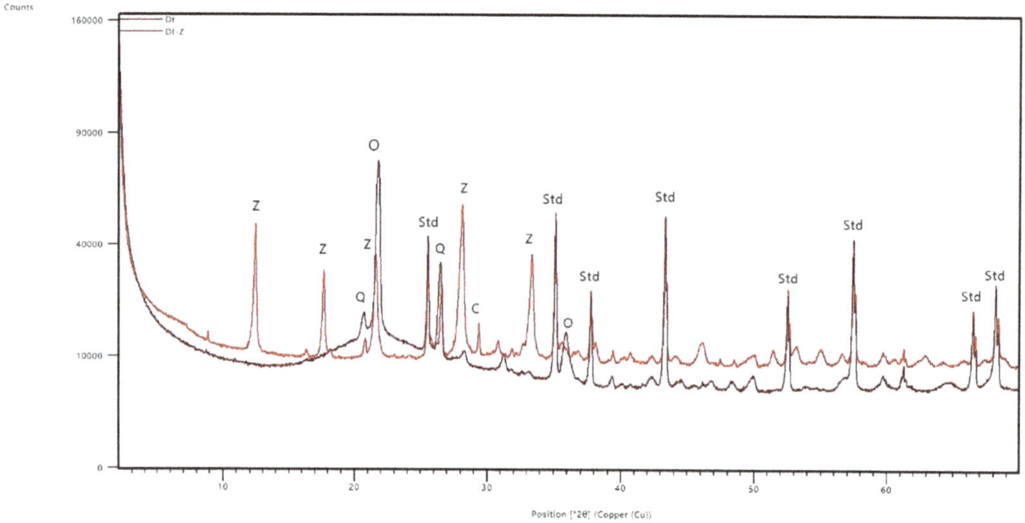


**Figure 1.** XRD diagrams of Dt and Dt-Z. Q = Quartz; O = Opal-CT; Std = Standard ($Al_2O_3$); C = Calcite; Z = Zeolite P.

**Table 2.** Quantitative mineralogical composition of the diatomite (Dt) and the material obtained after hydrothermal treatment (Dt-Z). Data in percentage. Δ % represent the percentage.

| | Amorphous | Opal-CT | Quartz | Calcite | Zeolite P |
|---|---|---|---|---|---|
| Dt | 81.2 | 11.1 | 7.7 | – | – |
| Dt-Z | 39.2 | – | 4.3 | 1.6 | 54.9 |
| Δ % | −52.8 | −100.0 | −44.2 | 100.0 | 100.0 |

Based on the database of zeolite structures published by the International Zeolite Association (IZA), zeolite P presents a GIS framework type [15–17], i.e., this is a zeolite topologically related to the mineral called gismondine, presenting a two-dimensional pore system with two intersecting eight-membered oxygen ring channels [15,16]. Although increasingly in disuse compared to other zeolites, due to this structure, zeolite P has potential applications as a separator and adsorbent both at the industrial level and in environmental technologies [7,9,18].

Considering that calcite does not contain silicon in its structure, the zeolite P developed in this investigation would not have been formed solely from the chemical dissolution of the amorphous phase, but also from opal-CT and quartz, minerals that in theory could present greater difficulties for it. In fact, according to the data in Table 2, the alkaline treatment has been totally effective in dissolving the opal-CT in its entirety, while the original amorphous phase has been wiped out by 52.8% and the quartz by 44.2%. These results refute what has been published by other authors, such as Du et al. [9], who synthesized zeolite P also from diatomite after previously removing all crystalline $SiO_2$ by considering that the latter could not intervene in zeolite formation, something that is not the case in this study. Although opal-CT could have contributed all its $SiO_2$ in the neoformation of zeolite P, its original proportion is relatively low, so the amorphous phase can be considered as the one that has released the most silicon involved in the zeolitization of the diatomite. However, on the other hand, it should be noted that the volume of sediment obtained through the hydrothermal synthesis was significantly lower than that of the raw material used. This would indicate that zeolitization has been restricted, a priori, by the low aluminum content of the diatomite and that, therefore, further investigation would be required including some source of aluminum to react with the dissolved silicon.

### 3.2. $N_2$ adsorption–Desorption Isotherms: Porosity and Surface Area

The graphs obtained from the $N_2$ adsorption–desorption isotherms test are shown in Figure 2. In the case of the Dt adsorption curve (Figure 2a), a slight increase in the amount of $N_2$ adsorbed is observed when the relative pressure is practically zero, and then the adsorption increases at a low rate until approximately P/Po = 0.7. Above this relative pressure, the adsorption rate increases up to about P/Po = 0.95, above which the slope of the curve is again somewhat reduced. In the case of the Dt-Z adsorption curve (Figure 2b), the pathway is similar to that of the Dt sample, with the largest differences observed when P/Po > 0.7, with a less pronounced rise in the rate of adsorbed amount and no final deceleration if compared with Dt. Despite this, during desorption a hysteresis loop is observed for both Dt and Dt-Z, decoupling from the adsorption curve from relative pressures of approximately 0.7 to 0.95 for Dt, and 0.5 to 0.95 for Dt-Z.

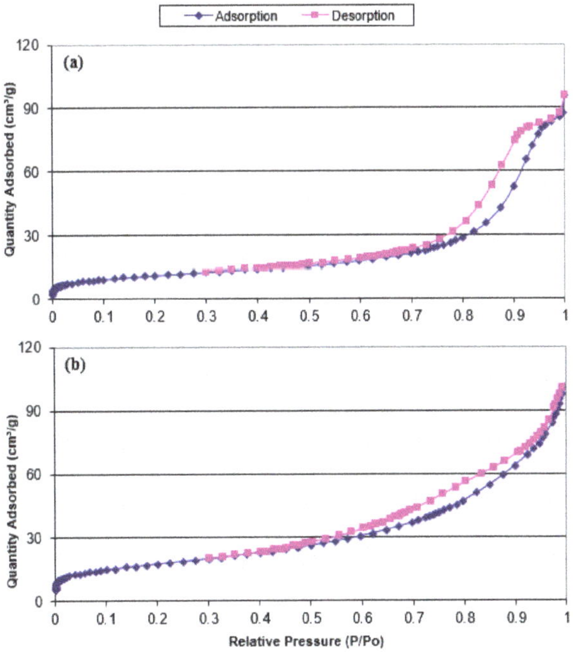


**Figure 2.** $N_2$ adsorption–desorption isotherms of: (**a**) initial diatomite, Dt; (**b**) the material obtained after the zeolitization, Dt-Z.

The International Union of Pure and Applied Chemistry (IUPAC) defines six possible types of adsorption isotherms, from which the type of the dominant porosity (if any) in the sample can be deduced [19–22]. The Dt and Dt-Z isotherms would be classified within type IV, typical of mesoporous materials, i.e., in theory, they would present a predominance of pores with sizes between 2 and 50 nm according to the IUPAC [22,23]. In the case of Dt, this type of isotherm (Figure 2a) differs from others for diatomites presented in the literature [8], which show type I isotherms, typical of non-porous or macroporous materials. Probably, the pretreatment at 550 °C may have helped to unclog the micro- and mesopores in Dt, hence their curves show a different trajectory than expected. From the four types of hysteresis loops classified by IUPAC [19,20,22], Dt presents the H1 type, which is typical of materials with well-defined cylindrical-type porous channels or agglomerates of more or less uniform spheres. For its part, Dt-Z exhibits an H3-type hysteresis loop, which is usually linked to slit-shaped pores and non-rigid aggregates of plate-like particles [22].

The main characteristics deduced from the $N_2$ isotherms test are summarized in Table 3. As can be seen, the alkaline hydrothermal treatment has produced a substantial increase in the specific surface area according to the BET method. $S_{BET}$ has increased from 39.4 $m^2/g$ for Dt to 62.6 $m^2/g$ for Dt-Z, which in percentage terms is almost a 60% increase in specific surface area. This specific surface area value is slightly higher than the maximum $S_{BET}$ obtained by Du et al. [9] in zeolites P also produced from diatomites (59.5 $m^2/g$) by using a more laborious method, which also included the addition of an aluminum source.

**Table 3.** Main properties deduced from $N_2$ adsorption–desorption isotherms.

| | $S_{BET}$ ($m^2/g$) | DFT Micro ($cm^3/g$) | BJH Cumulative Pore Volume ($cm^3/g$) | | | BJH Average Pore Width (Å) | | |
|---|---|---|---|---|---|---|---|---|
| | | | Adsorption | Desorption | Average | Adsorp. | Desorp. | Average |
| Dt | 39.38 | 0.002 | 0.11 | 0.13 | 0.12 | 121.8 | 114.0 | 117.9 |
| Dt-Z | 62.57 | 0.005 | 0.10 | 0.13 | 0.11 | 67.8 | 74.3 | 71.0 |
| Δ% | 58.9 | 124.6 | −10.9 | 0.5 | −4.9 | −44.4 | −34.8 | −39.7 |

*NOTE:* $S_{BET}$ = specific surface area calculated according to the BET theory; DFT micro = microporosity volume calculated according to the DFT theory, considering the cumulative pore volume < 20 Å. BJH results embrace both micro- and mesoporosity.

Returning to the results of the present investigation, although the microporosity (represented in this case as the accumulated volume of pores smaller than 2 nm) is not too high in either sample, an increase is again observed, from 0.002 $cm^3/g$ in Dt to more than double in Dt-Z (0.005 $cm^3/g$). However, the data estimated from the BJH method indicate that the total pore volume has hardly varied between both samples, even decreasing slightly (−4.9%) this property in the Dt-Z sample with respect to the untreated diatomite. The fact that the volume of accumulated porosity has hardly changed when applying the hydrothermal treatment, while the specific surface area has increased is an indicator that there has been a change in the pore size, so that in Dt-Z the pores would present a smaller size, being also more numerous than in Dt as a result of the zeolitization process (Figure 3). This is demonstrated not only in the increase in microporosity just mentioned, but also in the mean pore size estimated with the BJH theory (Table 3), showing a reduction in pore size of almost 40%, from 118 Å in Dt to 71 Å in Dt-Z (Figure 3).

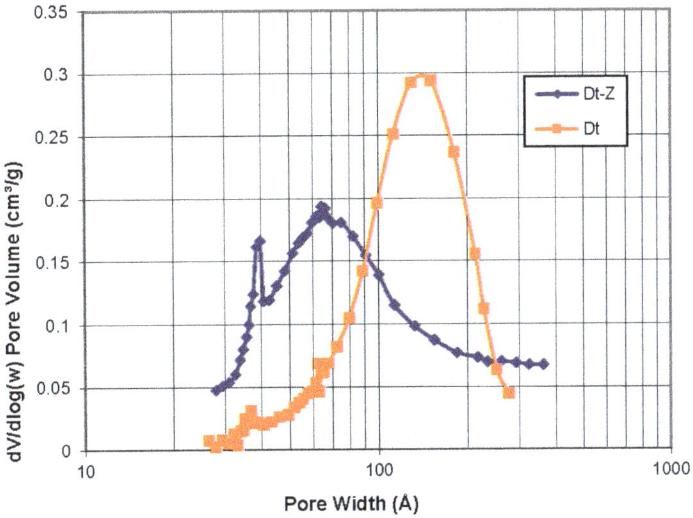


**Figure 3.** BJH Desorption dV/dlog(w) Pore Volume vs. Pore Width for Dt and Dt-Z.

Although there is a predominance of mesoporosity over microporosity, as explained in the previous section, according to the results of Figure 1 and Table 2, zeolitization of Dt has been effectively carried out. In fact, this is consistent with other works already published that show the obtaining of mesoporous zeolites from diatomite, such as that of Servatan et al. [11] or the previously cited of Du et al. [9]. However, it should be noted that the volume of sediment obtained was significantly lower than that of the raw material used. This would indicate that zeolitization has been restricted, a priori, by the low aluminum content of the diatomite and that, therefore, further investigation would be required including some source of aluminum to react with the dissolved silicon.

### *3.3. FTIR Results: Endorsing the Findings on Zeolitization*

The diagrams obtained by FTIR spectra are shown in Figure 4. Between 2950–3600 $cm^{-1}$ there is a slight variation in the absorbance of Dt-Z which is practically negligible for Dt. This band is related to hydration, namely to the H-O-H vibrations of stretching and bending of the free water molecules, as well as those absorbed in the pores and adsorbed on the particles' surface [24–26]. Similarly, at 1597 $cm^{-1}$ there is a band in Dt-Z, which is not observed in Dt, and is attributable to the bending of water molecules bound to the zeolite particles [2,27].

At 1044 $cm^{-1}$ a band of high intensity is observed in Dt, which is associated with asymmetric bending/stretching vibrations in T-O-(T) bonds, where T is Si or Al [11,28,29]. Considering the chemical composition of the Dt sample (Table 1), it is expected that T is mostly silicon. In the case of Dt-Z this band appears somewhat shifted towards smaller wavelengths, specifically at 957 $cm^{-1}$. According to Milkey [30], this displacement towards shorter wavelengths could be directly related to the inclusion of Al atoms in the T-O-T bonds, to give rise not only to Si-O-Si bonds, but also Si-O-Al ones during the formation of the zeolite P obtained. However, it must be taken into account that the sample used has a low percentage of aluminum (3.8%; Table 1) and that no reagent has been incorporated to increase its content, so that this type of Si-O-Al bonds would be, in theory, less representative.

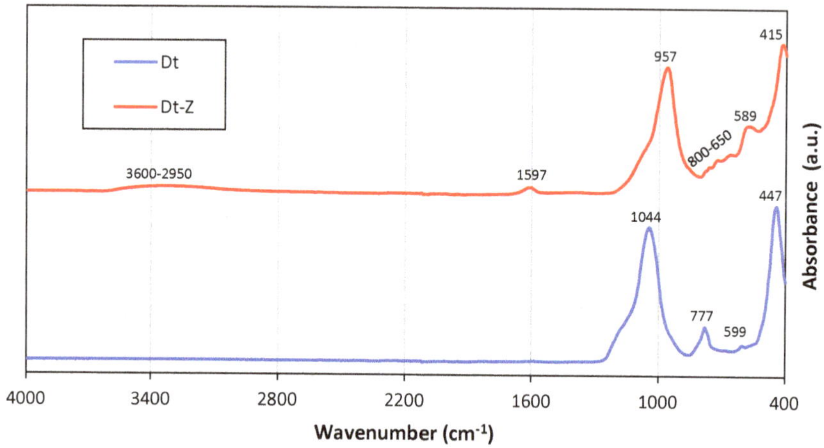


**Figure 4.** FTIR spectrum bands of the diatomite waste (Dt) and the material obtained after the hydrothermal synthesis (Dt-Z).

In the wavelength range between approximately 400 and 800 $cm^{-1}$, a series of peaks are seen that differ depending on the sample. In the non-hydrothermally treated diatomite, bands at 777 $cm^{-1}$ and 447 $cm^{-1}$ stand out, with the presence also of a very low intensity band at 599 $cm^{-1}$. In contrast, the zeolitization process in Dt-Z is mainly associated with a band at 589 $cm^{-1}$ (of much higher intensity than that observed in Dt at 599 $cm^{-1}$) and another at 415 $cm^{-1}$ (again shifted towards lower wavelengths with respect to that observed in Dt at 447 $cm^{-1}$), while between 650–800 $cm^{-1}$ there is a succession of small peaks of low intensity, with no evidence of the band at 777 $cm^{-1}$ observed in Dt. All these bands are related to symmetric stretching vibrations of the T-O-T bridge bonds described above, with the exception of the bands at 447 $cm^{-1}$ (Dt) and 415 $cm^{-1}$ (Dt-Z), which are due to bending vibrations in O-Si-O bonds, and to a lesser extent in O-Al-O (in the latter case, linked mainly to the lower wavelengths of Dt-Z) [11,26,28].

Therefore, the FTIR spectra shown in Figure 4 corroborate the changes in the bonds between atoms occurred during the alkaline hydrothermal treatment leading to the formation of zeolite P in the sediment obtained after treating the diatomite without any additional reagents other than NaOH.

## 4. Conclusions

A simple approach has been presented to investigate the potential of brewing industry diatomite-rich wastes with a Si/Al molar ratio of 17.4 as zeolite precursor. A 24 h hydrothermal treatment was applied using a 3M NaOH solution at low pressure and temperature (80 °C). The main conclusions drawn from this investigation are the following:

- It has been demonstrated that it is possible to synthesize zeolites (in this case zeolite P) from this type of raw materials without the addition of any aluminum supplying component. Despite this, the volume of sediment obtained was significantly lower than that of the raw material used, so further investigation would be required including some source of aluminum to improve the protocol.
- Quantitative XRD tests have shown that the zeolitization process has been effective, forming a material with 55% zeolite P from the hydrothermal transformation of both the amorphous phase, the opal-CT and, to a lesser extent, the quartz present in the original diatomite-rich residue. Such mineralogical transformations have been corroborated through changes in the bands observed in the FTIR spectra.
- The $N_2$ adsorption–desorption isotherms tests have revealed that although the starting diatomite presented a relatively high specific surface (39.4 $m^2/g$), the zeolitization treatment has involved a positive effect, increasing this property by almost 60% (62.6 $m^2/g$).

This specific surface area value is slightly higher than the maximum $S_{BET}$ obtained by Du et al. [9] in zeolites P also produced from diatomites by applying a more laborious method and with the addition of an Al source. Similarly, although the pore volume has hardly changed, the treatment described has favored a reduction in its size by almost 40%, from an average pore size of 118 Å to 71 Å. Therefore, the zeolite formed would be fundamentally mesoporous, like others previously reported in the literature.

- This study shows only a first approach to obtain zeolites from diatomite wastes. Therefore, the findings of this investigation could be useful for future research in order to improve certain properties, such as increasing the purity of the zeolite obtained, increasing its specific surface area and reducing the overall pore size.

**Author Contributions:** J.M.M.-M.: Conceptualization, investigation, methodology, visualization, writing original draft, writing—review and editing, funding acquisition. J.A.-A.: investigation, writing—review and editing, funding acquisition. M.R. and A.L.-D.: writing—review and editing. C.M.-G. and T.C.-P.: writing—review and editing, funding acquisition. All authors have read and agreed to the published version of the manuscript.

**Funding:** This research was funded by the Spanish Ministry of Science, Innovation and Universities and ERDF funds through the ECO-MET-AL Project, PID2019-109520RB-I00/AEI/10.13039/501100011033, "Can industrial and mining metalliferous wastes produce green lightweight aggregates? Applying the Circular Economy" framed in the "Ayudas a "Proyectos I+D+i" en el marco de los Programas Estatales de Generación de Conocimiento y Fortalecimiento Científico y Tecnológico del Sistema de I+D+i y de I+D+i orientada a los Retos de la Sociedad, Convocatoria 2019". This research was also funded by the Postdoctoral Contract granted through the Resolution of 21 May 2020, of the General Secretariat for Universities, Research and Technology of the Government of Andalusia (Spain) within the scope of the Andalusian Plan for Research, Development and Innovation (PAIDI 2020).

**Acknowledgments:** This research was conducted as a part of the ECO-MET-AL Project, PID2019-109520RB-I00/AEI/10.13039/501100011033, "Can industrial and mining metalliferous wastes produce green lightweight aggregates? Applying the Circular Economy" funded by the Spanish Ministry of Science, Innovation and Universities and ERDF funds, framed in the "Ayudas a "Proyectos I+D+i" en el marco de los Programas Estatales de Generación de Conocimiento y Fortalecimiento Científico y Tecnológico del Sistema de I+D+i y de I+D+i orientada a los Retos de la Sociedad, Convocatoria 2019". Thanks also to the SCAI of the University of Jaén, the University of Castilla-La Mancha and the University of Málaga for their services. The authors also acknowledge the Postdoctoral Contract granted through the Resolution of 21 May 2020, of the General Secretariat for Universities, Research and Technology of the Government of Andalusia (Spain) within the scope of the Andalusian Plan for Research, Development and Innovation (PAIDI 2020).

**Conflicts of Interest:** The authors declare no conflict of interest.

## References

1. Biswajit, G.; Dinesh, C.; Subhash, B. Synthesis of zeolite A from calcined diatomaceous clay: Optimization studies. *Ind. Eng. Chem. Res.* **1994**, *33*, 2107–2110. [CrossRef]
2. Chaisena, A.; Rangsriwatananon, K. Effects of thermal and acid treatments on some physico-chemical poroperies of Lampang diatomite. *Suranaree J. Sci. Technol.* **2004**, *11*, 289–299.
3. Fillaudeau, L.; Blanpain-Avet, P.; Daufin, G. Water, wastewater and waste management in brewing industries. *J. Clean. Prod.* **2006**, *14*, 463–471. [CrossRef]
4. Stafin, G.; Grzebielucka, E.C.; Antunes, S.R.; Ferreira Borges, C.P.; Chaves de Andrade, A.V.; Aparecida Alves, S.; Ferreira de Souza, E.C. Synthesis of zeolites from residual diatomite using a microwave-assisted hydrothermal method. *Waste Manag.* **2021**, *126*, 853–860. [CrossRef] [PubMed]
5. Lobo-Recio, M.A.; Rodrigues, C.; Jeremias, T.C.; Lapolli, F.R.; Padilla, I.; López-Delgado, A. Highly efficient removal of aluminum, iron, and manganese ions using Linde type-A zeolite obtained from hazardous waste. *Chemosphere* **2021**, *267*, 128919. [CrossRef]
6. Shan, W.; Zhang, Y.; Wang, Y.; Xia, J.; Tang, Y. Synthesis of Meso-/Macroporous Zeolite (Fe,Al)-ZSM-5 Microspheres from Diatomite. *Chem. Lett.* **2004**, *33*, 270–271. [CrossRef]
7. Wajima, T.; Haga, M.; Kuzawa, K.; Ishimoto, H.; Tamada, O.; Ito, K.; Nishiyama, T.; Downs, R.T.; Rakovan, J.F. Zeolite synthesis from paper sludge ash at low temperature (90 °C) with addition of diatomite. *J. Hazard. Mater.* **2006**, *132*, 244–252. [CrossRef] [PubMed]

8. Jia, Y.; Han, W.; Xiong, G.; Yang, W. A method for diatomite zeolitization through steam-assisted crystallization with in-situ seeding. *Mater. Lett.* **2008**, *62*, 2400–2403. [CrossRef]
9. Du, Y.; Shi, S.; Dai, H. Water-bathing synthesis of high-surface-area zeolite P from diatomite. *Particuology* **2011**, *9*, 174–178. [CrossRef]
10. Yao, G.; Lei, J.; Zhang, X.; Sun, Z.; Zheng, S. One-Step Hydrothermal Synthesis of Zeolite X Powder from Natural Low-Grade Diatomite. *Materials* **2018**, *11*, 906. [CrossRef]
11. Servatan, M.; Ghadiri, M.; Yazdi, M.K.; Jouyandeh, M.; Mahmodi, G.; Samadi, A.; Zarrintaj, P.; Habibzadeh, S.; Ganjali, M.R.; Saeb, M.R. Synthesis of Cost-Effective Hierarchical MFI-Type Mesoporous Zeolite: Introducing Diatomite as Silica Source. *Silicon* **2021**, *13*, 3461–3472. [CrossRef]
12. Danil de Namor, A.F.; El Gamouz, A.; Frangie, S.; Martinez, V.; Valiente, L.; Webb, O.A. Turning the volume down on heavy metals using tuned diatomite. A review of diatomite and modified diatomite for the extraction of heavy metals from water. *J. Hazard. Mater.* **2012**, *241–242*, 14–31. [CrossRef] [PubMed]
13. Bish, D.L.; Post, J.E. Quantitative mineralogical analysis using the Rietveld full pattern fitting method. *Am. Mineral.* **1993**, *78*, 932–940.
14. Moreno-Maroto, J.M.; González-Corrochano, B.; Alonso-Azcárate, J.; Rodríguez, L.; Acosta, A. Assessment of crystalline phase changes and glass formation by Rietveld- XRD method on ceramic lightweight aggregates sintered from mineral and polymeric wastes. *Ceram. Int.* **2018**, *44*, 11840–11851. [CrossRef]
15. IZA. International Zeolite Association (IZA). Database of Zeolite Structures. 2022. Available online: http://www.iza-structure.org/databases/ (accessed on 1 August 2022).
16. Albert, B.R.; Cheetham, A.K.; Adams, C.J. Investigations on P zeolites: Synthesis and structure of the gismondine analogue, highly crystalline low-silica CaP. *Microporous Mesoporous Mater.* **1998**, *21*, 127–132. [CrossRef]
17. Yang, S.; Lach-hab, M.; Vaisman, I.; Blaisten-Barojas, E.; Li, X.; Karen, V.L. Framework-Type Determination for Zeolite Structures in the Inorganic Crystal Structure Database. *J. Phys. Chem. Ref. Data.* **2010**, *39*, 033102. [CrossRef]
18. Baerlocher, C.; Meier, W.M. The crystal structure of synthetic zeolite Na-P 1, an isotype of gismondine. *Z. Für. Krist. -Cryst. Mater.* **1972**, *135*, 339–354. [CrossRef]
19. Broekhoff, J.C.P. Mesopore determination from nitrogen sorption isotherms: Fundamentals, scope, limitations. *Stud. Surf. Sci. Catal.* **1979**, *3*, 663–684.
20. Sing, K.S.W.; Everett, D.H.; Haul, R.A.W.; Moscou, L.; Pierotti, R.A.; Rouquerol, J.; Siemieniewska, T. Reporting physisorption data for gas/solid systems with special reference to the determination of surface area and porosity. *Pure Appl. Chem.* **1985**, *57*, 603–619. [CrossRef]
21. Shields, J.E.; Lowell, S.; Thomas, M.A.; Thommes, M. *Characterization of Porous Solids and Powders: Surface Area, Pore Size and Density*; Kluwer Academic Publisher: Boston, MA, USA, 2004; pp. 43–45.
22. Alothman, Z.A. A Review: Fundamental Aspects of Silicate Mesoporous. *Materials* **2012**, *5*, 2874–2902. [CrossRef]
23. Tangestaninejad, S.; Moghadam, M.; Mirkhani, V.; Baltork, I.M.; Ghani, K. Alkene epoxidation catalyzed by molybdenum supported on functionalized MCM-41 containing N–S chelating Schiff base ligand. *Catal. Commun.* **2009**, *10*, 853–858. [CrossRef]
24. Madejova, J.; Janek, M.; Komadel, P.; Herbert, H.J.; Moog, H.C. FTIR analyses of water in MX-80 bentonite compacted from high salinary salt solution systems. *Appl. Clay Sci.* **2002**, *20*, 255–271. [CrossRef]
25. Ilia, I.K.; Stamatakis, M.G.; Perraki, T.S. Mineralogy and technical properties of clayey diatomites from north and central Greece. *Open Geosci.* **2009**, *1*, 393–403. [CrossRef]
26. Moreno-Maroto, J.M.; Delgado-Plana, P.; Cabezas-Rodríguez, R.; Mejía de Gutiérrez, R.; Eliche-Quesada, D.; Pérez-Villarejo, L.; Galán-Arboledas, R.J.; Bueno, S. Alkaline activation of high-crystalline low-$Al_2O_3$ Construction and Demolition Wastes to obtain geopolymers. *J. Clean. Prod.* **2022**, *330*, 129770. [CrossRef]
27. Zainal Abidin, A.; Abu Bakar, N.H.H.; Ng, E.P.; Tan, W.L. Rapid Degradation of Methyl Orange by Ag Doped Zeolite X in the Presence of Borohydride. *J. Taibah Univ. Sci.* **2017**, *11*, 1070–1079. [CrossRef]
28. Mozgawa, W.; Król, M.; Barczyk, K. FT-IR studies of zeolites from different structural groups. *Chemik* **2011**, *65*, 667–674.
29. López-Delgado, A.; Robla, J.I.; Padilla, I.; López-Andrés, S.; Romero, M. Zero-waste process for the transformation of a hazardous aluminum waste into a raw material to obtain zeolites. *J. Clean. Prod.* **2020**, *255*, 120178. [CrossRef]
30. Milkey, R.G. Infrared spectra of some tectosilicates. *Am. Mineral.* **1960**, *45*, 990–1007.

applied sciences

MDPI

Article

# Investigation of Compressive Strength Characteristics of Hardfill Material and Seismic Stability of Hardfill Dams

**Soogeun Kim [1], Wonhyuk Choi [2], Yunhee Kim [3], Jaewoo Shin [3] and Bumjoo Kim [3,*]**

1 K-Water Ulsan Local Office, Korea Water Resources Corporation, Ulsan 44662, Republic of Korea
2 New Town Project Group, Incheon Housing and City Development Corporation, Incheon 21591, Republic of Korea
3 Department of Civil and Environmental Engineering, Dongguk University, Seoul 04620, Republic of Korea
* Correspondence: bkim1@dongguk.edu

**Abstract:** A hardfill dam can reduce the natural damage caused by the development of quarries and recycle submerged resources. However, the particle size distribution of the aggregate can result in large variations in the mix design, resulting in a wide range of strengths in the hardfill dam body. Therefore, quality control during construction is crucial, and the stability of the dam body after construction should be thoroughly examined in advance. This study investigated the strength characteristics of hardfill dam materials according to the particle size and mixing ratio through indoor uniaxial compressive strength tests using large specimens and uniaxial compressive strength tests for field compaction and collected cores. Furthermore, the stability of hardfill dams with three types of slopes during earthquakes was evaluated through a finite element analysis. The distributions of stress in the hardfill dam body and the strength required to stabilize the dam body were investigated. Except for a 1:0.6 inclined hardfill dam body to which artificial seismic waves were applied, the overall strength range calculated from the indoor specimens and field compaction cores exceeded the required hardfill strength, thereby ensuring stability in the event of an earthquake.

**Keywords:** hardfill; dam; aggregates; particle size distribution; strength; seismic stability

**Citation:** Kim, S.; Choi, W.; Kim, Y.; Shin, J.; Kim, B. Investigation of Compressive Strength Characteristics of Hardfill Material and Seismic Stability of Hardfill Dams. *Appl. Sci.* **2023**, *13*, 2492. https://doi.org/10.3390/app13042492

Academic Editor: José Manuel Moreno-Maroto

Received: 23 January 2023
Revised: 6 February 2023
Accepted: 13 February 2023
Published: 15 February 2023



## 1. Introduction

Despite numerous technological advancements in the field of dams, the demands for further cost reductions and eco-friendly technology developments persist. A hardfill dam is a relatively new type of dam developed in response to such demands. A hardfill dam is constructed by evenly spreading and roller compacting a material that is continuously mixed with water and a small amount of cement (hardfill) in a simple, temporary facility. The construction process uses materials (such as bed gravel and soil) readily available near the dam construction site with minimal processing. The hardfill dams that broaden the choices of material sources for dam construction are advantageous from both economic and environmental perspectives because they can reduce the damage to nature caused by the development of quarries while recycling submerged resources [1–4].

Hardfill dams are often compared with roller-compacted concrete (RCC) dams constructed by the roller compaction of concrete, such as rockfill dams. RCC dams allow for faster construction relative to block-type concrete gravity dams; for example, the latter have drawbacks concerning the various processes required for controlling the hydration heat caused by pouring block-shaped mass concrete, resulting in an increased construction time (decreased construction speed) [5,6]. In other words, hardfill and RCC dams are similar in that the main material sources are aggregates, water, and cement, and the construction speeds are faster than those of concrete dams owing to the roller compaction [7,8]. However, hardfill dams have further advantages relative to RCCs. In particular, an RCC dam follows a mixing design method for concrete during manufacturing. Thus, the range of material sources (and, correspondingly, the particle size distribution of the aggregates, cement content, and unit water content) is regulated according to the design goal. Accordingly, complex mechanical equipment is often required during manufacturing. In contrast, a hardfill dam uses the on-site aggregate at the planned dam construction site as-is without requiring any additional treatment other than the removal of particles exceeding the maximum particle size (80 mm in general). Therefore, the dams can be manufactured with simpler equipment after determining the cement content and unit water content according to the characteristics of the aggregate, thereby simplifying the construction. Furthermore, in terms of structural stability, hardfill dams with a trapezoidal cross-section have a larger dam body volume than RCC dams with a right-triangle cross-section but nevertheless provide high stability against overturning and sliding [3,9]. Despite the above advantages, however, since hardfill dams use almost the same riverbed materials as the construction materials, the range of variations in the mix design can be large according to the particle size distribution (PSD) of the aggregate; correspondingly, the range of changes in the strength of the hardfill dam body can also be large [9,10]. This makes it difficult to manage the quality of the material(s). Therefore, in terms of the stability of the dam body and quality control, the differences in behavior according to the characteristics of the materials should be thoroughly examined during the design phase.

Many construction cases of hardfill dams have been reported from several countries since the first development of a hardfill dam in the early 1990s [11–16]. Moreover, meanwhile, a number of studies have been made on the mechanical properties of hardfill material and the stability of hardfill dams through experimental or numerical investigations by many researchers [15–28]. However, it is difficult to find studies investigating the strength characteristics of hardfill materials in different aggregate PSDs and mixing ratios and the resulting stability of hardfill dams by applying the same hardfill materials through indoor tests, field test construction, and numerical analyses at a time.

In this study, the maximum (coarse) and the minimum (fine) PSDs of aggregates with a maximum particle size of 80 mm, considered to be a suitable aggregate PSD range for the construction of hardfill dams [11], were selected for analysis. The analysis was based on the results of PSD investigations of the riverbed aggregates for each major river system in South Korea presently without hardfill dam construction. After fabricating large hardfill specimens by mixing the aggregate, cement, and water, the strength characteristics of the hardfill were investigated through laboratory uniaxial compressive strength tests. In addition, test constructions were conducted on-site by applying the same mixing ratios and aggregates as those for hardfill samples in the laboratory tests. Then, the hardfill's constructability was not only investigated using the on-site compaction equipment, but also the strength range was obtained from uniaxial compressive strength tests on the cores collected from the test fills. Subsequently, by applying the properties of the hardfill obtained from the laboratory test results, a finite element analysis was performed for hardfill dams with three different cross-sectional slopes (i.e., 1:0.6, 1:0.8, and 1:1) and a height of 50 m, subjected to earthquake. The seismic stability of the hardfill dams was evaluated according to the changes in the strengths and cross-sections (slopes) of the hardfill dams.

## 2. Materials and Methods

### 2.1. Materials

As stated in the introduction, the aggregates of a hardfill are typically unclassified, and their particle size is not adjusted (except for eliminating oversized particles and conducting crushing for effective utilization). Therefore, the differences in the PSDs of aggregate may be significant even in the same aggregate collection area, making it difficult to maintain a consistent unit water content. A hardfill dam's strength naturally changes owing to the use of materials with variable particle size distributions and unit water content, even if the amount of cement remains constant. Accordingly, when designing a hardfill dam, the hardfill's target strength is typically set after investigating the range of the hardfill strength according to changes in the aggregate PSD, unit cement content, and unit water content.

This study excluded aggregates with a maximum particle size exceeding 80 mm, as these generally have low applicability as hardfill aggregates, based on previous studies [18,19]. Two types of PSDs, i.e., maximum PSD (coarse aggregates) and minimum PSD (fine aggregates), were selected based on previous investigations of the aggregate PSD range in major river basins in South Korea [11]. The aggregates used in the experiment were collected from a spoil pit within the Danyang Submerged Weir construction site and a nearby river basin in Danyang-gun, Chungcheongbuk-do, South Korea. They were based on quartzose sandstone and generally well-graded. The shape of the aggregate particles was generally sub-rounded to rounded. An oversize cut was performed by applying the maximum aggregate size of 80 mm using a backhoe and aggregate separator. Then, the screened aggregates were transferred indoors. Figure 1 shows the PSDs of the two types of aggregates used in the laboratory test of this study. The same materials were applied in the laboratory and field tests as described in Section 2.3.

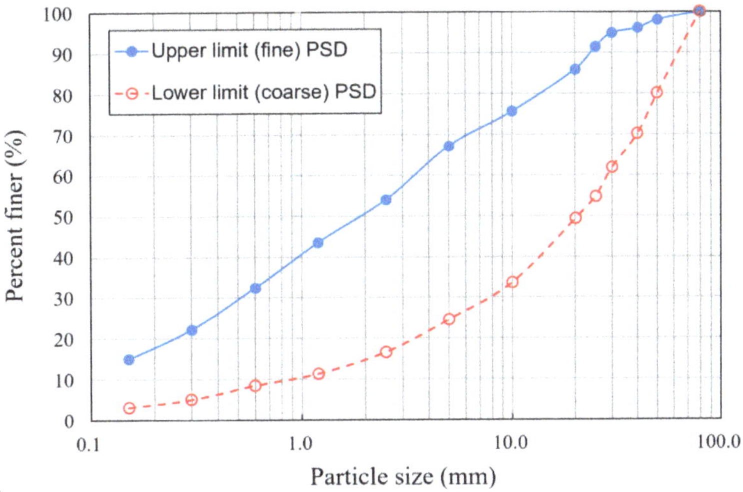


**Figure 1.** Particle size distribution of aggregate.

Prior to this study, standard specimens were fabricated using aggregates with a maximum particle diameter of 40 mm and particle size similar to that of the PSD collected from the same place shown in Figure 1, but with different quantities of cement. Then, compaction and uniaxial compressive strength tests were performed with the standard specimens [29–31]. Based on the results, a unit of cement content of 80 kg/m$^3$ and three units of water content of 70, 85, and 100 kg/m$^3$ were selected as the range of appropriate mixing ratios for the laboratory test investigation with large specimens using aggregates with a maximum particle diameter of 80 mm. Ordinary Portland cement (Type I) was used for the hardfill mix (Table 1).

**Table 1.** Physical properties of cement.

| Specific Gravity | Fineness (cm$^2$/g) | Setting Time (min) | | Compressive Strength (MPa) | |
|---|---|---|---|---|---|
| | | Initial | Final | 7 d | 28 d |
| 3.17 | 3475 | 205 | 295 | 44.4 | 59.3 |

*2.2. Laboratory Tests*

For the laboratory tests (as described above), uniaxial compressive strength tests were performed on cylindrical large specimens with a diameter of 300 mm and height of 600 mm by applying the mixing ratio shown in Table 2 (i.e., a unit cement content of 80 kg/m$^3$ and three unit water contents of 70, 85, and 100 kg/m$^3$) to aggregates of fine (upper-limit) and coarse (lower-limit) PSDs with a maximum particle diameter of 80 mm. This corresponded to the same range of aggregate sizes used in the construction of hardfill dams in the actual field. Three large specimens (samples) were fabricated for each mixing ratio for a total of 18 specimens.

**Table 2.** Mix ratios for large specimen.

| Aggregate | Cement Content per Unit Water Content (kg/m$^3$) | Water Content per Unit Water Content (kg/m$^3$) |
|---|---|---|
| Upper-limit (fine) PSD | 80 | 70, 85, 100 |
| Lower-limit (coarse) PSD | 80 | 70, 85, 100 |

For the fabrication of the large specimens, the aggregate, cement, and water were mixed according to a predetermined mixing ratio, and then a mold was filled in four layers to prevent material separation. After compacting 25 times with a compaction rod for each layer, the compaction time at which a compaction energy equal to that of the field compaction roller was transmitted was calculated; then, a compaction was performed using a vibrating tamper. The compaction energy was based on the results from performing two non-vibration and six vibration compactions with a vibration roller during a field test. When compacting the sample in the mold using a hammer drill, the compaction time for each layer was calculated as 82 s, as shown in Table 3, so that the same energy as the field compaction energy would be transmitted to the sample. Then, the specimens were fabricated by performing the compaction for the calculated compaction time.

**Table 3.** Estimation of compaction energy for large specimen.

| | |
|---|---|
| Compaction energy of vibration roller per unit time ($E_{0p}$, J/min) | $E_{0p} = 2a\left(W + \frac{F}{2}\right)f = 983,443\ \text{J/min}$ where, $a$: vibrating width (=0.00131 m), $W$: weight of vibration roller (=10.6 t), $F$: Average vibrating force (17,900 kgf) and $f$: Frequency (1920 Hz) |
| Compaction energy in field | $E_0 = E_{0p} n\, t_p = 354,040\ \text{J} = 354\ \text{kJ}$ where, $n$: number of roller passes (=6) and $t_p$: average time per compaction (=0.06 min) |
| Compaction energy of hammer drill per unit time ($E_{0l}$, J/min) | $E_{0l}$ = [single impact energy × number of impact per min] = 64,500 J/min |
| Compaction energy for large specimen | $E_l = E_{0l} L t_l$ where, $L$: number of compaction layer (=4) and $t_l$: average compaction time per layer |
| Compaction time per layer for large specimen | $t_l = \frac{E_l}{E_{0l} L} = 1.37\ \text{min} = 82\ \text{s}$ |
| Compaction energy of vibration roller per unit time ($E_{0p}$, J/min) | $E_{0p} = 2a\left(W + \frac{F}{2}\right)f = 983,443\ \text{J/min}$ where, $a$: vibrating width (=0.00131 m), $W$: weight of vibration roller (=10.6 t), $F$: Average vibrating force (17,900 kgf) and $f$: Frequency (1920 Hz) |

The compacted large specimens were sealed and cured for 28 d before performing the uniaxial compressive strength test. Each specimen for testing was continuously compressed to generate a compression strain of 1% per minute. Compression was terminated when more than 2% of the strain occurred after the maximum compressive force, when the maximum compressive force decreased by 2/3, or when the compressive strain reached 15%. Figure 2 shows the fabrication process of the large specimens and uniaxial compressive strength test for a specimen aged 28 d.

Figure 2. Preparation of a large specimen for unconfined compressive strength test. (**a**) Material placement; (**b**) material mixing; (**c**) sample compaction; (**d**) sample curing; (**e**) large specimen; (**f**) uniaxial compressive strength test.

### 2.3. Field Test Construction

The constructability of hardfill on-site was examined through a field test construction using field compaction equipment. The construction specifications (lift thickness, number of roller passes) were established, and the strength range for the hardfill was derived in preparation for actual future construction of a hardfill dam. Furthermore, field test results were compared with the indoor laboratory tests and numerical analysis results. The field test construction site was a spoil area in the Danyang underwater weir construction site, i.e., the same place where the aggregate was collected for the indoor laboratory tests. The hardfill field mixing was performed by applying three unit water contents of 70, 85, and 100 kg/m$^3$ and a unit cement content of 80 kg/m$^3$ to the aggregates with a maximum particle size of 80 mm with a PSD similar to the lower limit of the PSD in the indoor test. The compaction conditions for the vibration roller were eight roller passes (six vibrated passes and two non-vibrated passes). The lift thicknesses of one layer were set as 50 cm and 75 cm, respectively. As a result, a total of six cases were tested.

The planned scale of the test construction was a compaction height of one floor on a floor area with a total width of 19 m and length of 44 m. Construction was performed by dividing sections for each test case according to material mixing conditions and lift thickness. Once each section was compacted using the vibration roller, the degree of compaction was measured by the sand cone method [32]. After measuring the field compaction degree, the corresponding section was covered with curing cloth, and a sufficient volume of water was sprayed for the first 2 to 3 d. Then, the specimens were cured for 28 d. Once curing was completed, the cores were collected, and a uniaxial compressive strength test was performed [33,34].

Table 4 and Figure 3 show the field test construction conditions and process, respectively.

**Table 4.** Conditions of field test construction.

| **Material (Mixing Ratio)** | |
|---|---|
| Aggregate | PSD between upper- and lower-limit PSD with 80 mm max. particle size |
| Cement content per unit volume | 80 kg/m$^3$ |
| Water content per unit volume | 70, 85, 100 (kg/m$^3$) |
| **Hardfill Test Construction** | |
| Size | One fill lift with a floor area of 19 m width and 44 m length, which was partitioned into six parts for a total of six test cases |
| Lift thickness | 50, 75 (cm) |
| Compaction | Eight passes with a vibration roller (six vibrated passes and two non-vibrated passes) |

**Figure 3.** Field test construction and core collection. (**a**,**b**) Material transportation and placement; (**c**) compaction by vibration roller; (**d**) measurement of lift thickness; (**e**) measurement of field compaction (sand cone method); (**f**) curing after compaction; (**g**) core collection; (**h**) core aged 28 d.

## 2.4. Numerical Investigation

As described above, the seismic stability was evaluated by performing a finite element analysis for a hardfill dam body (Figure 4) with three types of slopes (1:0.6, 1:0.8, and 1:1) and a height of 50 m, as constructed on a soft rock foundation based on the properties of the hardfill determined through the indoor laboratory tests. In this analysis, the state of the freshwater above the planned flood level was applied by setting the most vulnerable condition for the stability of the dam body. The hardfill was modeled as an elastic body with a unit weight of 24.1 $kNg/m^3$, Young's modulus of 991 MPa, and Poisson's ratio of 0.29. Furthermore, the damping of the material was set in accordance with the first two eigenfrequencies of the dam and with an estimated damping coefficient equal to 5% for defining the Rayleigh's coefficients. The dam foundation ground was assumed according to the ground conditions near Wonjucheon Dam in South Korea, which was determined to be suitable as a candidate site for the hardfill dam construction. The static and dynamic geotechnical parameters for the dam foundation ground for the analysis were selected based on the results from subsurface investigations conducted in the area (Table 5).

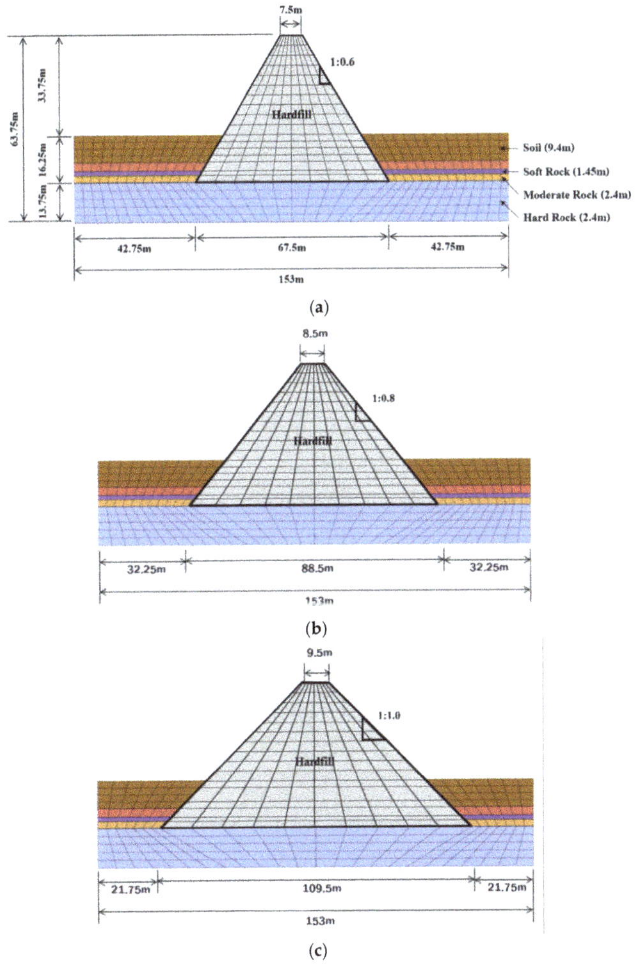


**Figure 4.** Cross-sections of dams for finite element (FE) analysis. (**a**) Dam slope gradient of 1:0.6 (vertical-to-horizontal ratio); (**b**) dam slope gradient of 1:0.8 (vertical-to-horizontal ratio); (**c**) dam slope gradient of 1:1.0 (vertical-to-horizontal ratio).

**Table 5.** Finite element method (FEM) model parameters for the ground.

| | Unit Weight (kN/m$^3$) | Cohesion (kPa) | Friction Angle (°) | Elastic Modulus (MPa) | Poisson's Ratio | P-Wave Velocity (m/s) | S-Wave Velocity (m/s) | Shear Modulus (MPa) | Dashpot Coefficient | |
|---|---|---|---|---|---|---|---|---|---|---|
| | | | | | | | | | $C_P$ (=$\rho V_p$) | $C_S$ (=$\rho V_s$) |
| Soil | 18.0 | 5 | 25 | 20 | 0.38 | 450 | 63.2 | 7.2 | 810 | 114 |
| Soft rock | 24.0 | 200 | 34 | 1500 | 0.27 | 2000 | 496.1 | 590.6 | 4800 | 1191 |
| Moderate rock | 25.0 | 350 | 37 | 4000 | 0.24 | 2300 | 803.2 | 1612.9 | 5750 | 2008 |
| Hard rock | 26.0 | 2000 | 44 | 8000 | 0.20 | 3000 | 1132.3 | 3333.3 | 7800 | 2944 |

A time history analysis method was applied for the dynamic analysis. Three types of seismic waves were used as the input seismic waves: Ofunato and Hachinohe waves, i.e., representative short-period and long-period seismic waves widely used in dynamic analysis, respectively, and artificial waves generated by reflecting the characteristics of the Wonjucheon dam site (Figure 5). In the analysis domain, the left and right sides were fixed in the horizontal direction, and the lower part was set in the vertical direction. A viscous boundary condition was applied to show a behavior similar to reality when a seismic wave is applied by absorbing the wave reaching the boundary.

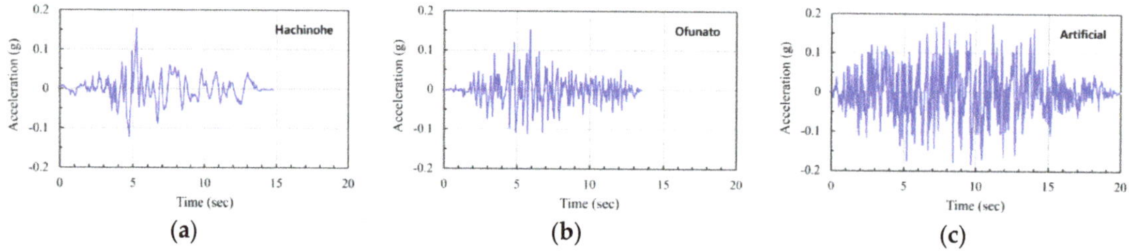


**Figure 5.** Input seismic waves used in FE analysis. (**a**) Hachinohe wave; (**b**) Ofunato wave; (**c**) artificial wave.

Lastly, from the numerical analysis results, the seismic stability of the hardfill dam body was evaluated by comparing the maximum stress generated inside the hardfill dam body with the "required hardfill strength" as applied with a safety factor of 1.5 under the seismic load condition [3].

## 3. Laboratory Test Results

### *3.1. Variations in Dry Unit Weight and Uniaxial Compressive Strength for Hardfill Mix Conditions with Different Aggregate Size Distributions and Water Content per Unit Volume*

Figures 6 and 7 show the relationship between the unit water content and dry unit weight and between the unit water content and uniaxial compressive strength for the fabricated large hardfill specimens, respectively. These specimens were fabricated by mixing aggregates with the upper-limit (fine) and lower-limit (coarse) particle size distributions as described in Section 2.1 with a unit cement content of 80 kg/m$^3$ and three unit water contents of 70, 85, and 100 kg/m$^3$, respectively. Since three specimens were fabricated and tested for one mixing condition, the average values for the corresponding mixing conditions are shown together with the measured values for each specimen.

The dry unit weight increases with the unit water content for both aggregate mixes with the upper-limit and lower-limit PSDs. The increase rate of dry unit weight is slightly higher in the aggregate mix with the upper-limit PSD condition than that with the lower-limit PSD condition (Figure 6). At the smallest unit water content (i.e., 70 kg/m$^3$), the dry unit weight of the aggregate mix with the lower-limit PSD slightly exceeds that of the aggregate mix with the upper-limit PSD (2410 vs. 2357 kg/m$^3$). However, the dry unit weights are similar at the largest unit water content (i.e., 100 kg/m$^3$) (2468 vs. 2470 kg/m$^3$). Meanwhile, the uniaxial compressive strength according to the unit water content change tends to increase gradually with the unit water content for both aggregate mixes with upper- and lower-limit PSDs (Figure 7). The ranges of the uniaxial compressive strengths according to the unit water content change are 3.152–4.106 MPa for the aggregate mix with the lower-limit PSD and 1.619–2.094 MPa for the aggregate mix with the upper-limit PSD. As a result, the uniaxial compressive strength of the aggregate mix with the lower-limit PSD exceeds that of the aggregate mix with the upper-limit PSD for the entire unit water content range.

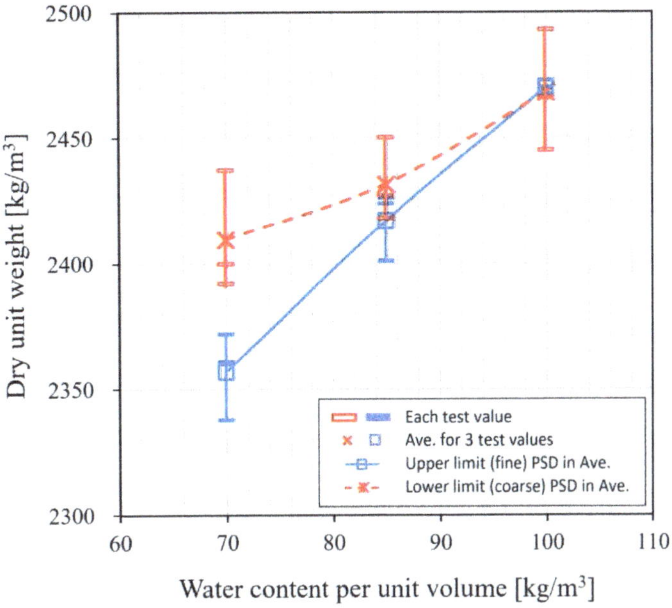


**Figure 6.** Change in dry unit weight according to hardfill unit water content.

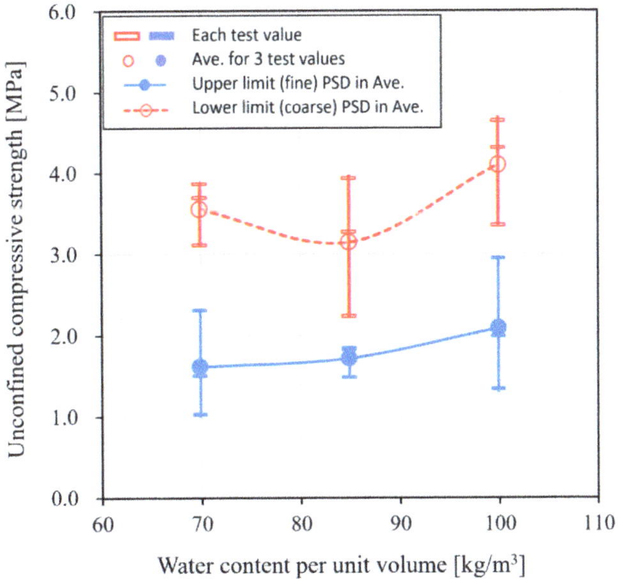


**Figure 7.** Change in hardfill's uniaxial compressive strength according to unit water content.

### 3.2. Elastic Properties of Hardfill

In designing a hardfill dam, in general, the strength in the elastic region is taken as the strength of the hardfill from the stress–strain behavior, and the hardfill dam is assumed to be a structure that behaves like an elastic body. For all samples subjected to the uniaxial compressive strength test, the maximum compressive strength and elastic modulus in the elastic region were calculated from the stress–strain relationship obtained from the uniaxial compressive strength test. For the range of hardfill mixes investigated in this study (i.e., mixes in the range of unit cement content of 80 kg/m$^3$ and unit water contents of 70–100 kg/m$^3$ for aggregates with upper- and lower-limit PSDs), the relationship between the hardfill's uniaxial compressive strength in the elastic range and elastic modulus is shown in Figure 8. For all mixing conditions, the hardfill's elastic modulus is in the range of approximately 220 to 609 times the uniaxial compressive strength (i.e., 220 q–609 q, where q is the uniaxial compressive strength). The elastic modulus of the aggregate mix with the upper-limit PSD is approximately 433 times (i.e., 433 q) the uniaxial compressive strength on average, and the elastic modulus of the aggregate mix with lower-limit PSD is approximately 314 times (i.e., 314 q) on average. Consequently, the elastic modulus of the hardfill of the aggregate mix with the upper-limit PSD is rather large. As shown in Figure 7, the peak value of the stress–strain relationship in the uniaxial compressive strength test (that is, the uniaxial compressive strength measured at the time of sample failure) shows a smaller range in the hardfill of the aggregate mix with the upper-limit PSD than that in the hardfill of the aggregate mix with the lower-limit PSD. However, the above-mentioned elastic modulus results indicate that, on average, the hardfill's stiffness, i.e., the elastic modulus (slope of the stress–strain curve) in the initial strain (elastic deformation) section is slightly larger in the hardfill of the aggregate mix with the upper-limit PSD than that in the hardfill of the aggregate mix with the lower-limit PSD.

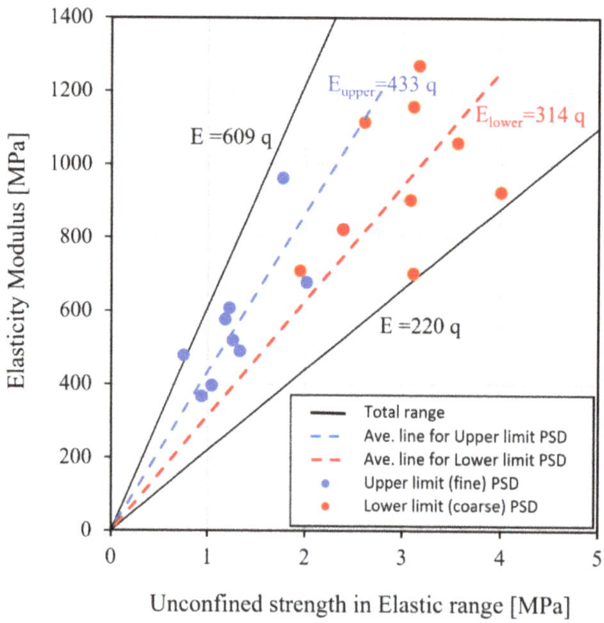


**Figure 8.** Relationship between unconfined compressive strength in elastic range and elastic modulus of hardfill.

## 4. Field Test Construction Results

Figure 9 shows the relationship between the unit water content and dry unit weight for field-compacted hardfills, i.e., hardfills mixed on-site using the same type of aggregates with the maximum particle diameter of 80 mm as used in the indoor laboratory test by matching the PSD and mixing ratio to the laboratory test conditions (i.e., a unit cement content of 80 kg/m$^3$ and three unit water contents of 70, 85, and 100 kg/m$^3$) and compacted on-site using a vibration roller. The change in the dry unit weight according to the unit water content shows that the dry unit weight increases in the case of the 50 cm lift compaction but decreases in the case of the 75 cm lift compaction. Thus, it shows a difference from the results from the indoor laboratory test, in which the values gradually increased in all mixing conditions. Regarding the range of changes in the unit water content, the average dry unit weight is in the range of 2232–2455 kg/m$^3$, i.e., generally similar to the indoor test results (2357–2468 kg/m$^3$) but shows a slightly larger range. In particular, the difference in the dry unit weight according to the lift thickness is larger with the unit water content of 70 kg/m$^3$ than with other unit water contents.

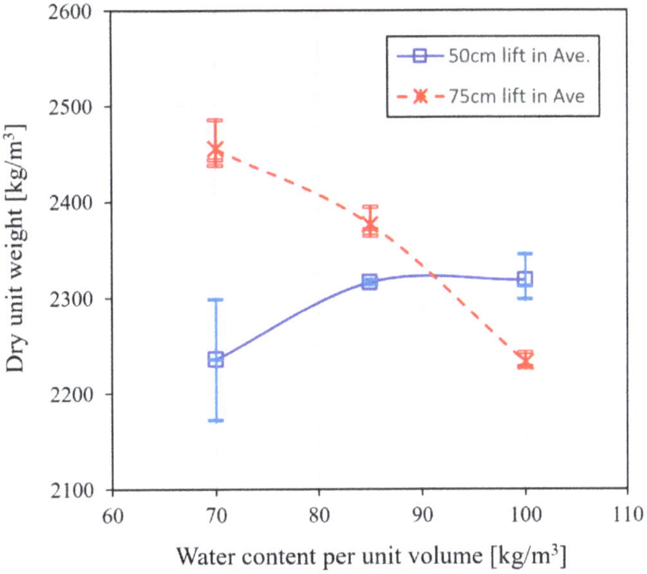


**Figure 9.** Compaction results for field-constructed hardfill (dry unit weight).

Prior to this study, a preliminary test was conducted for the same type of aggregate. In the preliminary test, the maximum dry unit weight was measured using a compaction test [3,30] on a standard specimen with an aggregate with a maximum allowable particle size of 37.5 mm. Figure 10 shows the calculations of the relative compaction using the indoor test results and field compaction measurement results in this study. For the three unit water contents, both the 50 cm and 75 cm lift compactions show relative compactions in the range of 101.7–108.3%. Thus, both exceed 100%. These results are owing to the fact that the size and compaction energy of the hardfill aggregate in the field compaction conditions are larger than those in the standard specimen compaction conditions.

Figure 11 shows the results from uniaxial compressive strength tests for 28-day cores collected after 28 d from field-compacted hardfills. For comparison, the test results for the large indoor specimens were plotted together. The uniaxial compressive strength of the field-compacted hardfill cores ranges from 5.3 to 8.2 MPa in the unit water content range of 70 to 100 kg/m$^3$ for the two lift thicknesses. The overall difference is more than doubles the range of 1.62–4.11 MPa in the indoor test results, and the largest strength value is observed at the unit water content of 85 kg/m$^3$. Thus, there is a rather large difference from the trend of the indoor test results. As can be seen from the error bars added in the data plot, there are larger variations in the indoor test results. The large variations in the indoor test results may be attributed to the fact that, for a large indoor specimen with large aggregate particles in a mold, uniform compaction tends to be hardly achieved by vibrating tamper, compared with a roller-compacted field core. As a result, it seems that less uniformly compacted indoor specimens exhibit lower strength. Even acknowledging such differences, the relatively large difference between the laboratory (indoor) and field tests suggests that the strength measurements from laboratory specimens under similar mixing conditions can generally be evaluated conservatively.

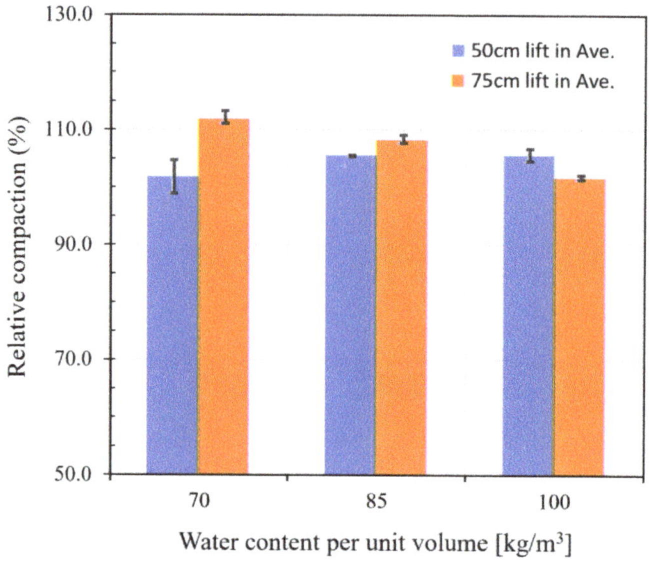


**Figure 10.** Relative compaction of field-constructed hardfill.

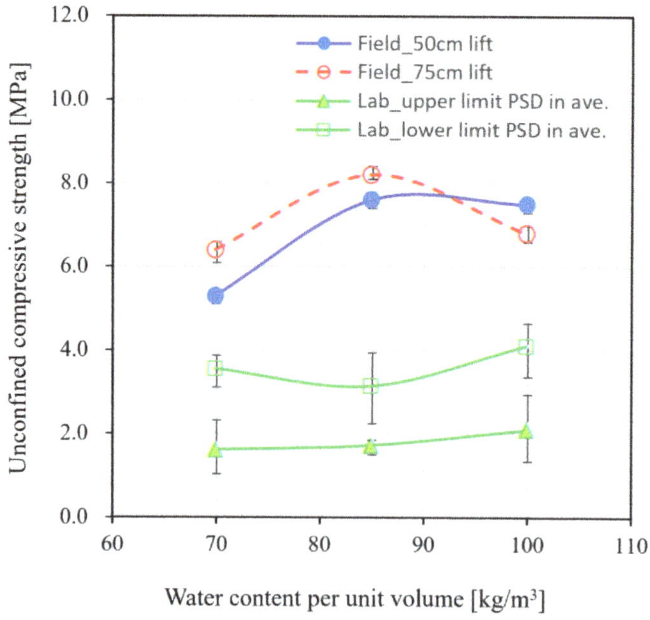


**Figure 11.** Uniaxial compressive strength of field-constructed hardfill (28 d of age).

## 5. Numerical Analysis Results

Figure 12 shows the finite element analysis results for a hardfill dam body with three slopes (1:0.6, 1:0.8, 1:1) and a height of 50 m. It shows the stress distribution of the cross-section of the dam body when the maximum compressive stress and maximum tensile stress occur in the dam body, and the three types of seismic waves in Figure 5 are applied. The location where the maximum compressive stress occurs in the dam body differs according to the type of seismic wave. The maximum compressive stress occurs at the lower part of the upstream side (left side) of the dam body with a 1:0.6 slope for the Hachinohe wave, at the upper part of the downstream side (right side) of the dam body with a 1:0.8 slope for the Ofunato wave, and at the upper side of the dam body with a 1:1.0 slope in the case of an artificial wave. Meanwhile, the maximum tensile stress occurs similarly at the lower part of the upstream side for all three types of seismic waves, as shown similarly in a previous study [13].

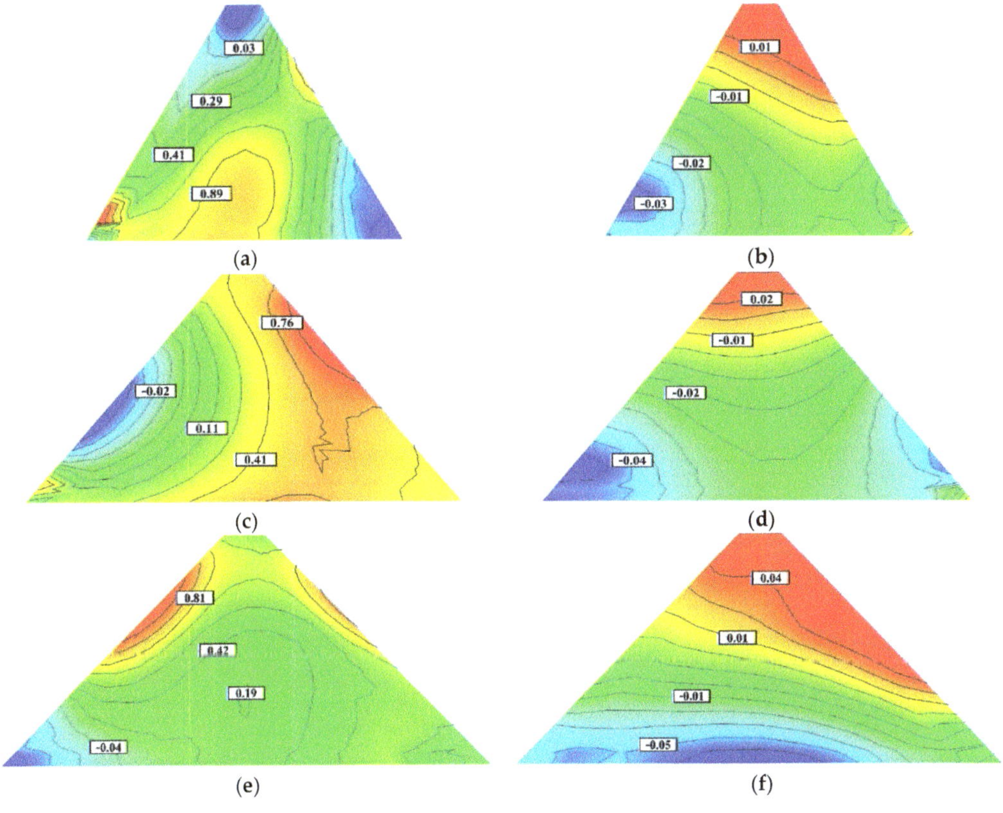


**Figure 12.** Results of FE analysis (stress distributions). (**a**) Compressive stress of Hachinohe wave (1:0.6); (**b**) tensile stress of Hachinohe wave (1:0.6); (**c**) compressive stress of Ofunato wave (1:0.8); (**d**) tensile stress of Ofunato wave (1:0.8); (**e**) compressive stress of artificial seismic wave (1:1.0); (**f**) tensile stress of artificial seismic wave (1:1.0).

Figure 13 shows the changes in the maximum compressive stress and maximum tensile stress according to the change in the slope of the dam body for the three types of seismic waves. The maximum compressive stress is the largest at 2.14 MPa in the hardfill dam body, with the steepest slope of 1:0.6 when the artificial wave with the largest maximum acceleration is applied. However, it rapidly decreases to 1.3 MPa when the slope is relaxed to 1:0.8. The maximum compressive stresses under the action of the Hachinohe and Ofunato waves are 1.07 MPa and 1.14 MPa in the hardfill dam body with a slope of 1:0.6, respectively, i.e., significantly lower than that of the artificial wave. Subsequently, all of them gradually decrease as the slope is relaxed. The largest value of the maximum tensile stress occurs in the hardfill dam body with a slope of 1:0.6 when the artificial wave is applied, as in the case of the maximum compressive stress. However, the magnitude is only 0.1 MPa, i.e., significantly lower than that of the compressive stress. Furthermore, the maximum tensile stress tends to decrease overall as the slope of the dam body is relaxed. However, the decrease is relatively insignificant or increases in some parts (the dam body with a slope of 1:0.8 in the Hachinohe wave).

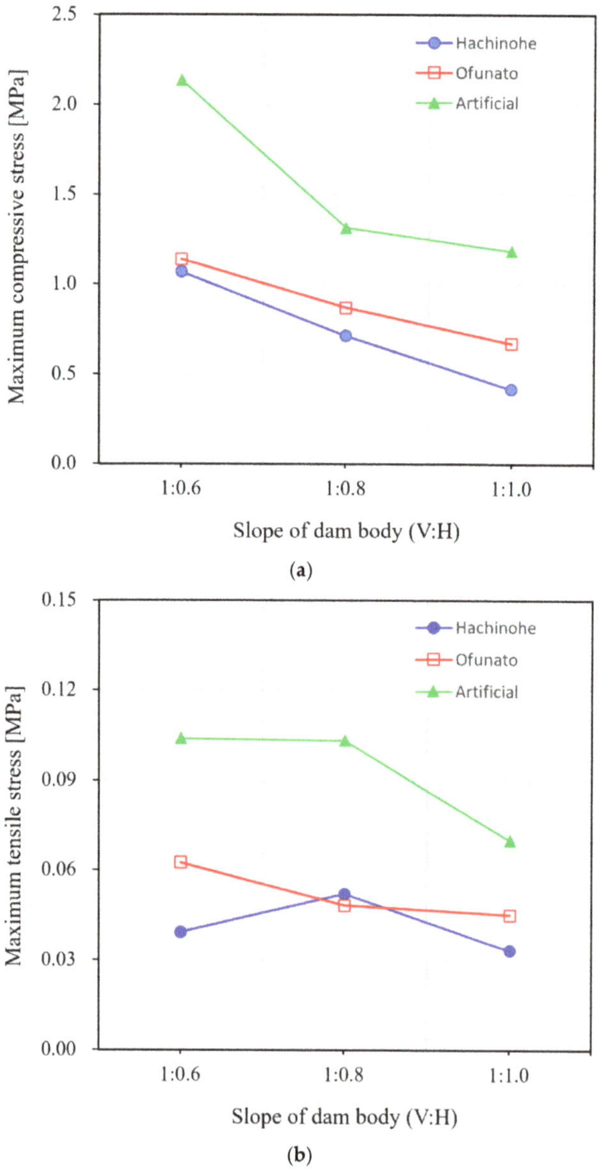


**Figure 13.** Results of FE analysis (maximum compressive and tensile stresses). (**a**) Maximum compressive stress. (**b**) Maximum tensile stress.

The hardfill comprising the dam body must have sufficient strength to resist the maximum compressive stress and maximum tensile stress occurring in the dam body. In this study, the stability of the hardfill dam body was evaluated by applying a safety factor of 1.5 to the maximum stress values shown in Figure 13. That is, the value obtained by multiplying the safety factor of 1.5 by the larger value of the maximum compressive stress and the compressive stress converted into seven times the maximum tensile stress was set as the minimum compressive strength of hardfill required to maintain the stability of the hardfill dam body (i.e., required hardfill strength) [3]. This was compared with the uniaxial compressive strengths of the large hardfill specimens through indoor laboratory tests described in Section 3 and the uniaxial compressive strengths of the field-compacted hardfill cores described in Section 4. The results are shown in Figure 14. The required hardfill strengths of the three inclined hardfill dam bodies for the three seismic waves are displayed as bar graphs. The upper and lower limits of the range of uniaxial compressive strength for the indoor and field samples in the unit water content range of 70 to 100 $kg/m^3$ in Figure 11 are indicated by solid lines. As can be seen, except for the case where the artificial wave with the largest maximum acceleration is applied to the hardfill dam body with the steepest slope of 1:0.6, the uniaxial compressive strength range of the large hardfill specimens in the laboratory tests mostly appears to exceed the required strength of the hardfill dam body. Moreover, the range of the uniaxial compressive strength of the field-compacted hardfill core exceeds the required hardfill strength relatively significantly (approximately 2.7 to 3.8 times), thereby ensuring seismic stability.

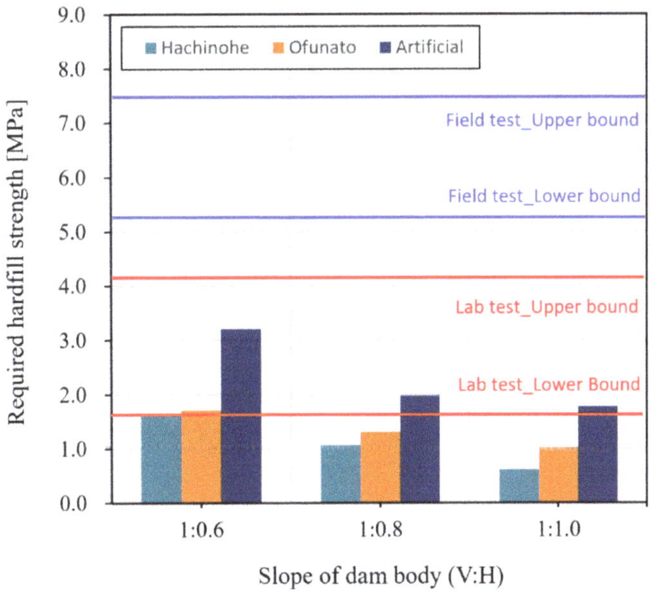


**Figure 14.** Results of FE analysis (required hardfill strength).

## 6. Conclusions

In this study, the strength characteristics of a hardfill material and the stability of hardfill dams during earthquakes were investigated by conducting uniaxial compressive strength tests on both large indoor hard fill specimens and field-compacted cores and by performing a finite element analysis, respectively:

1. From the results of the uniaxial compressive strength test on the large indoor specimens, the compressive strength tended to increase very gently with the unit water content in general for both types of aggregate PSDs, i.e., the upper limit (fine) and lower limit (coarse), mixed with a unit cement content of 80 $kg/m^3$ and three unit

water contents of 70, 85, and 100 kg/m$^3$, respectively. The strength range for the entire set of samples was 1.62–4.11 MPa, i.e., 3.15–4.11 MPa for the aggregate mix with the lower-limit PSD and 1.62–2.10 MPa for the aggregate mix with the upper-limit PSD. The uniaxial compressive strength of the aggregate mix with the lower limit PSD exceeded that of the aggregate mix with the upper limit PSD for the entire unit water content range. From the stress–strain relationship obtained from the uniaxial compressive strength test, the hardfill's elastic modulus was found to be in the range of approximately 220 to 609 times the uniaxial compressive strength.

2. Field compaction was performed with a vibration roller on the same type of aggregates as used in the indoor laboratory test by matching the PSD and mixing ratio to the laboratory test conditions. The results of the uniaxial compressive strength test on the cores collected from the field compacted hardfill after 28 d showed that the strength range for the entire core was 5.30–8.20 MPa, i.e., more than twice as large as the indoor test results, suggesting that the strength measurement of a hardfill through a laboratory test can conservatively evaluate the strength of a field compacted hardfill.
3. The results of the finite element analysis on the hardfill dams modeled using the parameters determined based on the laboratory test results showed that the largest value of the maximum tensile stress of 0.1 MPa occurs in the hardfill dam body with a slope of 1:0.6 when the artificial wave is applied. The maximum tensile stress tended to decrease overall as the slope of the dam body was relaxed. Furthermore, it was shown that the overall strength range obtained for the indoor laboratory specimens and field compaction cores exceeded the required strength of the hardfill, thereby securing seismic stability (except for the hardfill dam with a 1:0.6 slope to which artificial wave was applied).

**Author Contributions:** Formal analysis, W.C. and J.S.; Data curation, Y.K.; Writing—original draft, S.K.; Writing—review & editing, B.K. All authors have read and agreed to the published version of the manuscript.

**Funding:** This research was funded by the Korea Water Resources Corporation.

**Institutional Review Board Statement:** Not applicable.

**Informed Consent Statement:** Not applicable.

**Data Availability Statement:** Not applicable.

**Acknowledgments:** The authors wish to express their appreciation to the Korea Water Resources Corporation and the Expert Group for Earth and Environment Corporation for funding and supporting this work.

**Conflicts of Interest:** The authors declare no conflict of interest.

## References

1. Jia, J.; Lino, M.; Jin, F.; Zheng, C. The Cemented Material Dam: A New, Environmentally Friendly Type of Dam. *Engineering* **2016**, *2*, 490–497. [CrossRef]
2. Kim, K.Y.; Park, H.G.; Jeon, J.S. Strength Characteristics of Cemented Sand and Gravel. *J. Korean Geotech. Soc.* **2005**, *21*, 61–71. (In Korean)
3. Japan Dam Engineering Center. *Engineering Manual for Design, Construction, and Quality Control of Trapezoidal CSG Dam*; Japan Dam Engineering Center: Tokyo, Japan, 2012. (In Japan)
4. Xia, B.; Ding, T.; Xiao, J. Life cycle assessment of concrete structures with reuse and recycling strategies: A novel framework and case study. *Waste Manag.* **2020**, *105*, 268–278. [CrossRef] [PubMed]
5. Portland Cement Association. *Guide for developing RCC Specifications and Commentary*; Portland Cement Association: New York, NY, USA, 2000.
6. U.S. Army Corps of Engineers. *Engineering and Design Roller-Compacted Concrete*; U.S. Army Corps of Engineers: Washington, DC, USA, 2006.
7. Londe, P.; Lino, M. The faced symmetrical hardfill dam: A new concept for RCC. *Int. Water Power Dam Constr.* **1992**, *44*, 19–24.
8. Toshio, H.; Tadahiko, F.; Hideaki, K.; Takashi, S. Design concept of trapezoid-shaped CSG dam. In *Roller Compacted Concrete Dams*; Routledge: London, UK, 2018; pp. 457–464.

9. Cai, X.; Wu, Y.; Gui, X.; Ming, Y. Research review of the cement sand and gravel (CSG) dam. *Front. Struct. Civ. Eng.* **2012**, *6*, 19–24. [CrossRef]
10. Chen, J.; Liu, P.; Xu, Q.; Li, J. Seismic analysis of hardfill dams considering spatial variability of material parameters. *Eng. Struct.* **2020**, *211*, 110439. [CrossRef]
11. Yoon, J.S.; Park, W.Y.; Kim, K.T. Applicability analysis of Local Riverbed Aggregate for Hardfill Dam. *Korean Soc. Civ. Eng.* **2016**, *10*, 218–219. (In Korean)
12. Guillemot, T.; Lino, M. Design and Construction Advantages of Hardfill Symmetrical Dams-Case Study: Safsaf Dam In Eastern Algeria. In Proceeding of the 6th International Symposium on Roller Compacted Concrete (RCC) Dams, Zaragoza, Algeria, 23–25 October 2012.
13. Yuan, P.; He, Y.; Wang, Y. Study on seismic safety of Poshitougou flood intercepting dam. In *IOP Conference Series: Earth and Environmental Science*; IOP Publishing: Bristol, UK, 2021; Volume 861, pp. 072–073. [CrossRef]
14. Karimi, S.; Aghajani, H.F. A New Type of Cemented Sand-Gravel (CSG) Mixtures for Water-Tightening of Hardfill Dams. *Res. Sq. Platf. LLC* **2021**. [CrossRef]
15. Zheng, C. Progress on Cemented Material Dams in CHINA. In Proceedings of the International Symposium on the State-Of-The-Art Dam Construction Technology, Daejoen, Republic of Korea, 5–6 November 2015.
16. Ayagh, A.G.; Mohammadian, A. Optimum characteristic compressive strength for cmds (case study: Dasht-e-palang dam). In Proceedings of the 2017 International Conference on Long-Term Behaviour and Environmentally Friendly Rehabilitation Technologies of Dams, Tehran, Iran, 17–19 October 2017; pp. 243–250. [CrossRef]
17. Karimizad, N.; Teshnizi, E.S.; Mahdad, M.; Karimiazar, J. Investigating the design features of CSG dams. *JOJ Sci.* **2020**, *3*, 555588.
18. Kim, K.Y. The Factors Affecting the Unconfined Compressive Strength of C.S.G Materials. *J. Korean Geotech. Soc.* **2006**, *22*, 33–45. (In Korean)
19. Kim, K.Y.; Jeon, J.S.; Kim, Y.S. Laboratory Mix Design of C.S.G Method. *J. Korean Geotech. Soc.* **2006**, *22*, 27–37. (In Korean)
20. Omae, S.; Sato, N.; Oomoto, I. Dynamic properties of CSG. In Proceedings of the International Symposium on Roller Compacted Concrete (RCC) Dams, Madrid, Spain, 17–19 November 2003; pp. 511–518.
21. Liu, Z.; Jia, J.; Feng, W.; Ma, F.; Zheng, C. Shear Strength of Cemented Sand Gravel and Rock Materials. *Sains Malays.* **2017**, *46*, 2101–2108. [CrossRef]
22. Fujisawa, T.; Nakamura, A.; Kawasaki, H.; Hirayama, D.; Yamaguchi, Y.; Sasaki, T. Material Properties of CSG for the Seismic Design of Trapezoid-Shaped CSG Dam. In Proceedings of the 13th World Conference on Earthquake Engineering, Vancouver, BC, Canada, 1–6 August 2004; pp. 292–296.
23. Peng, Y.; Yuan, Y. Analysis on the Dynamic Properties of Hardfill Materials. *J. Appl. Sci. Eng. Innov.* **2014**, *1*, 54–58.
24. Ren, H.; Cai, X.; Wu, Y.; Jing, P.; Guo, W. A Study of Strength Parameter Evolution and a Statistical Damage Constitutive Model of Cemented Sand and Gravel. *Materials* **2023**, *16*, 542. [CrossRef] [PubMed]
25. Farshbaf Aghajani, H.; Soltani-Jigheh, H.; Salimi, M.; Karimi, S.; Estekanchi, V.; Ahari, R.A. Investigating the strength, hydraulic conductivity, and durability of the CSG (cemented sand-gravel) check dams: A case study in Iran. *SN Appl. Sci.* **2022**, *4*, 169. [CrossRef]
26. Guo, L.; Zhang, J.; Guo, L.; Wang, J.; Shen, W. Research on profile design criteria of 100 m CSG dams. *Case Stud. Constr. Mater.* **2022**, *16*, e01137. [CrossRef]
27. Peng, Y. Study on Structural Characteristics of CSG Dam. *Int. J. Sci.* **2014**, *1*, 70–74.
28. Zafeiratos, E.; Psycharis, I.; Papanicolaou, P. Dynamic behavior of faced symmetrical hardfill dams. In Proceedings of the 11th ICOLD European Club Symposium, Crete, Greece, 2–4 October 2019.
29. *Standard No. ASTM D698-12:2021*; Standard Test Method for Laboratory Compaction Characteristics of Soil Using Standard Effort (12,400 ft-lbf/ft3 (600 kN-m/m$^3$)). American Society for Testing and Materials (ASTM): West Conshohocken, PA, USA, 2021.
30. *Standard No. ASTM D1557-12:2021*; Standard Test Methods for Laboratory Compaction Characteristics of Soil Using Modified Effort (56,000 ft-lbf/ft3 (2700 kN-m/m$^3$)). American Society for Testing and Materials (ASTM): West Conshohocken, PA, USA, 2021.
31. *Standard No. KS F 2314:2018*; Standard Test Method for Unconfined Compression Test of Soils. Korean Standards Association: Seoul, South Korea, 2013. (In Korean)
32. *Standard No. KS F 2311:2016*; Standard Test Method for Density of Soil in Place by the Sand Cone Method. Korean Standards Association: Seoul, South Korea, 2016. (In Korean)
33. *Standard No. ASTM C42/C42M:2020*; Standard Test Method for Obtaining and Testing Drilled Cores and Sawed Beams of Concrete. American Society for Testing and Materials (ASTM): West Conshohocken, PA, USA, 2020.
34. *Standard No. ASTM C39/C39M:2021*; Standard Test Method for Compressive Strength of Cylindrical Concrete Specimens. American Society for Testing and Materials (ASTM): West Conshohocken, PA, USA, 2021.

applied sciences

MDPI

Article

# Consistency of Water Vapour Pressure and Specific Heat Capacity Values for Modelling Clay-Based Engineered Barriers

**Laura Asensio *, Gema Urraca and Vicente Navarro**

Geoenvironmental Engineering Group, Universidad de Castilla-La Mancha, Avda. Camilo José Cela 2, 13071 Ciudad Real, Spain
* Correspondence: laura.asensio@uclm.es

**Featured Application: Thermo-hydraulic modelling of engineering geomaterials consistently with vaporisation.**

**Abstract:** The aim of this study is to assess the consistency in the modelling of thermo-hydraulic problems in clay-based engineered barriers. This study focuses on two aspects: the modelling of vapour pressure as a function of temperature, and the specific heat capacities of liquid water and water vapour in relation to the enthalpy of vaporisation and the internal energy of liquid water and water vapour. Regarding the first aspect, several formulations of the saturated vapour pressure have been inspected, evaluating their accuracy and information provided in the temperature range from 0 to 150 °C. Regarding the second aspect, the enthalpy of vaporisation and the internal energy of water were used to assess the consistency of pairs of specific heat capacity values in the same temperature range. Values from the literature were also inspected. An accurate and simple enough expression for the saturated water vapour pressure with temperature has been identified as the optimal option for modelling. Recommendations on specific heat capacity constant values for liquid water and vapour are suggested to maximise consistency in the studied temperature range. However, the loss of accuracy in the enthalpy or internal energy of vaporisation associated with the inspected specific heat capacity pairs is limited.

**Keywords:** thermo-hydraulic modelling; vapour pressure; heat capacity; unsaturated soil; geomaterial; clay and soil; bentonite barrier

**Citation:** Asensio, L.; Urraca, G.; Navarro, V. Consistency of Water Vapour Pressure and Specific Heat Capacity Values for Modelling Clay-Based Engineered Barriers. *Appl. Sci.* **2023**, *13*, 3361. https://doi.org/10.3390/app13053361

Academic Editors: Daniel Dias and José Manuel Moreno-Maroto

Received: 28 October 2022
Revised: 23 February 2023
Accepted: 3 March 2023
Published: 6 March 2023



## 1. Introduction

To accurately model a thermo-hydraulic problem in clay-based engineered barriers, especially at temperatures significantly higher than standard temperature (20 °C), it is crucial to include accurate water vaporisation in the formulation. First, the water content of geomaterials is relevant for their thermal, hydraulic and mechanical properties. Second, vaporisation consumes energy and can affect the temperature of the system. In the problem of clay-based engineered barriers in deep geological repositories for high-level radioactive waste, the formulation used must be robust for temperatures up to 150 °C, a temperature that could be reached in the surface of the canisters that will be surrounded with clay barriers [1]. This applies in particular to the modelling of vapour pressure as a function of temperature and to a consistent use of the specific heat capacities of liquid water and water vapour.

The current international standard, as accepted by the International Association for the properties of Water and Steam (IAPWS), for saturated vapour pressure values is the equation of state in the IAPWS-95 formulation [2]. Tabulated values for temperatures between 0.01 and 150 °C can be found, for example, in the CRC Handbook of Chemistry and Physics [3]. In addition, the auxiliary equations of Wagner and Pruss [4] for saturated vapour pressure as a function of temperature are of comparable accuracy to the uncertainty of the experimental data. However, other simpler formulations are most frequently used in modelling. Some of them were formulated for meteorology. For instance, Buck [5] developed a set of nine equations "to be easily implemented on a calculator or computer". The nine equations are of different complexity and are optimised for different temperature ranges. In addition, for multiple purposes, Huang [6] developed an expression from the Clausius–Clapeyron equation, and fitted it to the IAPWS reference dataset for a temperature range between 0 and 100 °C. In the case of the thermo-hydraulic modelling of clay-based engineered barriers in deep geological repositories for high-level radioactive waste, other approximations based on exponential functional structures are common. In this way, Thomas and He [7] use an expression of this kind in a general model for deformable unsaturated soil. Gens et al. [8] and Dupray et al. [9] both used an equation with an exponential structure to model the full-scale FEBEX in situ test (Grimsel Test Site, Switzerland) for high-level nuclear waste disposal. Nowak et al. [10] modelled two full-scale in situ experiments analysing the behaviour of radioactive waste repositories (Buffer/Container Experiment and Isothermal Test, conducted at the Whiteshell Underground Research Laboratory, Manitoba, Canada, by Atomic Energy of Canada Limited) using an exponential function for the saturated vapour pressure. Wang et al. [11] also used an expression of this kind to model a long-term laboratory heating and hydration test on bentonite pellets simulating the behaviour of a repository buffer (HE-E cells, conducted by CIEMAT in Madrid, Spain). Likewise, Abed and Sołowski [12] presented a framework for modelling unsaturated soils where an exponential expression is used. While all these works [5–12] include an exponential function in their formulation of the saturated water vapour pressure, the arguments of the exponential are, in turn, different functions (polynomials, fractions) of temperature, and include a different number of parameters (two to five). These differences make the accuracy of these equations to be variable for different temperature ranges. While many may be acceptable for standard temperature, the suitability of a particular equation should be inspected for temperatures up to 150 °C. In addition, it would be advisable to use a model in which the number of parameters is justified by the additional information provided by them (parsimony principle).

Further, the values used for the specific heat capacity of water in liquid and vapour forms also play a role in the modelling of water vaporisation. These specific heat capacities are also a function of temperature, as can be seen, for instance, in the National Institute of Standards and Technology (NIST) Chemistry WebBook [13]. However, this temperature variability is not usually introduced in the modelling of clay-based engineered barriers. The most common approach is to use constant specific heat capacity values, as was done, for instance, by Collin et al. [14] in their model for clay barriers in nuclear waste deep geological disposal. Jussila and Ruokolainen [15] also used constant values when modelling compacted bentonite in the context of geological spent fuel disposal. Gens [16] reviewed a formulation for unsaturated soils to model soil-environment interactions, in which constant heat capacity values were also assumed. Like in the mentioned works, Zheng et al. [17] used constant values to model the full-scale FEBEX in situ test. Moreover, the works in the previous paragraph giving specific heat capacity values [7,12] also use constant values for them. Constant values are an approximation that should be used consistently for the temperature range expected. This consistency can be evaluated with regard to the reference specific heat capacity values, and with regard to the enthalpy of vaporisation or the internal energy of liquid water and vapour.

The aim of this work is to analyse the consistency in the modelling of the thermo-hydraulic behaviour of clay-based engineered barriers for deep geological repositories. To this end, the formulation of the saturated vapour pressure and the values used for the specific heat capacities of liquid water and vapour will be inspected for repository temperatures. The accuracy of saturated vapour pressure formulations will be assessed using a relative error, and their parsimony will be assessed using the Akaike [18] and Bayesian [19] information criteria for model selection. Regarding specific heat capacities, the liquid water and vapour values will be assessed comparing it to reference values. In addition, taking into account the energy balance equation solved for soils, the difference between liquid water and vapour values will be evaluated so that it can reproduce as accurately as possible the evolution of the enthalpy of vaporisation and the internal energy difference with temperature. Recommendations for modelling will be proposed accordingly.

## 2. Materials and Methods

### *2.1. Reference Information from the Literature*

#### 2.1.1. Saturated Water Vapour Pressure

Seven expressions from the literature for the saturated water vapour pressure $p_{V0}$ as a function of temperature were inspected. All the equations and their reference work are included in Table 1. The equation of Wagner and Pruss [4], Equation (1), is taken as a reference. Given its high accuracy, it was used to develop the IAPWS-95 formulation [2]. The first inspected expression, Equation (2), is taken from Buck [5]. From among Buck's expressions, equation $e_{w6}$ is used in the present work, since it is the one with the smallest relative error in the temperature interval 0 to 100 °C. It involves temperature fractions as an argument of an exponential expression, and includes four parameters. Next, Huang's [6] expression, Equation (3), includes temperature fractions multiplying and as an argument of an exponential expression, and five parameters.

**Table 1.** Expressions for saturated water vapour pressure as a function of temperature $T$ from the literature. $R$: universal gas constant (8.3144 J/mol/K), $MW$: molar mass of water (0.018016 kg/mol).

| Reference | Expression | Coefficients and Auxiliary Expressions | Equation No. |
|---|---|---|---|
| Wagner and Pruss [4] | $p_{V0} = p_c exp\left[\frac{T_c}{T}\left(c_1\theta + c_2\theta^{1.5} + c_3\theta^3 + c_4\theta^{3.5} + c_5\theta^4 + c_6\theta^{7.5}\right)\right]$ | $c_1 = -7.85951783$, $c_2 = 1.84408259$, $c_3 = -11.7866497$, $c_4 = 22.6807411$, $c_5 = -15.9618719$, $c_6 = 1.80122502$. $\theta = 1 - \frac{T}{T_c}$, $T_c = 647.096$ K, $p_c = 22064$ kPa. | (1) |
| Buck [5] | $p_{v0} = c_1\, exp\left[\frac{\left(c_2 - \frac{T}{c_3}\right)T}{T + c_4}\right]$ | $c_1 = 0.61121$ kPa, $c_2 = 18.564$, $c_3 = 254.4$ °C, $c_4 = 255.57$ °C. | (2) |
| Huang [6] | $p_{v0} = \frac{1\text{ Pa}}{(T + c_1)^{c_2}} exp\left[c_3 + \frac{c_4}{T + c_5}\right]$ | $c_1 = 105$, $c_2 = 1.57$, $c_3 = 34.494$, $c_4 = -4924.99$ °C, $c_5 = 237.1$ °C. $T$ in °C. | (3) |
| Thomas and He [7] | $p_{v0} = c_1 \frac{RT}{MW} exp[c_2(T - 273\text{ K}) + c_3(T - 273\text{ K})^2]$ | $c_1 = \frac{1}{194.4}$ kg/m$^3$, $c_2 = 0.06374$ K$^{-1}$, $c_3 = -1.634 \times 10^{-4}$ K$^{-2}$. | (4) |
| Gens et al. [8] | $p_{v0} = c_1\, exp\left(\frac{c_2}{T + 273\text{ C}}\right)$ | $c_1 = 1.36075 \times 10^8$ kPa, $c_2 = -5239.7$ °C. | (5) |
| Dupray et al. [9] | $p_{v0} = c_1\, exp\left(\frac{c_2}{T}\right)$ | $c_1 = 1.12659 \times 10^8$ kPa, $c_2 = -5192.74$ K. | (6) |
| Nowak et al. [10], Wang et al. [11] | $p_{v0} = 10^{-3}\text{kg/m}^3 \frac{RT}{MW} exp\left(c_1 + \frac{c_2}{T}\right)$ | $c_1 = 19.84$, $c_2 = -4975.9$ K. | (7) |
| Abed and Sołowski [12] | $p_{v0} = 10^{-3}\text{kg/m}^3 \frac{RT}{MW} exp\left(c_1 + \frac{c_2}{T}\right)$ | $c_1 = 19.891$, $c_2 = -4974.0$ K. | (8) |

Then, expressions used to model thermal problems in unsaturated soils and clay-based engineered barriers were inspected. First, Thomas and He [7] use an expression for $p_{V0}$, Equation (4), that includes a quadratic polynomial of temperature $T$ inside an exponential function, and three parameters. Gens et al. [8], Equation (5), and Dupray et al. [9], Equation (6), use the same expression structure for $p_{V0}$, an exponential of the inverse of temperature with two parameters, although their values are slightly different. Finally, Nowak et al. [10] and Wang et al. [11], Equation (7), and Abed and Sołowski [12], Equation (8), use the same expression structure for $p_{V0}$, an exponential function involving the inverse of temperature with three parameters. The two former works use exactly the same parameters, while the latter uses somewhat different figures. The expressions and their reference work are included in Table 1.

2.1.2. Specific Heat Capacity of Water

Specific heat capacity values of water in liquid and vapour forms used for modelling unsaturated soils and clay-based engineered barriers from six works in the literature were selected for inspection. Their values are given in Table 2. Thomas and He [7] and Collin et al. [14] use isobaric specific heat capacities in their formulations, and give isobaric values for liquid water and vapour. Although Jussila and Ruokolainen [15] use isochoric specific heat capacities in their free energy model, they formulate its application to their thermo-hydro-mechanical model in terms of the isobaric specific heat capacities. Then, Jussila and Ruokolainen [15] also give isobaric values for liquid water and vapour. Gens [16] states that isochoric specific heat capacities should be used with the formulation it presents, because it is "established in terms of internal energy". It gives isobaric values for liquid water and vapour, and it explains how to obtain isochoric values from isobaric ones (for vapour, treating it as an ideal gas [20], by substracting $R/MW$, while for liquid water, treating it as incompressible [20], the difference can be disregarded). Table 2 contains both the given isobaric values and the calculated isochoric values. Zheng et al. [17] and Abed and Sołowski [12] do not state if the specific heat capacities given for liquid water and vapour are isobaric or isochoric. The values of specific heat capacities as a function of temperature in the NIST Chemistry WebBook [13] for the minimum pressure between saturated vapour pressure and 100 kPa will be used as a reference for comparison.

**Table 2.** Specific heat capacity of water in liquid and vapour forms given in the literature for modelling unsaturated soils and clay-based engineered barriers.

| Reference | $c^L$ (J/kg/K) | $c^V$ (J/kg/K) |
|---|---|---|
| Thomas and He [7] | $4180^{\mathrm{p}}$ | $1870^{\mathrm{p}}$ |
| Collin et al. [14] | $4180^{\mathrm{p}}$ | $1900^{\mathrm{p}}$ |
| Jussila and Ruokolainen [15] | $4180^{\mathrm{p}}$ | $1870^{\mathrm{p}}$ |
| Gens [16] | $4180^{\mathrm{p,\,vol}}$ | $1890^{\mathrm{p}}$, $1428^{\mathrm{vol}}$ |
| Zheng et al. [17] | $4202^{\mathrm{ns}}$ | $1620^{\mathrm{ns}}$ |
| Abed and Sołowski [12] | $4180^{\mathrm{ns}}$ | $1900^{\mathrm{ns}}$ |

Superscripts p: isobaric, vol: isochoric, ns: not specified.

*2.2. Model Analysis and Selection for Saturated Water Vapour Pressure*

The drift of each saturated water vapour pressure model presented in Section 2.1.1 with respect to the reference of Wagner and Pruss [4] can be evaluated with its relative error $RE_{\mathrm{pv0},i}$

$$RE_{\mathrm{pv0,i}} = \left(p_{v0,i} - p_{v0,(1)}\right) / p_{v0,(1)} \tag{9}$$

where $i$ denotes each of the formulations in Table 1, and $p_{\mathrm{v0},(1)}$ is the reference saturated water pressure of Wagner and Pruss [4], Equation (1).

However, in model selection, in addition to the similarity to a reference, the number of parameters used is relevant, and parsimony has to be taken into account. Model selection is then a balance between under- and overfitted models [21]. Two of the most usual selection criteria are the Akaike information criterion (AIC) and the Bayesian information criterion (BIC).

The AIC [18] is an estimator of the expected Kullback–Leibler information loss for each model. In the case of assuming least squares fitting with normally distributed errors for the models, the AIC can be written as [21]

$$\mathrm{AIC} = n \log\left(\frac{\sum \varepsilon^2}{n}\right) + 2k \tag{10}$$

where $n$ is the sample size, $k$ is the number of parameters of the model, and $\varepsilon$ are the residuals from the model. The value of the AIC for each model should be interpreted in terms relative to the values of the other models compared. From a set of $S$ models, the one obtaining the minimum AIC will define $\mathrm{AIC}_{\mathrm{min}}$. A normalised measure of the relative likelihood of each model $i$ is given by the Akaike weights $Aw$ [21]

$$Aw_i = \frac{\exp[(\mathrm{AIC}_{\mathrm{min}} - \mathrm{AIC}_i)/2]}{\sum_{s=1}^{S} \exp[(\mathrm{AIC}_{\mathrm{min}} - \mathrm{AIC}_s)/2]} \tag{11}$$

$Aw_i$ can be interpreted as the probability that model $i$ is the best model for the data in terms of Kullback–Leibler information among the $S$ models analysed. The BIC [19] is based on Bayes factors. In the case of assuming least squares fitting with normally distributed errors for the models, the BIC can be written as

$$\mathrm{BIC} = n \log\left(\frac{\sum \varepsilon^2}{n}\right) + k \log(n) \tag{12}$$

As with the AIC, BIC values should be interpreted in relative terms. From a set of $S$ models, posterior model probabilities $Bp$ can be computed for each model $i$ as [21]

$$Bp_i = \frac{\exp[(\mathrm{BIC}_{\mathrm{min}} - \mathrm{BIC}_i)/2]}{\sum_{s=1}^{S} \exp[(\mathrm{BIC}_{\mathrm{min}} - \mathrm{BIC}_s)/2]} \tag{13}$$

where $\mathrm{BIC}_{\mathrm{min}}$ is the minimum BIC obtained among the models in the set. $Bp_i$ is the inferred probability that model $i$ is the most parsimonious model closest to the reference in the model set.

The model obtaining the $Aw$ and $Bp$ values closest to 1 will be selected following AIC and BIC model selection criteria.

### 2.3. Thermodynamic Basis for Analysing the Consistency of the Formulations

The energy balance equation for the thermal problem in soils can be written in terms of the specific internal energy $u$ of each phase $i$ (solid, liquid, gas) and species $j$ (soil, water, air) composing the mixture of the soil [8,11,12,16,22,23]

$$\sum_{i,j} \frac{\partial}{\partial t}\left(\rho^{ij}\theta^{ij}u^{ij}\right) = \sum_{i,j} -\nabla\cdot\left(-\mathbf{K}\cdot\nabla T + \rho^{ij}\theta^{ij}u^{ij}\mathbf{v}^{ij}\right) + g \tag{14}$$

where $t$ is time, $\nabla$ is the gradient operator, $\nabla\cdot$ is the divergence operator, $\mathbf{K}$ is the thermal conductivity tensor, $g$ is an energy source term, and $\rho^{ij}$, $\theta^{ij}$ and $\mathbf{v}^{ij}$ are, respectively, the density, the specific volume and the velocity of component ($i$, $j$). However, it is usual to write this balance equation in terms of the specific enthalpies $h^{ij}$ of each component [7,9,10,14,17,23]

$$\sum_{i,j} \frac{\partial}{\partial t}\left(\rho^{ij}\theta^{ij}h^{ij}\right) = \sum_{i,j} -\nabla\cdot\left(-\mathbf{K}\cdot\nabla T + \rho^{ij}\theta^{ij}h^{ij}\mathbf{v}^{ij}\right) + g \tag{15}$$

The total differentials d$u$ and d$h$ can, in a general case, be written in terms of the isochoric $c_v$ and isobaric $c_p$ specific heat capacities as [20]

$$du = c_v \, dT + \left[ T \left( \frac{\partial P}{\partial T} \right)_v - P \right] dv \tag{16}$$

$$dh = c_p \, dT + \left[ -T \left( \frac{\partial v}{\partial T} \right)_p + v \right] dP \tag{17}$$

where $P$ is pressure, v is volume, and $()_v$ and $()_p$ indicate derivative under constant volume and pressure, respectively. Although specific heat capacities are functions of temperature and pressure, they are usually taken as constants in the modelling of engineering geomaterials. In addition, in such modelling, the differentials d$u$ and d$h$ are usually formulated as a function of a constant specific heat capacity for each component and temperature only, see [7–12,14,16,17,23]. Then, for the components water vapour V and liquid water L, the internal energy and enthalpy terms can be written as

$$u_V - u_{V0} = c_v^V (T - T_0) \tag{18}$$

$$u_L - u_{L0} = c_v^L (T - T_0) \tag{19}$$

$$h_V - h_{V0} = c_p^V (T - T_0) \tag{20}$$

$$h_L - h_{L0} = c_p^L (T - T_0) \tag{21}$$

where $u_{V0}$ and $u_{L0}$ are the internal energies and $h_{V0}$ and $h_{L0}$ are the enthalpies of vapour and liquid water, respectively, at reference temperature $T_0$.

The enthalpy of vaporisation is an energy sink in the energy balance equation in soils. At a given temperature, the enthalpy of vaporisation $\Delta h_{LV}$ is equal to the difference between water vapour and liquid water enthalpies, which following Equations (20) and (21) gives

$$\Delta h_{LV} = h_V - h_L = \left[ \Delta h_{LV,T0} - \left( c_p^V - c_p^L \right) T_0 \right] + \left( c_p^V - c_p^L \right) T \tag{22}$$

where $\Delta h_{LV,T0}$ is the enthalpy of vaporisation at $T_0$. Given that $\Delta h_{LV,T0}$ and $T_0$ are constant, if $c^V{}_p$ and $c^L{}_p$ are constant too, the enthalpy of vaporisation should respond to a linear equation

$$\Delta h_{LV} = a + b\,T \tag{23}$$

where $a$ and $b$ are constant parameters. An equation such as Equation (23) can be fitted to reference data of the enthalpy of vaporisation to find these parameters. Reference data for $\Delta h_{LV}$ will be taken from the NIST Chemistry WebBook [13] for the minimum pressure between saturated vapour pressure and 100 kPa. Comparing Equations (22) and (23), for the pairs of constant heat capacities to be consistent with the enthalpy of vaporisation, their difference ($c^V{}_p - c^L{}_p$) should be close to the value of parameter $b$ identified to fit Equation (23). Ideally, the difference should be equal to $b$, but there are two simplifications in deriving the equations that may make the two values differ. The first one is disregarding the last term on the right hand side of Equation (17) to derive Equations (20) and (21). The second one is considering $c^V{}_p - c^L{}_p$ as constant with temperature. To assess the effect of constant specific heat capacities, the values of $c_p$ for temperatures 0–150 °C in the NIST Chemistry WebBook [13] will be used as a reference for comparison.

Analogously to Equation (22), the difference in internal energy $\Delta u_{LV}$ between vapour, Equation (18), and liquid water, Equation (19), can be expressed as

$$\Delta u_{LV} = u_V - u_L = \left[ \Delta u_{LV,T0} - \left( c_v^V - c_v^L \right) T_0 \right] + \left( c_v^V - c_v^L \right) T \tag{24}$$

where $\Delta u_{LV,T0}$ is the internal energy difference at $T_0$. Given that $\Delta u_{LV,T0}$ and $T_0$ are constant, if $c^V{}_v$ and $c^L{}_v$ are constant too, $\Delta u_{LV}$ should also respond to a linear equation

$$\Delta u_{LV} = a' + b'\,T \tag{25}$$

Comparing Equations (24) and (25), the difference $(c^V{}_v - c^L{}_v)$ should now be close to the value of parameter $b'$. The values of specific heat capacities $c_v$ for temperatures 0–150 °C in the NIST Chemistry WebBook [13] will again be used as a reference for comparison.

## 3. Results and Discussion

### *3.1. Saturated Water Vapour Pressure*

The saturated water vapour pressure values obtained with the formulations in Table 1 are plotted as a function of temperature in Figure 1 for a temperature range between 0 and 150 °C.

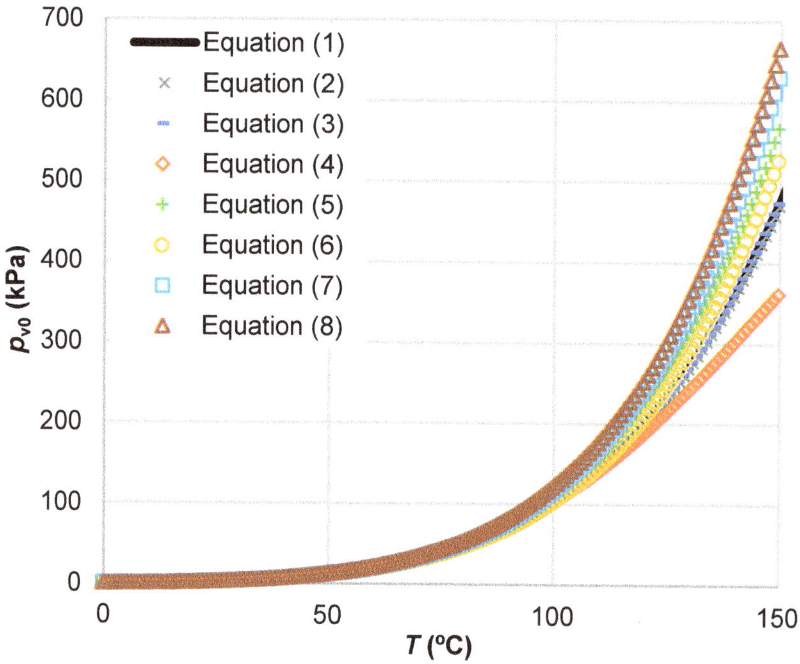


**Figure 1.** Saturated water vapour pressure as a function of temperature in the temperature range 0–150 °C for the different formulations analysed, Equations (1)–(8) [4–12]. Note that the results of Equations (2) and (3) practically overlap with the reference values of Equation (1).

Figure 1 shows that all the formulations inspected obtain similar values of the saturated water vapour pressure for the lower temperatures. However, the formulations diverge increasingly with temperature, with notable differences at 150 °C. To illustrate the drift of each formulation, their relative error $RE_{pv0}$, Equation (9), with respect to the reference of Equation (1) [4] is plotted in Figure 2.

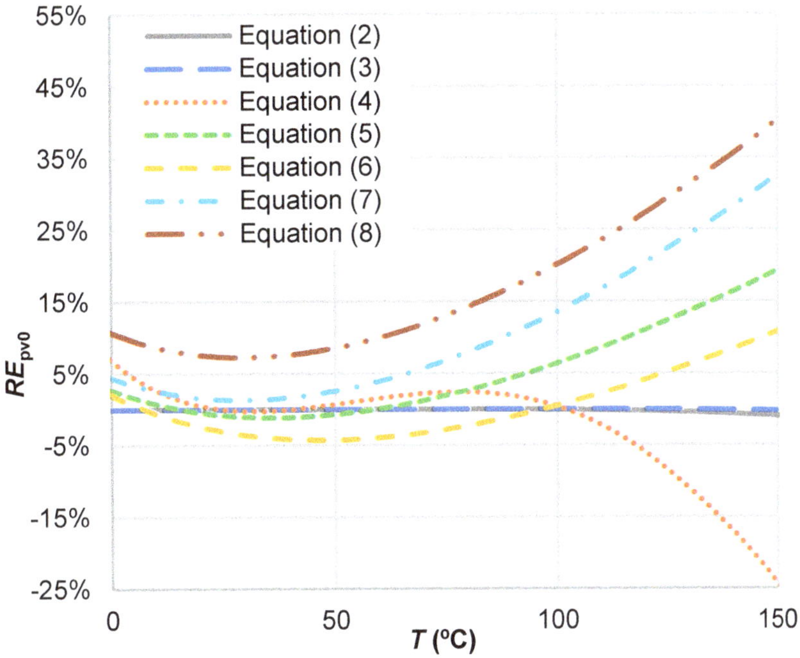


**Figure 2.** Relative error in the saturated water vapour pressure, $RE_{pv0}$, with respect to the reference, Equation (1) [4], as a function of temperature for the different formulations analysed. Equations (2)–(8) [5–12].

Analysing Figure 2, Huang's [6] equation, Equation (3), is the saturated water vapour pressure formulation with the least error in the temperature range analysed, 0–150 °C, and also in the more limited temperature range, 0–100 °C. The relative error of Equation (3) is kept within ±0.2% in the temperature range 0–150 °C, and is up to ±0.006% in the range 0–100 °C. Among the other formulations, only Buck's [5] equation, Equation (2), limits its relative error under ±1.0% in the temperature range 0–150 °C. On the side of the lower accuracy, Equation (7), Nowak et al. [10] and Wang et al. [11], and Equation (8), Abed and Sołowski [12], both show relative errors higher than 10% for temperatures higher than 90 °C.

Equation (3) is able to reproduce the reference saturated water pressure values with a notably lower error than Equations (2) and (4)–(8). This result should not be surprising, since Equation (3) is the one containing the highest number of parameters from the equations in Table 1 after the reference of Wagner and Pruss [4], Equation (1), (5 compared to 2–4 parameters). Then, it should be analysed if the increase in number of parameters is justified or if, on the contrary, Equation (3) is overfitted. To this end, the AIC and the BIC were applied to the models, and the values of Akaike weights $Aw$ from the AIC, Equation (11), and posterior model probabilities $Bp$ from the BIC, Equation (13), were computed for all of them. For a number of observations equidistant between 0.01 and 150 °C of $n = 3$, $Aw$ and $Bp$ are plotted for Equations (2)−(8) in Figure 3. In the computations, the number of parameters $k$ of Equations (2)−(8) are (4, 5, 3, 2, 2, 2, 2) (Table 1). The residual $\varepsilon$ has been computed as the difference between the pressure obtained with each formulation and that of Equation (1). For any $n$, both information criteria for model selection yield the same result: the selected model for the data is Equation (3), the formulation by Huang [6]. It obtains $Aw > 0.977$ and $Bp > 0.985$ for any $n \geq 3$, and $Aw > 0.999$ and $Bp > 0.999$ for any $n \geq 5$. The rest of models, Equations (2) and (4)–(8), obtain $Aw < 0.023$ and $Bp < 0.015$ for any $n \geq 3$, and tend to 0 with increasing $n$. Then, the increase in number of parameters for

Equation (3) with respect to the rest of equations is justified with the increase in accuracy (information) they provide.

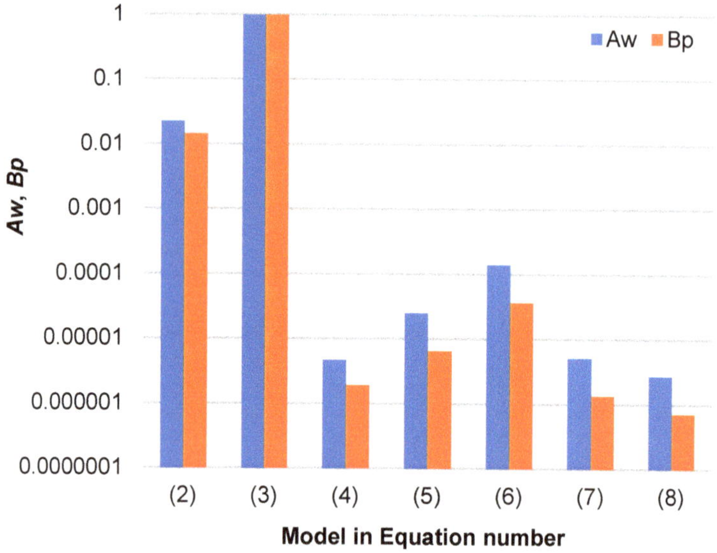


**Figure 3.** Model selection results from information criteria for $n$ = 3. Akaike weights $Aw$ from the Akaike information criterion (blue), and posterior model probabilities $Bp$ from the Bayesian information criterion (orange).

### 3.2. *Vapour and Liquid Water Heat Capacities*

Figure 4 plots the evolution of the reference enthalpy of vaporisation of water (Figure 4a) and the difference in internal energy between vapour and liquid water (Figure 4b) with temperature for the minimum pressure between saturated vapour pressure and 100 kPa. The source of the data is the NIST Chemistry WebBook [13]. The figure also plots the best linear fits $\Delta h_{LV} = a + bT$, Equation (23), and $\Delta u_{LV} = a' + b'T$, Equation (25), to the data in the mentioned range, where the fitting parameters have been found to be $a$ = 2500.9 kJ/kg, $b$ = − 2.3989 kJ/kg/K, $a'$ = 2374.8 kJ/kg and $b'$ = − 2.8415 kJ/kg/K. The squared Pearson correlation coefficient $R^2$ obtained is nearly 1 for both $\Delta h_{LV}$ and $\Delta u_{LV}$ (Figure 4), which shows they are well represented with a linear fit. As mentioned, the difference in specific heat capacity between vapour and liquid water in isobaric conditions, $c^V{}_p - c^L{}_p$, Equation (22), should be close to the identified coefficient $b$, while the difference in isochoric conditions, $c^V{}_v - c^L{}_v$, Equation (24), should be close to the identified coefficient $b'$.

To illustrate the evolution of the specific heat capacities of liquid water and vapour with temperature, the values given by the NIST [13] for the range 0–150 °C are plotted in Figure 5. The figure includes values for isochoric and isobaric conditions at the minimum pressure between vapour saturation pressure and 100 kPa. Please note that, although the specific heat capacity of liquid water shows practically no dependence on pressure, that of water vapour does. For pressure conditions other than atmospheric, the corresponding specific heat capacity values can be obtained from the NIST [13], and an expression of the specific heat capacity of vapour as a function of pressure is found, for instance, in Vestfálová and Šafařík [24].

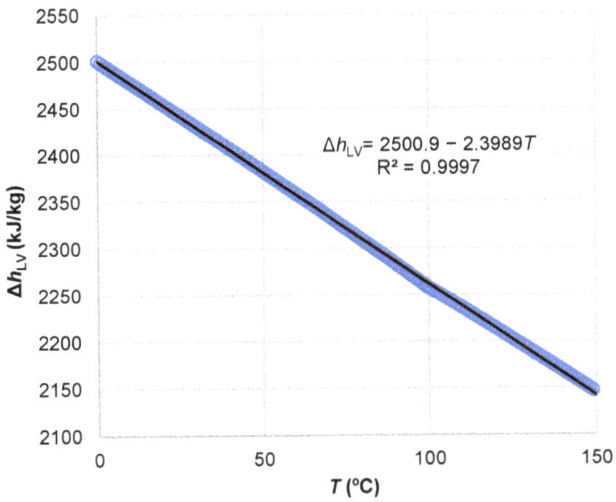


(a)

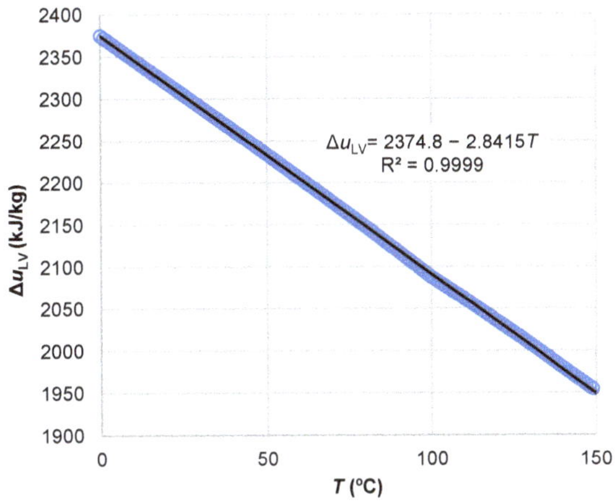


(b)

**Figure 4.** (**a**) Enthalpy of vaporisation $\Delta h_{LV}$, and (**b**) difference in internal energy between vapour and liquid water $\Delta u_{LV}$, for temperatures between 0 and 150 °C. Markers: reference values from [13]. Line and equation: best linear fit.

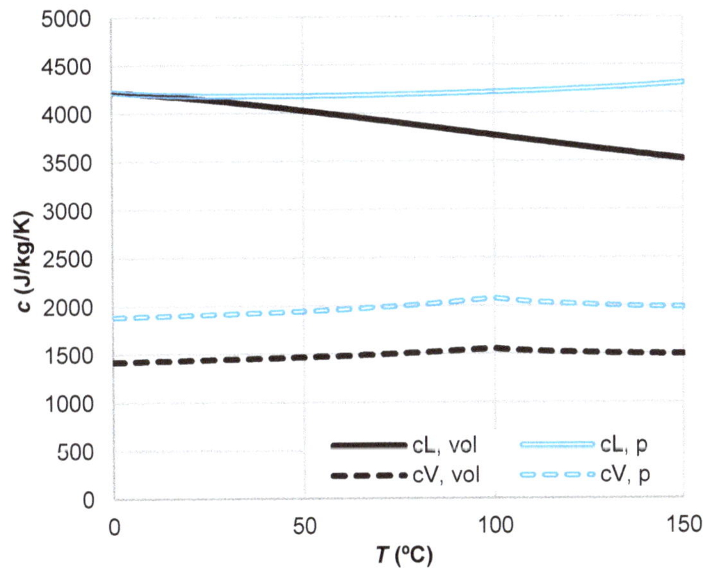


**Figure 5.** Specific heat capacity $c$ of liquid (L, solid lines) water and water vapour (V, dashed lines) for the temperature range 0–150 °C, at the minimum between saturation pressure and 100 kPa, for isochoric (vol, black) and isobaric (p, blue outline) conditions. All values from the NIST [13].

The temperature (and pressure) conditions corresponding to the specific heat capacity values used for modelling are not always reported in the literature. Then, a comparison of the literature and reference values is useful to identify such conditions and assess their consistency with the modelling conditions. The literature values in Table 2 are consistent with the reference values represented in Figure 5. The two values used for liquid water (4180 and 4202 J/kg/K) correspond to the reference specific heat capacities at 30 and 7 °C, respectively, for isobaric conditions, and at 13 and 6 °C, respectively, for isochoric conditions. The values used for water vapour in isobaric conditions (between 1870 and

1900 J/kg/K) correspond to the reference values between 0 and 15 °C. The vapour value specified for isochoric conditions (1428 J/kg/K) corresponds to the reference value at 11 °C. The vapour value given by Zheng et al. [17] (1620 J/kg/K) corresponds to the reference value in isochoric conditions at 121 °C although at saturation pressure. The vapour value given by Abed and Sołowski [12] (1900 J/kg/K) corresponds to the reference value in isobaric conditions at 15 °C.

However, the main interest in this study of the consistency of the formulation with the modelling of water vaporisation is the difference between the specific heat capacities of liquid water and vapour. The difference between the reference values in Figure 5 as a function of temperature from 0 to 150 °C is plotted in Figure 6, for both isobaric (Figure 6a) and isochoric (Figure 6b) conditions. In addition, the fit coefficients $b$ and $b'$ identified from Figure 4 and the differences obtained from the literature values in Table 2, which are constant with temperature, are included for comparison.

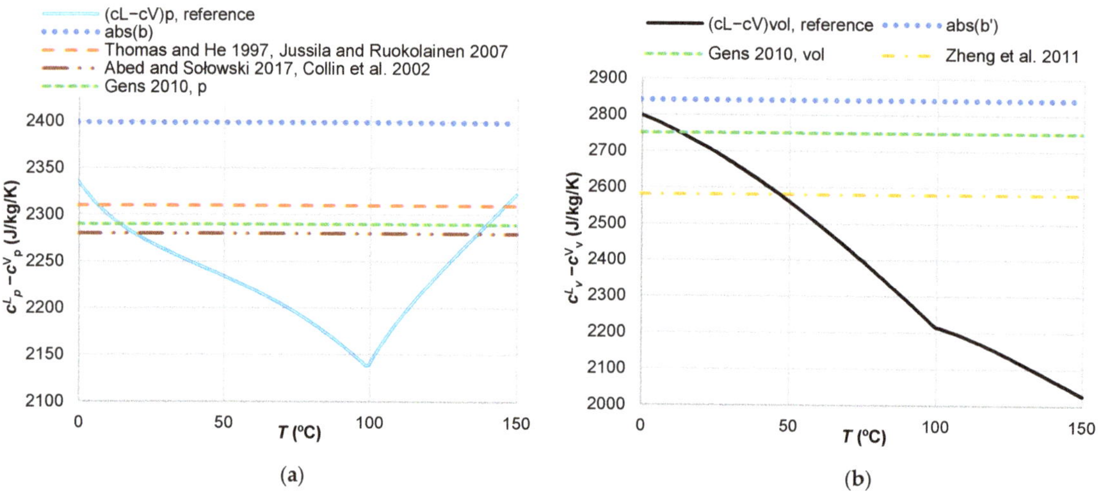


**Figure 6.** Difference in specific heat capacity between liquid water and vapour as a function of temperature. Solid lines: reference values [13] for (**a**) isobaric and (**b**) isochoric conditions; dotted and dashed lines: identified $b$ and $b'$ coefficients and the literature values in Table 2 [7,12,14–17].

Regarding the literature values (Table 2), all those given for isobaric conditions (Thomas and He [7], Collin et al. [14], Jussila and Ruokolainen [15] and Gens [16] (p)), or corresponding to such conditions according to the values in Figure 5 (Abed and Sołowski [12]), are consistent with the difference between isobaric reference values, since they are contained in the range of reference values for the studied temperatures (Figure 6a). The literature values for isochoric conditions (Gens [16] (vol), Zheng et al. [17]) are also consistent with the isochoric reference values (Figure 6b).

The absolute value of $b$, 2398.9 J/kg/K, is within a maximum relative error of 12% from the literature values and isobaric reference values. However, $b$ is greater than reference values in all the temperature range (Figure 6a), which show a maximum of 2335.5 J/kg/K at 0 °C. Something similar occurs with the absolute value of $b'$, 2841.5 J/kg/K, which is greater than the maximum isochoric reference value, 2799.0 J/kg/K at 0 °C (Figure 6b). This result shows that using Equations (18)–(21) with a constant pair of specific heat capacities for liquid water and vapour equal to reference values at the same temperature (Figure 5), for any temperature between 0 and 150 °C, leads to underestimating the slope of the linear laws of Equations (22) and (24) plotted in Figure 4. Thus, it implies underestimating the decrease in $\Delta h_{LV}$ or $\Delta u_{LV}$ with temperature in the modelling. Assuming that the enthalpy of vaporisation, or the internal energy difference between vapour and liquid water, for the reference temperature, $\Delta h_{LV,T0}$ or $\Delta u_{LV,T0}$, is accurately set in the model, $\Delta h_{LV}$ or $\Delta u_{LV}$ will be accurately computed at $T_0$ and lose accuracy as temperature deviates from $T_0$.

To inspect this loss of accuracy, the evolution of $\Delta h_{LV}$ and $\Delta u_{LV}$ computed with Equations (22) and (24), respectively, in a case that would maximise this loss, has been studied and plotted in Figure 7. $T_0$ is set at one extreme of the temperature range considered, 0 °C, and $\Delta h_{LV,T0}$ and $\Delta u_{LV,T0}$ have been accurately taken as the reference values for 0 °C. Then, the maximum error will be made at the other extreme of the range, 150 °C, as a consequence of using a too small slope. To study the magnitude of the error at 150 °C, two values of the slopes ($c^V{}_p - c^L{}_p$) and ($c^V{}_v - c^L{}_v$) have been used. First, the values in the inspected literature furthest from the identified values for $b$ and $b'$, from references [12,14] for $\Delta h_{LV}$ (Figure 6a) and [17] for $\Delta u_{LV}$ (Figure 6b), have been used (solid grey lines in Figure 7). In this case, the relative error at 150 °C is 0.7% for $\Delta h_{LV}$ and 1.9% for $\Delta u_{LV}$. Second, the reference specific heat capacity pairs in Figure 6 furthest from $b$ and $b'$ have been used: 2138 J/kg/K at 99 °C for isobaric values and 2027.3 J/kg/K at 150 °C for isochoric values (dashed orange lines in Figure 7). The relative error at 150 °C then increases to 1.7% for $\Delta h_{LV}$ and 6.1% for $\Delta u_{LV}$.

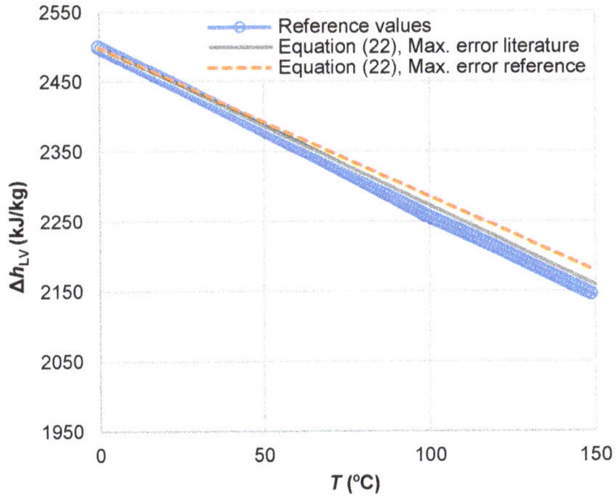


(a)

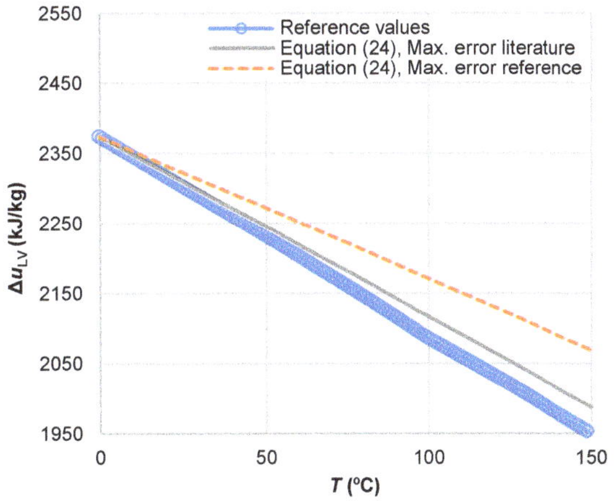


(b)

**Figure 7.** Values of (**a**) $\Delta h_{LV}$ and (**b**) $\Delta u_{LV}$ computed with Equations (22) and (24), respectively, for specific heat capacity pairs that maximise error, from the literature (solid grey lines, values from [12,14], $\Delta h_{LV}$, and [17], $\Delta u_{LV}$) and from reference values (dashed orange lines). Reference $\Delta h_{LV}$ and $\Delta u_{LV}$ [13] in blue markers.

Using the same approach, the relative error at 150 °C can, in addition, be computed for specific heat capacity pairs from the reference values [13] corresponding to all the temperature range between 0 and 150 °C (Figure 8). All errors are below the maximum errors mentioned in the previous paragraph (1.7% for $\Delta h_{LV}$ and 6.1% for $\Delta u_{LV}$). If the relative error is to be further limited, the reference pairs used should be limited to lower temperatures. For instance, to limit the relative error to a 1%, for $\Delta h_{LV}$, any ($c^V{}_p$, $c^L{}_p$) pair from the literature values inspected would be valid, but the pairs of reference values for the same temperature should be limited to temperatures below 50 °C. Analogously, for $\Delta u_{LV}$, the pair of ($c^V{}_v$, $c^L{}_v$) from reference [16] would be valid, and the pairs of reference values for the same temperature should be limited to temperatures below 25 °C.

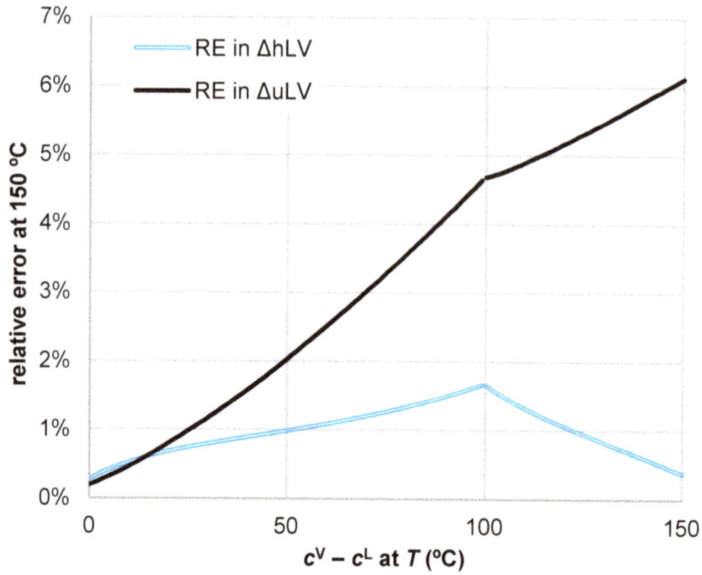


**Figure 8.** Relative error in $\Delta h_{LV}$ and $\Delta u_{LV}$ at 150 °C using specific heat capacity pairs from reference values [13] at temperatures 0–150 °C.

## 4. Conclusions

Two aspects relevant for a consistent thermo-hydraulic modelling of clay-based engineered barriers for deep geological repositories for radioactive waste have been studied: the saturated water vapour pressure and the specific heat capacity of liquid water and vapour. The temperature range of interest has been set between 0 and 150 °C.

Several expressions for the saturated water vapour pressure have been inspected, including reference expressions and those most applied in works modelling unsaturated soils and clay-based barriers. This has led to successfully identifying an accurate (relative error not greater than ±0.2%) and simple enough expression (ranked as the selected model using both the Akaike information criterion and the Bayesian information criterion), Huang's [6] equation, Equation (3), as the optimal option for modelling in the temperature range between 0 and 150 °C.

Further, the enthalpy of vaporisation of water and the internal energy difference between liquid water and vapour as a function of temperature have been studied to derive the consistency of pairs of constant specific heat capacity values for liquid water and water vapour. Several pairs of values from the literature used for modelling unsaturated soils and clay-based barriers have been inspected. Their difference has been compared to reference values and to coefficients identified from the enthalpy of vaporisation and the difference of internal energy. It has been found that using constant specific heat capacity pairs corresponding to reference values at the same temperature can lead to underestimating the decrease in the enthalpy of vaporisation and of the internal energy difference with increasing temperature. To minimise this loss of accuracy, a model should use specific heat capacity pairs as close as possible to the fitted coefficient 2398.9 J/kg/K for isobaric values and 2841.5 J/kg/K for isochoric values. In the case of using reference specific heat capacity pairs for the same temperature, this means using pairs for a temperature as close to 0 °C as possible. However, the relative error associated with this loss of accuracy is limited, being below 2% for all the specific heat capacity pairs from the literature inspected, and up to a 6% for the reference specific heat capacity pairs for the same temperature between 0 and 150 °C.

**Author Contributions:** Conceptualisation, V.N. and L.A.; methodology, V.N. and L.A.; software, V.N. and L.A.; validation, L.A. and V.N.; formal analysis, V.N. and L.A.; investigation, V.N. and L.A.; data curation, V.N. and L.A.; writing—original draft preparation, L.A.; writing—review and editing, V.N., G.U. and L.A.; visualisation, V.N., L.A. and G.U.; supervision, V.N.; project administration, V.N. and L.A.; funding acquisition, L.A. and V.N. All authors have read and agreed to the published version of the manuscript.

**Funding:** This research was funded by MCIN/AEI/10.13039/501100011033, grant number PID2020-118291RB-I00. The APC was funded by MDPI open access publishing in Basel/Switzerland.

**Institutional Review Board Statement:** Not applicable.

**Informed Consent Statement:** Not applicable.

**Data Availability Statement:** Data were obtained from the references cited throughout the text, and are available from those documents.

**Conflicts of Interest:** The authors declare no conflict of interest. The funder had no role in the design of the study; in the collection, analyses, or interpretation of data; in the writing of the manuscript; or in the decision to publish the results.

## References

1. Johnson, L.H.; Niemeyer, M.; Klubertanz, G.; Siegel, P.; Gribi, P. *Calculations of the Temperature Evolution of a Repository for Spent Fuel, Vitrified High-Level Waste and Intermediate Level Waste in Opalinus Clay*; Technical Report 01-04; Nagra: Wettingen, Switzerland, 2002.
2. Wagner, W.; Pruss, A. The IAPWS formulation 1995 for the thermodynamic properties of ordinary water substance for general and scientific use. *J. Phys. Chem. Ref. Data* **2002**, *31*, 387–535. [CrossRef]
3. Lemmon, E.W. Vapor pressure and other saturation properties of water. In *CRC Handbook of Chemistry and Physics*, 97th ed.; Haynes, W.M., Ed.; CRC Press: Boca Raton, FL, USA, 2017.
4. Wagner, W.; Pruss, A. International equations for the saturation properties of ordinary water substance—Revised according to the international temperature scale of 1990. *J. Phys. Chem. Ref. Data* **1993**, *22*, 783–787. [CrossRef]
5. Buck, A.L. New Equations for Computing Vapor Pressure and Enhancement Factor. *J. Appl. Meteorol. Climatol.* **1981**, *20*, 1527–1532. [CrossRef]
6. Huang, J. A Simple Accurate Formula for Calculating Saturation Vapor Pressure of Water and Ice. *J. Appl. Meteorol. Climatol.* **2018**, *57*, 1265–1272. [CrossRef]
7. Thomas, H.R.; He, Y. A coupled heat-moisture transfer theory for deformable unsaturated soil and its algorithmic implementation. *Int. J. Numer. Methods Eng.* **1997**, *40*, 3421–3441. [CrossRef]
8. Gens, A.; Sanchez, M.; Guimaraes, L.D.N.; Alonso, E.E.; Lloret, A.; Olivella, S.; Villar, M.V.; Huertas, F. A full-scale in situ heating test for high-level nuclear waste disposal: Observations, analysis and interpretation. *Geotechnique* **2009**, *59*, 377–399. [CrossRef]
9. Dupray, F.; François, B.; Laloui, L. Analysis of the FEBEX multi-barrier system including thermoplasticity of unsaturated bentonite. *Int. J. Numer. Anal. Methods Geomech.* **2013**, *37*, 399–422. [CrossRef]
10. Nowak, T.; Kunz, H.; Dixon, D.; Wang, W.; Görke, U.-J.; Kolditz, O. Coupled 3-D thermo-hydro-mechanical analysis of geotechnological in situ tests. *Int. J. Rock Mech. Min. Sci.* **2011**, *48*, 1–15. [CrossRef]
11. Wang, X.; Shao, H.; Wang, W.; Hesser, J.; Kolditz, O. Numerical modeling of heating and hydration experiments on bentonite pellets. *Eng. Geol.* **2015**, *198*, 94–106. [CrossRef]
12. Abed, A.A.; Sołowski, W.T. A study on how to couple thermo-hydro-mechanical behaviour of unsaturated soils: Physical equations, numerical implementation and examples. *Comput. Geotech.* **2017**, *92*, 132–155. [CrossRef]
13. Lemmon, E.W.; Bell, I.H.; Huber, M.L.; McLinden, M.O. Thermophysical Properties of Fluid Systems. In *NIST Chemistry WebBook, NIST Standard Reference Database Number 69*; Linstrom, P.J., Mallard, W.G., Eds.; National Institute of Standards and Technology: Gaithersburg, MD, USA, 2022. [CrossRef]
14. Collin, F.; Li, X.L.; Radu, J.P.; Charlier, R. Thermo-hydro-mechanical coupling in clay barriers. *Eng. Geol.* **2002**, *64*, 179–193. [CrossRef]
15. Jussila, P.; Ruokolainen, J. Thermomechanics of porous media—II: Thermo-hydro-mechanical model for compacted bentonite. *Transp. Porous Media* **2007**, *67*, 275–296. [CrossRef]
16. Gens, A. Soil-environment interactions in geotechnical engineering. *Geotechnique* **2010**, *60*, 3–74. [CrossRef]
17. Zheng, L.; Samper, J.; Montenegro, L. A coupled THC model of the FEBEX in situ test with bentonite swelling and chemical and thermal osmosis. *J. Contam. Hydrol.* **2011**, *126*, 45–60. [CrossRef] [PubMed]
18. Akaike, H. Information Theory as an Extension of the Maximum Likelihood Principle. In *Second International Symposium on Information Theory*; Petrov, B.N., Csaki, F., Eds.; Akademiai Kiado: Budapest, Hungary, 1973.
19. Schwarz, G. Estimating the Dimension of a Model. *Ann. Stat.* **1978**, *6*, 461–464. [CrossRef]
20. Çengel, Y.A.; Boles, M.A. *Thermodynamics: An Engineering Approach*; McGraw-Hill: New York, NY, USA, 1989.

21. Burnham, K.; Anderson, D.R. Multimodel Inference: Understanding AIC and BIC in Model Selection. *Sociol. Methods Res.* **2004**, *33*, 261–304. [CrossRef]
22. Pinder, G.F.; Shapiro, A. Physics of Flow in Geothermal Systems. *Geol. Soc. Am. Spec. Pap.* **1982**, *189*, 25–30. [CrossRef]
23. Navarro Gámir, V. Modelo del Comportamiento Mecánico e Hidráulico de Suelos No Saturados en Condiciones No Isotermas. Ph.D. Thesis, Universitat Politècnica de Catalunya, Barcelona, Spain, 1998.
24. Vestfálová, M.; Šafařík, P. Dependence of the isobaric specific heat capacity of water vapor on the pressure and temperature. *EPJ Web Conf.* **2016**, *114*, 02133. [CrossRef]

applied sciences

MDPI

Article

# Analysis of the Effect of Capillary Water Absorption on the Resistivity of Cementitious Materials

Xiangyu He, Xiaohui Zeng *, Rongzhen Dong and Jiangfan Yang

School of Civil Engineering, Central South University, Changsha 410075, China
* Correspondence: hxz9732@foxmail.com

**Abstract:** In a humid environment, the stray current generated by subway operation will corrode the underground pipe network and the internal steel structure of the track plate, which will seriously affect the service life of the subway track plate and increase the maintenance cost later. Groundwater in subway projects mainly enters the concrete interior through capillary action and affects the concrete resistivity, and there is a lack of research on the mechanism of groundwater effect on concrete resistance. In this study, the variation of cement mortar resistivity with capillary water absorption time for cement mortar with different amounts of mineral admixtures (fly ash, ground-granulated blast furnace slag, and silica fume) was measured by the four-electrode method, and the mechanism of the influence of the electrical properties of cementitious materials under the effect of capillary water absorption was analyzed based on the mercury-pressure method (MIP) and thermogravimetric method (TG-DTG). The results show that with the increase in capillary water absorption time, the change curve of cement mortar resistivity can be divided into two stages; in the first stage, capillary water absorption leads to gel pores and transition pores quickly connecting to capillary pores and other large pores to form a water-saturated conductive pathway, resulting in a rapid decrease in resistivity, when the gel pores and excessive pores have a greater impact on resistivity. The second stage is that of capillary water absorption, to a certain extent, after the specimen's internal water upward development rate slows down; at this time, the formation speed of the conductive pathway decreases, resulting in the rate of change in resistivity decreasing and gradually stabilizing. The incorporation of silica fume can effectively improve the resistivity of cementitious materials under the action of capillary water absorption, in which the resistivity of specimens incorporated with 15% silica fume after 36 days of capillary water absorption is 10.39 times that of the reference group, which is mainly due to its lower porosity and a higher percentage of gel pores.

**Keywords:** electro-chemical corrosion; capillary water absorption; resistivity; mineral admixtures; cement mortar; pore structure

**Citation:** He, X.; Zeng, X.; Dong, R.; Yang, J. Analysis of the Effect of Capillary Water Absorption on the Resistivity of Cementitious Materials. *Appl. Sci.* **2023**, *13*, 3562. https://doi.org/10.3390/app13063562

Academic Editor: José Manuel Moreno-Maroto

Received: 21 February 2023
Revised: 8 March 2023
Accepted: 9 March 2023
Published: 10 March 2023



## 1. Introduction

Cement-based material is a typical porous material whose resistivity is closely related to its pore structure and pore water saturation [1–3]. In practical engineering, decreases in the resistivity of cementitious materials can lead to several problems, such as subway stray currents [4]. Studies have shown that the track insulation performance is excellent at the beginning of the metro opening [5], but with the dirty, wet, iron powder overlay or the aging of the surface of the operational insulation fasteners, the walking track current will leak into the surrounding soil and structures through the concrete roadbed, causing serious electrochemical corrosion of the main structure reinforcement and metal pipes along the line [6,7]. In addition, the main structure of the subway is often in a groundwater-rich environment, and the pore water saturation of the concrete roadbed rises rapidly under the invasion of moisture, and the resistivity decreases sharply, increasing the hazard of stray currents [8]; the subway, as an underground, hidden and complex project, is more difficult to renovate or repair [9]. Therefore, studying the preparation of highly insulating cementitious materials and their application in metro rail engineering is crucial.

There are various methods to improve the insulation (resistivity) of cementitious materials, of which the most common application is the incorporation of mineral admixtures. Research on mineral admixtures to improve the resistivity of cementitious materials has made some progress in recent years; Ehtesham et al. [10] found that the resistivity of concrete could be increased by a factor of 2.2 after replacing cement with a mass of 30% fly ash. Baweja et al. found that the resistivity of concrete after replacing cement with 25% fly ash equivalent mass underwent a large increase [11]. Bagheri et al. [12] found an eight-fold increase in 180d resistivity after replacing cement with 10% silica fume of equal quality. Dotto et al. [13] replaced cement with 6% and 12% silica fume of equal mass at 0.5 water-cement ratios, and found that the mortar replacement rate was increased by 2.5 and 5 times for 6% and 12% silica fume replacement, respectively. Yang et al. showed that the resistivity of concrete after replacing cement with 26% silica fume equivalent mass can be increased by 13.4 times and 17 times after 28 d and 60 d of curing, respectively [14]; Justnes et al. [15] found that calcined marl as a supplementary cementitious material was able to increase the resistivity of the mortar in a water-filled state. Liang, by comparing the effect of fly ash, slag, and partial high territory on the resistivity of cement mortar, found that the partial high territory has the best effect on the resistivity of cement mortar [13]. You found that the resistivity of concrete can be increased by 2.67 times after replacing cement with 15% silica fume of equal quality in experiment [16]. Cai et al. [17] found that electrical resistivity increased with the increase in the slag replacement ratio, since the addition of slag would consume more conductive ions. Li et al. [18] tested the resistivity of fly ash mixed in the early stage by a non-contacting electrical resistivity meter, and found that concrete mixed with 25–50% had high early stage resistivity. It is worth noting that most scholars currently test the resistivity of cementitious materials in a water-saturated state, but in actual working conditions, due to the specificity of the environment in which the subway is located, the pore water saturation of cementitious materials will fluctuate greatly, especially in rainy weather or groundwater-rich areas. Water mainly enters the interior of cementitious materials through capillary action so that its pore water saturation increases rapidly [19,20] and its resistivity decreases significantly, which is extremely unfavorable for the prevention and control of stray currents, and there are few reports on the changes in the resistivity of cementitious materials under different water-saturated states due to capillary water absorption.

Therefore, in this paper, three typical mineral admixtures are selected: fly ash, mineral powder, and silica fume, and changes in cement mortar resistivity with capillary water absorption time and different admixtures are measured by the four-electrode method. The influence mechanism is analyzed by combining microscopic testing techniques, and the obtained research results can provide a reference for the prevention and control of stray currents in metro projects.

## 2. Materials and Methods

### 2.1. Materials

The cementitious material adopts P.I 42.5 cement (C) produced by Fushun Aosier Technology Co., Ltd. (Fushun, China), first-grade fly ash (FA), S95 ground granulated blast furnace slag (SL), and silica fume (SF) produced by Henan Hengyuan New Materials Co. (Xinxiang, China). The chemical composition of the cementitious material and the results of particle size tests (average particle size $D_{50}$) are shown in Table 1. The fine aggregate is ordinary river sand (S) with a fineness modulus of 2.72; the mixing water is ordinary tap water (W).

**Table 1.** The chemical compositions of raw materials (calculated by mass fraction %).

| | $SiO_2$ | $Al_2O_3$ | $Fe_2O_3$ | CaO | MgO | $SO_3$ | $Na_2O_{eq}$ | $TiO_2$ | f-CaO | $D_{50}$ |
|---|---|---|---|---|---|---|---|---|---|---|
| C | 20.58 | 4.97 | 3.76 | 63.57 | 2.29 | 2.00 | 0.53 | - | 0.75 | 17.763 |
| FA | 55.7 | 28.39 | 4.695 | 3.424 | 0.751 | 0.885 | 0.949 | 1.15 | - | 7.911 |
| SL | 30.48 | 15.46 | 0.322 | 37.45 | 8.12 | 2.36 | 0.46 | 1.16 | - | 11.895 |
| SF | 93.7 | 0.329 | 0.0957 | 0.478 | 0.248 | 1.35 | 0.893 | 0.017 | - | 14.860 |

### 2.2. Mixture Proportion Design

This test mainly considers the type and dosage of mineral admixtures and adopts a control variable method to design four groups of specimens; they are reference group (JZ group), fly ash group (FA group), ground-granulated blast furnace slag group (SL group), and silica fume (SF group), respectively. The water–binder ratio of the specimen was fixed at 0.40, and the comb–sand ratio was 1:2. The content of mineral admixture is the percentage of the mass of the admixture in the mass of cement, and the specific test mix design is shown in Table 2.

**Table 2.** Mix proportions of cement mortar.

| Group | | Cement | FA | SL | SF |
|---|---|---|---|---|---|
| JZ | | 100 | - | - | - |
| FA | FA-1 | 95 | 5 | - | - |
| | FA-2 | 90 | 10 | - | - |
| | FA-3 | 80 | 20 | - | - |
| | FA-4 | 70 | 30 | - | - |
| SL | SL-1 | 95 | - | 5 | - |
| | SL-2 | 90 | - | 10 | - |
| | SL-3 | 80 | - | 20 | - |
| | SL-4 | 70 | - | 30 | - |
| SF | SF-1 | 95 | - | - | 5 |
| | SF-2 | 90 | - | - | 10 |
| | SF-3 | 80 | - | - | 20 |

### 2.3. Test Method

(1) Test material preparation: Test specimens of mortar strength and resistivity were made according to the Chinese standard: GB/T17671-2021 "Test Method for Cement mortar Strength (ISO Method)". The size of cement mortar is 40 mm × 40 mm × 160 mm, and the strength test and resistivity test specimens are made, respectively. The electrical electrode of the resistivity test specimen is ordinary stainless steel mesh, and extends 20 mm out of the mortar surface for testing. The spacing between the mesh is 40 mm, and the distance between the two ends of the mesh is 20 mm from the end of the specimen. In the process of specimen pouring, four electrodes were inserted vertically into the bottom of the mold, and then the specimen was placed on the shaking table to vibrate and compact. To avoid water loss, the surface of the specimen was covered with a layer of plastic wrap immediately after the completion of vibration, and the specimen was demolded after being cured in the laboratory environment at 20 °C for 24 h, before being put into the saturated lime water solution at 20 ± 2 °C for curing to the corresponding age. The molded specimen is shown in Figure 1.

According to the mix ratio given in Table 2, six test pieces (three strength test pieces, and three resistivity test pieces) were made for each mixing ratio of cement mortar at the same time. The strength and resistivity test pieces of three strength test pieces and resistivity test pieces with the same ratio were tested. The average values of the three pieces were taken as the strength or resistivity test results of the test pieces under the mix ratio.

**Figure 1.** Resistivity test specimen.

(2) Compressive strength test: The compressive strength of the cement mortar specimens of corresponding age was tested according to the Chinese standard: GB/T17671-2021 "Test Method for Cement mortar Strength (ISO Method)".

(3) Resistivity test: At present, the commonly used methods of concrete resistance testing are the Wenner method [21], double-probe method [22], embedded four probe method [23], impedance spectroscopy [24], and ASTM C1760-12 method, etc. The four-electrode method is widely used because of its simple test method, uniform test electric field, and small error [21–23]. Therefore, this paper chooses the four-electrode method using ordinary stainless steel mesh as the electrode; a DC voltage of 24 V was used during the resistivity test. At the same time, to avoid the influence of temperature on the resistivity test [25,26], the whole test process was carried out in a laboratory environment of 20 °C.

(4) Capillary water absorption test: The capillary water absorption test was carried out according to ASTM C1585-13 standards. First, the 28 day aged mortar resistivity test specimen was removed from the saturated lime water solution, and the surface moisture of the specimen was dried with dust-free paper. The specimen was placed in a vacuum drying oven at 50 ± 2 °C, and the mass change was no more than 0.1% after 24 h. After that, silicone rubber and plastic film were used to seal all surfaces except the absorbent surface to ensure that capillary absorbent is only carried out on the absorbent surface [27], and to eliminate the influence of air humidity and carbonization, among which the absorbent surface is the opposite of the specimen-forming surface. After that, the specimen was placed at room temperature of 20 °C and stood for 1 d. After the temperature of the specimen remained constant, its mass (accurate to 0.01 g) was weighed and its resistivity was tested. Then, the specimen was placed in the water absorption device (as shown in Figure 2), and the depth of immersion was about 3 mm. Before testing, the specimen was first wiped with dust-free paper to remove any excess water from the suction surface, and then quickly weighed (±0.01 g). Its resistivity was then tested (control within 30 s to complete), and the specimen was placed in the suction device to continue the capillary suction test. The specific test time is shown in Table 3.

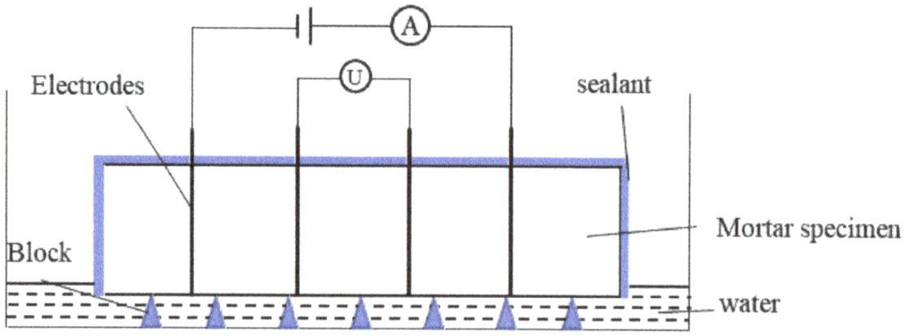


**Figure 2.** Schematic diagram of capillary water absorption and resistivity testing of mortar specimens [28,29].

**Table 3.** Timing and tolerance of capillary water absorption quality tests.

| Time | 60 s | 5 min | 10 min | 20 min | 30 min | 60 min | Every Hour up to 6 h | 12 h | 24 h | Once a Day up to 36 Days |
|---|---|---|---|---|---|---|---|---|---|---|
| Tolerance | 2 s | 10 s | 2 min | 2 min | 2 min | 2 min | 5 min | 5 min | 1 h | 2 h |

(5) TG: Thec water absorption test specimen at 28 d age was crushed into small pieces first. We took the pieces < 1.18 mm and immersed them in anhydrous alcohol for 7 days to terminate the hydration process, and then we put them in a vacuum oven for seven days. Finally, the specimens were ground into powder and 20 mg was selected for the TG test after passing through a 0.075 mm aperture sieve. The test temperature range was room temperature ~1000 °C, the test gas atmosphere was nitrogen, and the heating rate was 10 °C/min.

(6) MIP: Mortar was extracted from capillary water absorption test specimens of corresponding age for testing. The pore structure of the sample was tested by the automatic mercury injection instrument PoreMaster 33, produced by Konta company (Boynton Beach, FL, USA), and the pore size range was 5~106 nm.

## 3. Results and Discussion

### 3.1. Compressive Strength

Figure 3 shows the compressive strength test results of cement mortar after 28 d curing. It can be seen from the figure that compared with the specimens of the JZ group, the 28 d compressive strength of the specimens increased to different degrees after adding different amounts of mineral admixtures. The compressive strength of the FA group increased by 2.6%, 4.1%, 7.02%, and 8.3%, respectively. The compressive strength of the SL group increased by 5.2%, 6.89%, 7.72%, and 9.4%, respectively. The compressive strength of the SF group increased by 16.68%, 30.85%, and 45.81%, respectively, and the growing range increased with the increase in the admixtures. This is because the admixtures can promote the secondary hydration reaction of the system and react with cement hydration products to generate more C-S-H gels. At the same time, the content of $Ca(OH)_2$ with low strength in the system is reduced, so the pore diameters of specimens in each group doped with admixtures are refined compared with those in the JZ group, and the structural system is more compact. This is evidenced in the Section 3.3 and 3.4 below. Among the three kinds of admixtures, SF has the best effect on enhancing strength, because SF has high chemical activity and can better promote the secondary hydration reaction of cement [30]. This led to a significant decrease in the porosity of the specimen along with the refinement of the pore size, resulting in a significant increase in strength. As shown in the Section 3.3, it can be seen that more gel pores and transition pores can lead to strength improvement in the case of similar porosity, and lower porosity has a significant effect on strength improvement at the point of large difference in porosity [31–33].

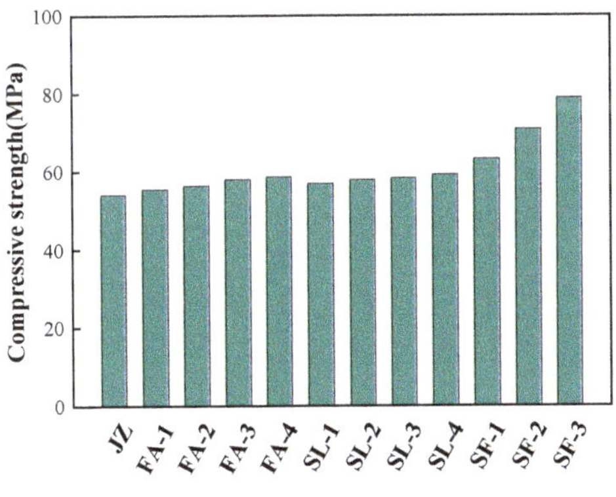


**Figure 3.** 28 d compressive strength of cement mortar.

### 3.2. Effect of Capillary Water Absorption on the Resistivity of Cement Mortar

Because in an underground environment, groundwater erosion on concrete through capillary action takes a long time, a longer capillary absorption test time of 36 d is chosen in this paper to restore the actual engineering situation as much as possible. Figure 4a,b show the resistivity of the cement mortar in the dry state and after 36 d of capillary water absorption, respectively. It can be seen in Figure 4a that in the dry state, the resistivity of the specimens in the JZ group is the highest, the resistivity of the specimens with admixture decreases to different degrees, and the resistivity of the specimens in the same group decreases with the increase of admixture. It can be seen in Figure 4b that after 36 d capillary water absorption, the resistivity of the specimens in the JZ group is the lowest, and the resistivity of the specimens doped with admixtures increases to different degrees. The resistivity of the specimens in the same group increases with the increase in the amount of admixture, and SF has the best effect on the resistivity of the specimens after improving water absorption. It can also be obtained from Figure 4 that after 36 d of capillary water absorption, the resistivity of specimens in the SF group decreases between 70% and 90%, while the resistivity of specimens in the other three groups decreases more than 90%, and the resistivity of specimens in the JZ group decreases the most, by 99.08%.

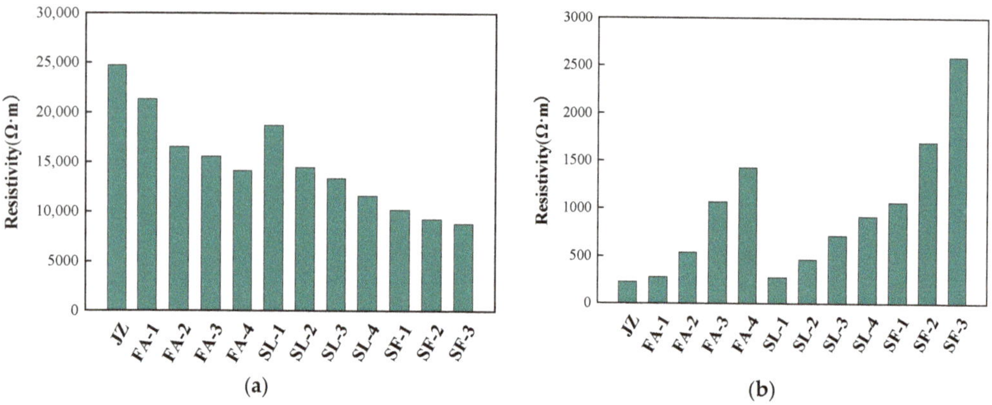


**Figure 4.** The resistivity of cement mortar: (**a**) Resistance in dry conditions; (**b**) Resistance after 36 d of capillary water absorption.

The above phenomenon is related to the influence of mineral admixtures on the pore structure of cement mortar. Cementitious material is a typical gas–solid–liquid three-phase composite, and its electrical conductivity is mainly determined by the solid phase formed by the aggregate and cementitious material, and the gas phase and liquid phase in the pores, of which the conductivity of the gas phase is the worst, and the conductivity of the solid phase is also poor; their electrical conductivity is negligible, so the liquid phase in the pores is a key factor in determining the electrical conductivity of cementitious materials [16]. In the dry state, the pores inside the specimens are almost free of water and are all gas phase, and their connectivity is poor. The number of pores in the JZ group specimens is larger and larger (as shown in the Section 3.3); thus, their resistivity is the largest, and after mixing with mineral admixtures, the pore size of the specimens is refined, the number of large pores is reduced, and the resistivity decreases. This phenomenon is more obvious with the increase in admixture, wherein the specimens mixed with SF have the lowest porosity; thus, their resistivity is the lowest. Under the effect of capillary absorption, the pores inside the specimen are partially connected by water, the liquid phase replaces part of the gas phase, the electrical conductivity is greatly enhanced, the resistivity drops rapidly, the internal tortuosity of the specimen in JZ group is worse than the other three groups, the difficulty of forming the conductive pathway is reduced, and the number is higher; thus, its resistivity after water absorption is the lowest after the addition of mineral admixtures, as is the tortuosity of the pores inside the specimen. After the addition of the mineral admixtures, the degree of tortuosity of the internal pores of the specimens increased, the difficulty of forming conductive pathways increased, the number decreased, and the resistivity increased compared with that of JZ group.

Figures 5–7 show the change curves of capillary water absorption and resistivity with time for cement mortar with different FA, SL, and SF contents, respectively. It can be seen from the figure that the changes in the capillary water absorption resistivity of each group of specimens show two stages of rapid decrease and slow decrease, and tend to be stable; in stage I, in the early stage of capillary water absorption, a saturated area appeared at the bottom of the specimen, forming a large number of conductive paths, and the resistivity decreased rapidly. In stage II, after capillary water absorption reaches a certain level, water diffuses to the unsaturated region on the upper part of the specimen. At this time, capillary water absorption has little influence on the pore connectivity of the specimen, and it is difficult to form more conductive channels, so the resistivity decreases slowly until it is stable [34]. (Figure 8 shows the cross-section of the specimen after 36 days of capillary water absorption).

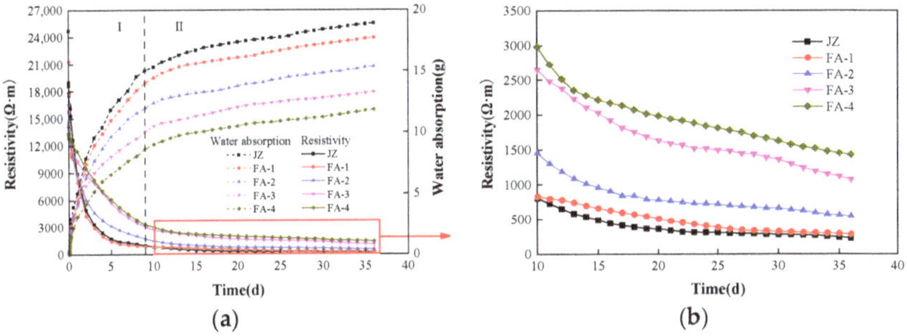


**Figure 5.** Changes in water absorption and resistivity of cement mortar with different FA contents over time: (**a**) Macroscopic damage; (**b**) Partially enlarged drawing.

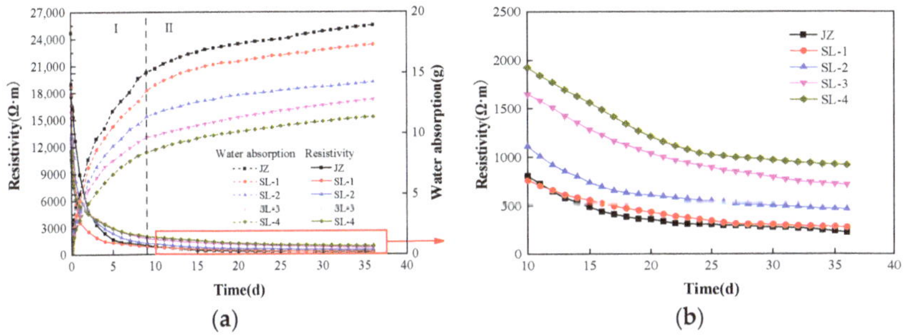


**Figure 6.** Changes in water absorption and resistivity of cement mortar with different SL contents over time:(**a**) Macroscopic damage; (**b**) Partially enlarged drawing.

In addition, by comparing Figures 5–7, it can be observed that after FA or SL incorporation, the capillary water absorption and resistivity of the specimens with water absorption time are similar to those of the JZ group, while the specimens with SF incorporation are different. The resistivity of the specimens in the SF group decreased relatively gently with the increase in capillary water absorption time. The stage I process was significantly shorter than that of the other three groups. The resistivity of the samples doped with 5%, 10%, and 20% SF in the dry state was 41.3%, 37.5%, and 35.7% of that of the JZ group, respectively. However, after 36 d capillary water absorption, the resistivity of the SF group was 3.69, 6.45, and 10.39 times higher than that of the JZ group, respectively. The resistivity of specimens in the SF group was significantly higher than that in the FA group and SL group after capillary water absorption. The capillary water absorption mass (g) and $t^{1/2}$ ($s^{1/2}$) of each group were linearly fitted according to the capillary water absorption data measured in the previous 6 h, and the fitted equations and the capillary water absorption coefficient data of each group were obtained, as shown in Table 4. It can be seen that the capillary water absorption coefficient of each group decreased after the addition of mineral admixtures compared with the reference group. With the increase in mineral admixture, the secondary hydration effect causes the internal pore structure of the specimen to be refined continuously, the degree of tortuosity increases, the internal pore connectivity becomes more difficult, the capillary absorption coefficient decreases continuously, and the capillary absorption coefficient of the SF group is the smallest. This explains the relatively gentle change in the second stage of capillary absorption in the SF group, and with the increase in the mineral admixture substitution, the SF group, being due to higher reactivity. The porosity of the specimens was significantly reduced (as shown in the Section 3.3), and the number of internal conductive pathways decreased significantly compared to other groups. Finally, the resistivity changes during the water absorption process are more gentle compared to the other groups. To a certain extent, it can be considered that the larger the initial water absorption rate, the lower the resistivity after capillary water absorption.

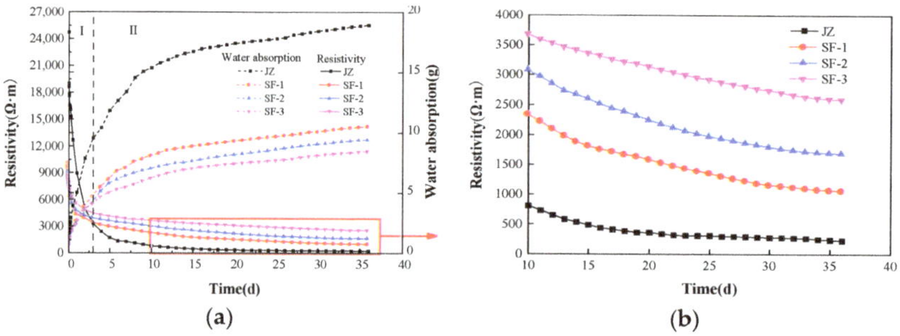


**Figure 7.** Changes in water absorption and resistivity of cement mortar with different SF contents over time: (**a**) Macroscopic damage; (**b**) Partially enlarged drawing.

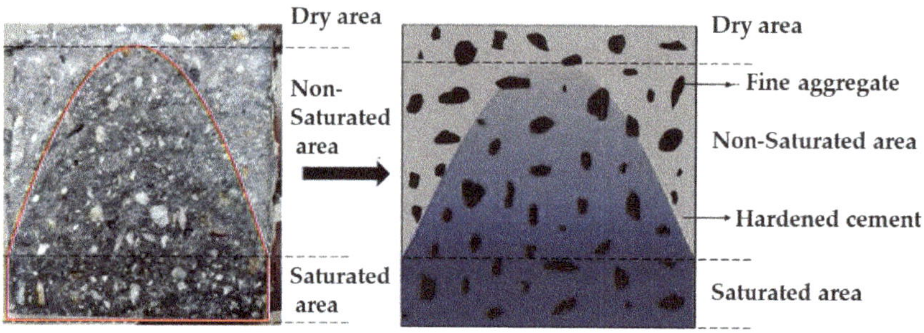


**Figure 8.** The capillary water distribution in the section of 1/2 length of the specimen (after 36 days of capillary water absorption).

Based on the above test results, the pore structure diagram of the cement mortar in the dry state and after capillary water absorption is given in Figure 9. In the dry state, the pore of the specimen is equivalent to the capacitance, which cuts off the conductive path, there is no connected pore inside the specimen, and the cement matrix is almost non-conductive, so the resistivity of the specimen is large. Under the action of capillary water absorption, the specimen can be divided into three areas according to the distribution of water inside the specimen: the saturated area, non-saturated area, and dry area. The bottom of the specimen is a saturated area, and the pores are connected by water to form a conductive path, which leads to a significant decrease in the resistivity of the specimen. However, the pores in the dry zone are hardly conductive.

**Table 4.** Capillary water absorption coefficient fitting equation and parameters.

| Group | Equation | Correlation Coefficient | Capillary Absorption Coefficient |
|---|---|---|---|
| JZ | $m = -0.0345 + 0.01984\ t^{1/2}$ | 0.9976 | 0.01984 |
| FA-1 | $m = 0.00953 + 0.01644\ t^{1/2}$ | 0.993 | 0.01644 |
| FA-2 | $m = -0.00631 + 0.01462\ t^{1/2}$ | 0.999 | 0.01462 |
| FA-3 | $m = -0.0122 + 0.01267\ t^{1/2}$ | 0.9917 | 0.01267 |
| FA-4 | $m = 0.01224 + 0.00992\ t^{1/2}$ | 0.9956 | 0.00992 |
| SL-1 | $m = -0.00323 + 0.01568\ t^{1/2}$ | 0.9957 | 0.01568 |
| SL-2 | $m = -0.00718 + 0.01402\ t^{1/2}$ | 0.9987 | 0.01402 |
| SL-3 | $m = 0.03619 + 0.01105\ t^{1/2}$ | 0.99551 | 0.01105 |
| SL-4 | $m = -0.00607 + 0.00983\ t^{1/2}$ | 0.991 | 0.00983 |
| SF-1 | $m = 0.01867 + 0.00938\ t^{1/2}$ | 0.9941 | 0.00938 |
| SF-2 | $m = -0.03766 + 0.00928\ t^{1/2}$ | 0.9917 | 0.00928 |
| SF-3 | $m = 0.01172 + 0.0089\ t^{1/2}$ | 0.99213 | 0.0089 |

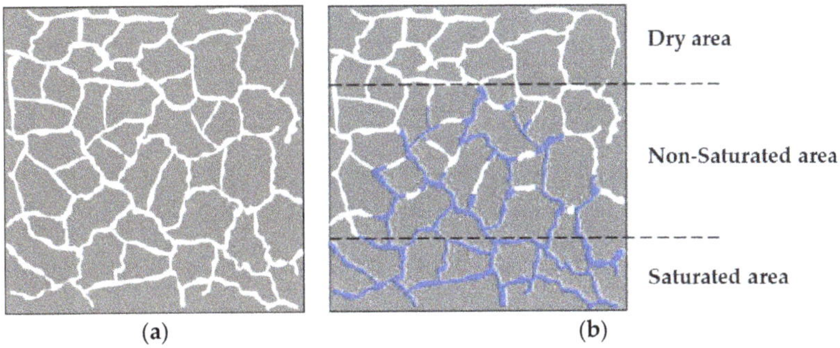


**Figure 9.** Diagram of a cement mortar opening pore: (**a**) Opening pore structure under dry conditions. (**b**) Opening pore structure under capillary water absorption.

### *3.3. Pore Structure Feature*

Figure 10 shows the MIP test results of the pore volume of the specimen, which can be divided into gel pore (5–10 nm), transition pore (10–100 nm), capillary pore (100–1000 nm), and macropore (>1000 nm) according to pore size [35].

It can be seen from Figure 10 that compared with the JZ group, the pore volumes of the specimens in the FA group and the SL group after 28 days of curing are not very different, while the number of gel pores and transition pores is significantly increased, and the number of capillary pores and macropores is significantly reduced. Combined with the analysis in the Section 3.4, this is because FA and SL have lower activity and a lesser secondary hydration reaction, so their pore volume is very much different from that of the JZ group. Due to its physical filling effect and certain secondary hydration reaction, it has a certain refinement effect on the pores. This increases the tortuous degree of pores in the specimen to a certain extent, so the number of conductive pathways formed in the specimen under the effect of capillary water absorption is reduced and the resistivity of the specimen after capillary water absorption is increased; this phenomenon is more obvious with the increase in the doping amount. Under the same doping amount, the specimen doped with FA is stronger than that doped with SL group due to its secondary hydration effect (as can be seen from the Section 3.4, under the same doping amount, 28 d $Ca(OH)_2$ in FA group is lower than that in SL group), which leads to more gel pores and transition pores in FA specimens compared with SL group, and a better effect of the refining pore structure. Therefore, the overall resistivity of FA specimens after capillary absorption is higher than that of SL group. In addition, combined with the test results of the resistivity of the specimens in the JZ group, the FA group, and the SL group, it can be found that when there is little difference in the pore volume, the resistivity of the specimens in the dry state increases with the increase in the number of capillary pores and macropores. According to this phenomenon, it is speculated that capillary pores and macropores contribute more to the resistivity of the specimens in the dry state.

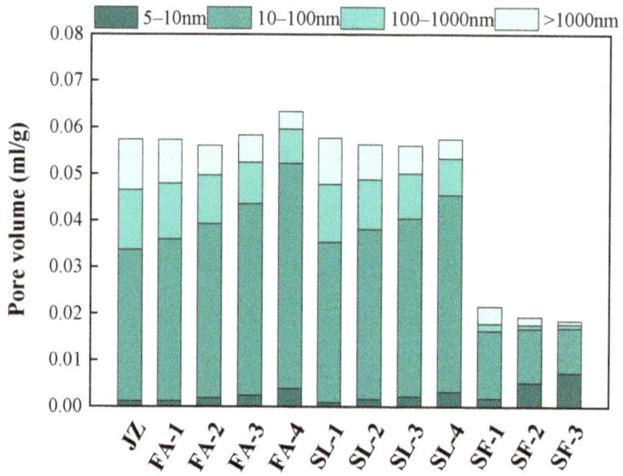


**Figure 10.** Pore volume distribution of cement mortar.

It can also be seen from Figure 10 that compared with the JZ group, the pore volume of the specimen in the SF group was reduced by more than half. This is mainly related to the higher activity of silica fume; the stronger secondary hydration leads to a significant reduction in its pore volume. which greatly reduced the number of conductive channels formed in the specimen under the action of capillary water absorption, and greatly improved the resistivity of the specimen after capillary water absorption. In addition, with the increase in SF content, the pore volume of the SF group specimen decreased, but the difference was not large. With the same pore volume, the increase in pore-specific surface area increases the degree of tortuosity of pores within the specimen [36], while the significant increase in the number of gel pores and the significant decrease in the number of transition pores, capillary pores, and macropores increases the comparative area of their internal pores, which results in a significant increase in the degree of tortuosity of pores within the specimen. A more substantial increase was observed in the resistivity of the specimen after capillary absorption under the dual influence of the reduction in pore volume and the increase in internal tortuosity.

### 3.4. Hydration Products

In order to make the test results more prominent, the samples of JZ group and the three samples with the largest admixture content (FA-4, SL-4, SF-3) were, respectively selected to analyze the hydration products. Figures 11 and 12 show the TG-DTG curve of cement paste. It can be seen from the figure that when the DTG curve has an endothermic peak, the TG curve has a relatively obvious mass loss. The position of the endothermic peak of the slurry is the same, and there are three peak intervals: 70–350 °C (dehydration and decomposition of hydration products such as C-S-H, Ettringite, and Monosulfate (AFM)), 350–470 °C (dehydration and decomposition of hydration products $Ca(OH)_2$), and 600–730 °C (decomposition of carbonate) [37]. The content of chemically bound water and

$Ca(OH)_2$ in the slurry can be quantitatively analyzed based on the TG-DTG curve [38], and the specific method is as follows.

$$m(H_2O) = m_A + m_B + m_C \frac{M_{H_2O}}{M_{CO_2}} \quad (1)$$

$$m(Ca(OH)_2) = m_B \frac{M_{Ca(OH)_2}}{M_{H_2O}} + m_C \frac{M_{Ca(OH)_2}}{M_{CO_2}} \quad (2)$$

In the above equation, $m_A$, $m_B$, $m_C$ are the weight loss rate (%) of the sample in the range of 70–350 °C, 350–470 °C, and 600–730 °C, respectively; m($H_2O$ is the content of chemically bound water (%); m($CaOH_2$) is the content of $Ca(OH)_2$ (%); M is the relative molecular mass, and the relative molecular masses of $H_2O$, $CO_2$ and $CaOH_2$ are 18, 44 and 74, respectively.

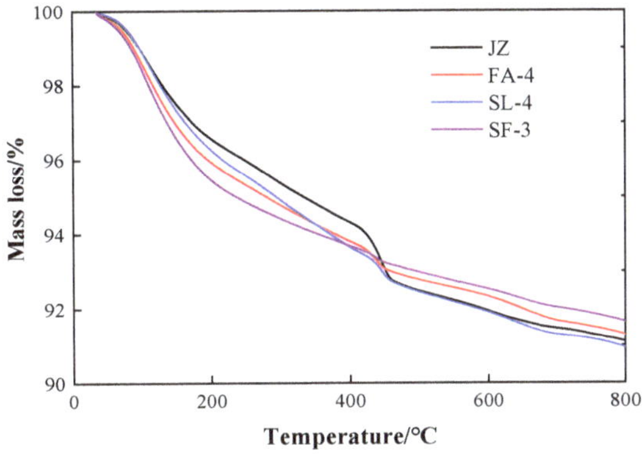


**Figure 11.** TG curve of cement mortar slurry.

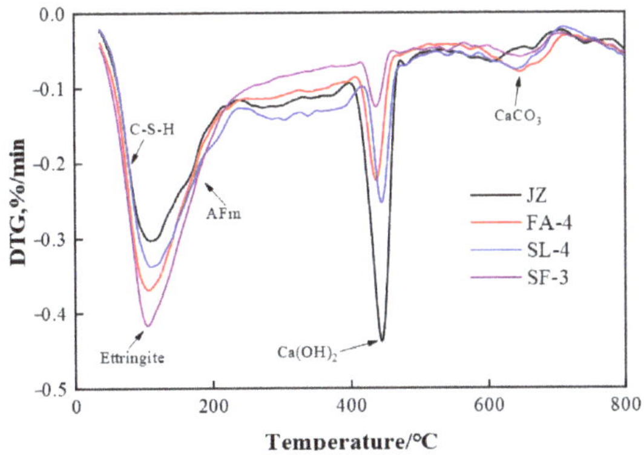


**Figure 12.** DTG curve of cement mortar slurry.

Figure 13 shows the calculated results of $Ca(OH)_2$ and the chemically bound water content of the sample. It can be seen from the figure that the content of $Ca(OH)_2$ in the four groups of samples is significantly different: JZ > SL-4 > FA-4 > SF-3. This well confirms the analysis in the Section 3.3, because the mineral admixture mixed with the cement hydration product ($Ca(OH)_2$)'s secondary hydration reaction, the consumption of $Ca(OH)_2$ content. As a result, the content of the mineral admixture mixed group $Ca(OH)_2$ is lower than in the JZ group. At the same time, the secondary hydration reaction and the filling effect of mineral admixtures together increased the number of gel pores or transition pores, reduced the number of large and capillary pores, and increased the tortuosity of pores, so that the resistivity of the specimen after capillary absorption in the mineral admixture group was increased compared with that in the JZ group. The SF reaction activity was the highest among the three mineral admixtures, and the secondary hydration effect was the most significant, resulting in porosity of the SF group decreasing significantly. The percentage of gel pores and transition pores increased (as shown in Figure 10). In addition, the chemical binding water content of the sample with mineral admixtures decreased compared with that of the JZ group. This is the result of the comprehensive action of two factors. First, the content of cement clinker is reduced after cementing material replaces cement, resulting in a substantial reduction in the content of chemically combined water generated by cement clinker [39]. Second, compared with the JZ group, although the total amount of chemically combined water generated by cement clinker is greatly reduced after the addition of mineral admixture, the formation space of the cement clinker phase hydrate after the addition of mineral admixture is larger than that in the JZ group, and the surface of mineral admixture particles also provides a place for the non-uniform precipitation and generation of cement clinker hydration products, which promotes the hydration of the remaining cement clinker [40–42]. In addition, secondary hydration of mineral admixtures will also generate a portion of chemically combined water [43]. Combining the above two factors, the chemically combined water content of the mineral admixture group decreased compared with that of the JZ group, but there was little difference.

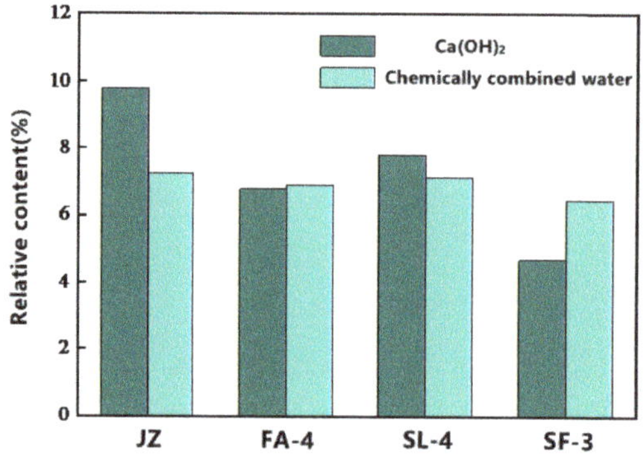


**Figure 13.** Content of chemically combined water and $Ca(OH)_2$ in cement mortar slurry.

## 4. Conclusions and Perspectives

### *4.1. Conclusions*

(1) The secondary hydration of mineral admixture can increase the number of gel pores and transition pores in the specimen. The lower activity and certain filling effects of fly ash and SL can have a certain refining effect on the pore size. The silica fume can significantly reduce the porosity of the specimen and increase the proportion of gel pores and transition pores.

(2) In the dry state, capillary pores and large pores contribute to the resistivity; the more capillary pores and large pores in the specimen, the higher the resistivity, the highest resistivity was found in the dry state of JZ group cement mortar, and the resistivity of cement mortar mixed with mineral admixture had different degrees of decline.

(3) With the increase in the amount of mineral admixture and the increase in the number of gel pores and transition pores inside the specimen, the degree of curvature inside the specimen also increased, leading to a continuous decrease in the capillary water absorption coefficient of the specimen. The resistivity of the specimen after capillary water absorption increased with the decrease in the capillary water absorption coefficient. Among the three admixtures, the addition of silica fume has the best effect on improving the resistivity of cement mortar after 36 days of capillary water absorption, compared with the JZ group.

(4) With the increase in capillary water absorption time, the change curve of the resistivity of cement mortar can be divided into two stages: rapid decrease and slow decrease. It tends to be stable, while the resistivity of cement mortar with silica fume decreases relatively slowly with capillary water absorption. This is mainly due to the low porosity and high proportion of gel pores of the silica fume formation, which lead to a weakening of its water absorption capacity, a significant reduction in the number of conductive paths, and a slow change in resistivity after capillary water absorption.
(5) Under the action of capillary water absorption, the specimen can be divided into a saturated zone, an unsaturated zone, and a dry zone. The saturated zone causes the resistivity of the specimen to decrease greatly, while the unsaturated zone and dry zone have little effect on the decrease in the resistivity of the specimen.
(6) To improve the resistivity of cement-based materials after capillary water absorption, two aspects can be considered. One is to increase the number of gel pores (5–10 nm) and transition pores (10–100 nm) when the pore volume is not very different, because the contribution of gel pores to the resistivity after capillary water absorption is much higher than that of transition pores; the other is to greatly reduce the volume of pores. This is helpful for improving the resistivity of concrete elements in subways and reducing stray current corrosion.

### 4.2. Prospectives

Due to the complexity and diversity of the subway operating environment, there are often multiple factors acting together on concrete members in actual projects, and the current research is mainly focused on single-factor mechanism research. In future, studying the resistivity change mechanism of concrete members under multi-factor coupled environments is a possible research direction, and the evaluation of the durability performance of members under long-term current passage is also very important.

**Author Contributions:** Conceptualization, X.H.; Methodology, X.H.; Resources, X.Z.; Writing—original draft, X.H. and X.Z.; Supervision, X.Z.; Validation, X.Z. and R.D.; Investigation, X.H. and R.D.; Date curation, X.H. and J.Y.; Writing—review and editing, X.H. and J.Y. All authors have read and agreed to the published version of the manuscript.

**Funding:** This research received no external funding.

**Institutional Review Board Statement:** Not applicable.

**Informed Consent Statement:** Not applicable.

**Data Availability Statement:** The data presented in this study are available on request from the corresponding author.

**Acknowledgments:** This research did not receive any specific grant from funding agencies in the public, commercial, or not-for-profit sectors.

**Conflicts of Interest:** The authors declare no conflict of interest.

## References

1. Li, Q.H.; Xu, S.L.; Zeng, Q. The effect of water saturation degree on the electrical properties of cement-based porous material. *Cem. Concr. Compos.* **2016**, *70*, 35–47. [CrossRef]
2. Zeng, X.H.; Chen, L.; Zheng, K.R. Electrical resistivity and capillary absorption in a mortar with styrene-acrylic emulsion and air-entrained agent: Improvement and correlation with pore structure. *Constr. Build. Mater.* **2020**, *255*, 119287. [CrossRef]
3. Liang, K.; Zeng, X.H.; Zhou, X.J. Investigation of the capillary rise in cement-based materials by using electrical resistivity measurement. *Constr. Build. Mater.* **2018**, *173*, 811–819. [CrossRef]
4. Shi, J.J.; Sun, W. Recent research on steel corrosion in concrete. *J. Chin. Ceram. Soc.* **2010**, *38*, 1753–1764.
5. Jin, H.; Yu, S. Effect of DC stray current on rebar corrosion in the cracked segment of shield tunnel. *Constr. Build. Mater.* **2021**, *272*, 121646. [CrossRef]
6. Tang, K.K. Stray current induced steel fibre reinforced concrete corrosion. *Cem. Concr. Res.* **2017**, *100*, 445–456. [CrossRef]
7. Zhang, Y.S.; Huang, R.; Yang, Y.G. Corrosion Behavior of Reinforced Concrete under Stray Current-Brine Coupling Effects. *J. Build. Mater.* **2017**, *20*, 7. [CrossRef]

8. Lu, X.; Chen, X.; Bai, G.Q. Study on the influence of coupling action of stray current and chloride solution on reinforced concrete. *New Build. Mater.* **2018**, *045*, 100–103.
9. Ding, H.T. The Research on the Subway Concrete of High Impedance and Impervious. Master's Thesis, Wuhan University of Technology, Wuhan, China, 2008.
10. Ehtesham, S.; Rasheeduzzafar, H. Corrosion resistance performance of fly ash blended cement concrete. *ACI Mater. J.* **1994**, *91*, 264–271.
11. Baweja, D.; Roper, H.; Sirivivatnanon, V. Corrosion of steel in marine concrete: Long-term half-cell potential and resistivity data. In Proceedings of the Third CANMET/ACI International Conference on Performance of Concrete in Marine Environment, St. Andrew, NB, Canada, 4–9 August 1996.
12. Bagheri, A.; Zanganeh, H.; Alizadeh, H. Comparing the performance of fine fly ash and silica fume in enhancing the properties of concretes containing fly ash. *Constr. Build. Mater.* **2013**, *47*, 1402–1408. [CrossRef]
13. Dotto, J.M.R.; Abreu, A.G.D.; Dal Molin, D.C.C. Influence of silica fume addition on concretes physical properties and on corrosion behaviour of reinforcement bars. *Cem. Concr. Compos.* **2004**, *26*, 31–39. [CrossRef]
14. Yang, W.S. Experimental Investigation on Effects of Mineral Admixtures on Electrical Receptivity of Cement-Based Materials. Master's Thesis, Hunan University, Changsha, China, 2010.
15. Justnes, H.; Østnor, T.A. Durability and microstructure of mortar with calcined marl as supplementary cementing material. In Proceedings of the XIII International Conference on Durability of Building Materials and Components—XIII DBMC, Sao Paulo, Brazil, 2–5 September 2014; Curran Associates, Inc.: Red Hook, NY, USA, 2015; pp. 752–761.
16. Liang, K. Experimental Study on Electric Conductivity Characteristics and Influential Factors of Concrete. Ph.D. Thesis, Southwest Jiaotong University, Chengdu, China, 2018.
17. Cai, J.; Pan, J.; Li, X. Electrical resistivity of fly ash and metakaolin based geopolymers. *Constr. Build. Mater.* **2020**, *234*, 117868. [CrossRef]
18. Li, Z.J.; Xiao, L.Z.; Wei, X.S. Study on the Properties of Young Concrete with Fly Ash Using Electrical Measurement. In *International RILEM Symposium on Concrete Science and Engineering: A Tribute to Arnon Bentur*; RILEM Publications SARL: Champs-sur-Marne, France, 2004.
19. Yang, L.; Liu, G.J.; Gao, D.Y. Experimental study on water absorption of unsaturated concrete: W/c ratio, coarse aggregate, and saturation degree. *Constr. Build. Mater.* **2020**, *272*, 121945. [CrossRef]
20. Hong, S.X.; Yao, W.Q.; Guo, B.W. Water distribution characteristics in cement paste with capillary absorption. *Constr. Build. Mater.* **2020**, *240*, 117767. [CrossRef]
21. Minagawa, H.; Miyamoto, S.; Hisada, M. Relationship of Apparent Electrical Resistivity Measured by Four-Probe Method with Water Content Distribution in Concrete. *J. Adv. Concr. Technol.* **2017**, *15*, 278–289. [CrossRef]
22. Dehghanpour, H.; Yilmaz, K. The relationship between resistances measured by two-probe, Wenner probe and C1760-12 ASTM methods in electrically conductive concretes. *SN Appl. Sci.* **2020**, *2*, 10. [CrossRef]
23. Zhou, Z.; Xie, N.; Cheng, X.; Feng, L.; Hou, P.; Huang, S.; Zhou, Z. Electrical properties of low dosage carbon nanofiber/cement composite: Percolation behavior and polarization effect. *Cem. Concr. Compos.* **2020**, *109*, 103539. [CrossRef]
24. Chiarello, M.; Zinno, R. Electrical conductivity of self-monitoring CFRC. *Cem. Concr. Res.* **2005**, *27*, 463–469. [CrossRef]
25. Liu, Y.B.; Francisco, J.; Presuel, M. Normalization of Temperature Effect on Concrete Resistivity by Method Using Arrhenius Law. *ACI Mater. J.* **2014**, *111*, 433–442. [CrossRef]
26. Hope, B.B.; Ip, A.K.; Manning, D.G. Corrosion and electrical impedance in concrete. *Cem. Concr. Res.* **1985**, *15*, 525–534. [CrossRef]
27. *ASTM C1585*; Standard Test Method for Measurement of Rate of Absorption of Water by Hydraulic-Cement Concretes. ASTM International: West Conshohocken, PA, USA, 2004.
28. Qian, J.S.; Xu, S.S.; Li, M. The Measurement and Application of Resistivity for Concrete. *J. Shandong Univ. Sci. Technol.* **2010**, *29*, 37–42.
29. Tian, D.M.; Deng, D.H.; Huang, B. Research on the Capillary Absorption of the Cement-Emulsified Asphalt Mortar. *China Railw. Sci.* **2010**, *31*, 32–37.
30. Ye, B.T.; Jiang, J.Y.; Wang, W.H. Influence of Mineral Admixtures on Strength Development of Basalt Fiber Reinforced Cement Mortar. *J. Build. Mater.* **2014**, *17*, 5.
31. Santos, A.R.; Veiga, M.D.R.; Silva, S.A. Evolution of the microstructure of lime based mortars and influence on the mechanical behaviour: The role of the aggregates. *Constr. Build. Mater.* **2018**, *187*, 907–922. [CrossRef]
32. Theoulakis, P.; Moropoulou, A. Microstructural and mechanical parameters determining the susceptibility of porous building stones to salt decay. *Constr. Build. Mater.* **1997**, *11*, 65–71. [CrossRef]
33. Theoulakis, P.; Moropoulou, A. Salt crystal growth as weathering mechanism of porous stone on historic masonry. *Porous Mater.* **1999**, *6*, 345–358. [CrossRef]
34. Zeng, X.H.; Pan, Z. Influence of Capillary Water Absorption on Resistivity of Cement Mortar. *J. Build. Mater.* **2018**, *21*, 18–23.
35. Guo, J.F. The Theoretical Research of the Pore Structure and Strength of Concrete. Master's Thesis, Zhejiang University, Hangzhou, China, 2004.
36. Bourbatache, M.K.; Bennai, F.; Zhao, C. Determination of geometrical parameters of the microstructure of a porous medium: Application to cementitious materials. *Int. Commun. Heat Mass* **2020**, *117*, 104786. [CrossRef]
37. Shi, C.J.; Yuan, Q. *Analytical Methods for Cementitious Material Testing*; China Architecture & Building Press: Beijing, China, 2018.

38. Yang, K.; Tang, Z.; Cheng, Z.Q. Mechanical properties of ultra-high strength cement-based materials (UHSC) incorporating metal powders and steel fibers. *Constr. Build. Mater.* **2020**, *318*, 125926. [CrossRef]
39. Li, X.; Ruhan, A.; Yan, P.Y. Research on Hydration Degree of Cement-Fly Ash Complex Binders. *J. Build. Mater.* **2010**, *13*, 5.
40. Scrivener, K.L.; Juilland, P.; Monteiro, P.J.M. Advances in understanding hydration of Portland cement. *Cem. Concr. Res.* **2015**, *78*, 38–56. [CrossRef]
41. Kocaba, V. *Development and Evaluation of Methods to Follow Microstructural Development of Cementitious Systems Including Slags*; Thesis EPFL No 4523; EPFL: Lausanne, Switzerland, 2009; 263p.
42. Bishnoi, S.; Scrivener, K.L. Studying nucleation and growth kinetics of alite hydration using μic. *Cem. Concr. Res.* **2009**, *39*, 849–860. [CrossRef]
43. Jensen, O.M.; Hansen, P.F. Influence of temperature on autogenous deformation and relative humidity change in hardening cement paste. *Cem. Concr. Res.* **1999**, *29*, 567–575. [CrossRef]

applied sciences

MDPI

Article

# Use of Foundry Sands in the Production of Ceramic and Geopolymers for Sustainable Construction Materials

**Caterina Sgarlata [1,2], Maria Camila Ariza-Tarazona [1], Enrico Paradisi [1], Cristina Siligardi [1,2] and Isabella Lancellotti [1,2,*]**

1 Department of Engineering "Enzo Ferrari", University of Modena and Reggio Emilia, 41125 Modena, Italy; caterina.sgarlata@unimore.it (C.S.); mariacamila.arizatarazona@unimore.it (M.C.A.-T.); enrico.paradisi@unimore.it (E.P.); cristina.siligardi@unimore.it (C.S.)

2 National Interuniversity Consortium of Materials Science and Technology (INSTM), 50121 Florence, Italy

* Correspondence: isabella.lancellotti@unimore.it

**Abstract:** The aim of this research was to evaluate the possibility of reusing waste foundry sands derived from the production of cast iron as a secondary raw material for the production of building materials obtained both by high-temperature (ceramic tiles and bricks) and room-temperature (binders such as geopolymers) consolidation. This approach can reduce the current demand for quarry sand and/or aluminosilicate precursors from the construction materials industries. Samples for porcelain stoneware and bricks were produced, replacing the standard sand contained in the mixtures with waste foundry sand in percentages of 10%, 50%, and 100% by weight. For geopolymers, the sand was used as a substitution for metakaolin (30, 50, 70 wt%) as an aluminosilicate precursor rather than as an aggregate to obtain geopolymer pastes. Ceramic samples obtained using waste foundry sand were characterized by tests for linear shrinkage, water absorption, and colorimetry. Geopolymers formulations, produced with a Si/Al ratio of 1.8 and Na/Al = 1, were characterized to evaluate their chemical stability through measurements of pH and ionic conductivity, integrity in water, compressive strength, and microstructural analysis. The results show that the addition of foundry sand up to 50% did not significantly affect the chemical-physical properties of the ceramic materials. However, for geopolymers, acceptable levels of chemical stability and mechanical strength were only achieved when using samples made with 30% foundry sand as a replacement for metakaolin.

**Keywords:** foundry sands; waste; ceramic products; bricks; geopolymers

**Citation:** Sgarlata, C.; Ariza-Tarazona, M.C.; Paradisi, E.; Siligardi, C.; Lancellotti, I. Use of Foundry Sands in the Production of Ceramic and Geopolymers for Sustainable Construction Materials. *Appl. Sci.* **2023**, *13*, 5166. https://doi.org/10.3390/app13085166

Academic Editor: José Manuel Moreno-Maroto

Received: 22 March 2023
Revised: 13 April 2023
Accepted: 19 April 2023
Published: 21 April 2023



## 1. Introduction

Foundry sands consist primarily of clean, uniformly sized, high-quality silica sand used in the production of molds for both ferrous (iron and steel) and nonferrous (copper, aluminum, and brass) metal casting industries.

Molding sands are obtained by adding binding agents to virgin silica sands. Silica sand is mainly used because of its thermal conductivity. It can absorb and transmit heat while allowing the gases generated during the thermal degradation of binders to pass through the grains. In general, molding sand is extensively used because it is easy to use, economical, has high-temperature resistance, and due to its ability to bind with other binders and organic materials, outperforming natural sand [1].

According to their composition, foundry sands are classified into two types: green sands and chemical foundry sands. Green sands are composed of 85–95% silica sand, 4–10% bentonite clay as a binder, 2–10% of carbonaceous additive, and water (2–5%); they are particularly suitable for geotechnical applications, such as structural fills and base courses. Chemical foundry sands are composed of 93–99% silica sand and 1–3% chemical binder; they are used as the "cores" in castings to withstand the heat of molten metal and as molds for nonferrous castings [2].

In the casting process, molding sands are recycled and reused multiple times, but after going through many production cycles, they become less uniform and lose their cleanliness. The sand grains begin to break down because of heat and mechanical abrasion, and therefore new sand must be continually added to the system. When waste foundry sands (WFS) are no longer suitable for the manufacturing process, they are removed from the system and discarded at foundry landfills or offsite municipal landfills [2].

It is estimated that approximately 100 million tons of sand are used in annual production, and 6–10 million tons are discarded annually and available to be recycled into other products and industries [3]. Further, about one ton of foundry sand is used for each ton of iron or steel casting produced [4]. Meanwhile, the foundry industry in Italy produces 1.5 million tons of raw casting and generates about one million tons of waste, most of which (around 80%) is sand. The automotive industries are the major generators of foundry sand (about 95% of the estimated WFS) [2].

As millions of tons of waste foundry sands are generated each year, their disposal has become an environmental challenge [5]. Foundry sand can be classified as a hazardous or non-hazardous material, depending on its source. Silica-based spent foundry sands, specifically from iron, steel, and aluminum foundries, are considered non-hazardous, while spent foundry sands from leaded brass and bronze foundries are often regulated as hazardous waste because of the presence of highly toxic organic pollutants that can contaminate the atmosphere or condense in the sand. These pollutants include phenols and inorganic elements such as lead, chromium, cadmium, iron, and zinc [6]. Even though the analysis of the leachate obtained from waste foundry sands has found that the concentration of most hazardous substances is below the acceptable limits, hazardous polyaromatic hydrocarbons have been found in all types of waste foundry sands [6]. Waste foundry sands should be dumped into controlled landfills; however, because of the growing waste foundry sands production and the need for landfill monitoring, the landfilling cost has been continuously increasing in the past years, reaching $135–657 million in the U.S.A. [7], making unviable to continue with this practice.

Consequently, recycling waste foundry sands into several applications has become an attractive opportunity to decrease the economic and environmental impact of their disposal [8]. Although natural materials are often preferred in terms of quality, their sources are depleting gradually over the years, making the use of waste materials viable [7,9,10].

Silica-based WFSs are classified as non-hazardous residues and can be used in other industries as secondary raw materials. For example, they have been used in the sustainable building sector, specifically in cold consolidated materials such as cementitious conglomerates, asphalts, concrete, and cement [11]. Ceramic materials regulations allow the use of secondary residual raw materials to replace sands normally used as inert material in the production of ceramic tiles and bricks, therefore contributing to the reduction of the environmental impact of their activities [12]. Other types of products suitable for exploiting foundry sands are geopolymers, materials obtained through the cold alkaline activation of suitable precursors. In geopolymers, foundry sands are potentially interesting as precursors and/or as fillers or aggregates to obtain geopolymer mortars [12,13]. In this work, we investigated the potential application of WFS, obtained from different Italian metal casting industries, in geopolymers and ceramic tiles formulations.

## 2. Materials and Methods

### 2.1. Waste Foundry Sands

The waste foundry sand (WFS) used for ceramic tile (porcelain stoneware) and bricks application is named 31A. For the geopolymers application, foundry sand with higher content of aluminum was chosen and is named 17A. The chemical composition of the sands was analyzed through an XRF spectrometer (Bruker S4 Pioneer), and it is reported in Table 1. As expected, $SiO_2$ is present in high quantities in both samples, but 31A has a higher amount of it compared to 17A. Furthermore, 31A has small amounts of Al, Fe, and Zr. In comparison, 17A sand has a higher amount of aluminum, which is useful for alkali activation, Fe, and Zr, and presents an organic fraction that is evident from both carbon content (4.350%) and LOI (6.2%). Given that the geopolymers were obtained at room temperature, the organic content does not represent a problem with respect to samples that need firing. Traces of sulfur are also present.

**Table 1.** Chemical composition of sands 17A and 31A.

| XRF | $SiO_2$ | $Al_2O_3$ | $TiO_2$ | $Fe_2O_3$ | CaO | MgO | $K_2O$ | $Zr_2O$ | L.O.I. | C | S |
|---|---|---|---|---|---|---|---|---|---|---|---|
| | (%) | (%) | (%) | (%) | (%) | (%) | (%) | (%) | (%) | (%) | (%) |
| 17A | 81.85 | 2.83 | 0.15 | 3.88 | 0.13 | 0.48 | 0.32 | 4.15 | 6.20 | 4.350 | 0.385 |
| 31A | 93.00 | 2.03 | 0.37 | 1.48 | 0.12 | 0.16 | 0.71 | 1.89 | 0.24 | 0.177 | 0.046 |

### *2.2. Particle Size Distribution Analysis*

The size and distribution of particles were determined using a Mastersizer 2000 Light Diffractometer. The sands were initially sieved using a 1 mm sieve and then were measured using the light diffractometer.

### *2.3. X-ray Diffraction (XRD)*

X-ray diffractions were recorded by a PW3710 diffractometer (Philips, Almelo, The Netherlands). Specimens were scanned in the 0–70° 2θ angular range on powdered samples, using Cu-Kα radiation from a conventional X-ray source operated at 35 kV, 35 mA, and a scan step of 0.02°. The patterns were analyzed using the HighScore Plus (PANAlytical) software.

### *2.4. Samples Formulation and Preparation*

The materials used for ceramic tiles are clays and feldspars of different origins and standard sand. The standard sand was replaced in varied percentages with the foundry sand, as reported in Table 2. For brick materials, different types of clays and basalt were used. The basalt, as for the sand in ceramics tiles, was replaced with the foundry sand, as reported in Table 3.

**Table 2.** Formulations of porcelain stoneware samples.

| Material \ Composition (%) | STD | SF10 | SF50 | SF100 |
|---|---|---|---|---|
| Clay 1 | 26 | 26 | 26 | 26 |
| Clay 2 | 14 | 14 | 14 | 14 |
| Feldspar 1 | 20 | 20 | 20 | 20 |
| Feldspar 2 | 28 | 28 | 28 | 28 |
| Sand | 12 | 10.8 | 6 | 0 |
| Foundry sand | 0 | 1.2 | 6 | 12 |

**Table 3.** Formulations of brick samples.

| Material \ Composition (%) | FSTD | F.10 | F.50 | F.100 |
|---|---|---|---|---|
| Clay (Vicenza) | 33.30 | 33.30 | 33.30 | 33.30 |
| Marl (Possagno) | 23.40 | 23.40 | 23.40 | 23.40 |
| Clay Modena 1 | 16.65 | 16.65 | 16.65 | 16.65 |
| Clay Modena 2 | 16.65 | 16.65 | 16.65 | 16.65 |
| Basalt fine | 10.00 | 9.00 | 5.00 | 0 |
| Foundry sand | 0 | 1.00 | 5.00 | 10.00 |

For tile samples, the standard mixture shown in Table 2 was used. Raw materials were mixed with water in a 2:1 ratio, and a fluidizer was added to improve the pourability of the dough. Then the dried dough was pulverized and humidified to 6% before being pressed to obtain the samples and cooked up to 1205 °C.

For brick mixtures, the raw materials previously dried and sieved under 500 µm were mixed with around 27% water to obtain a smooth and malleable dough suitable for extrusion. Then, samples were extruded and cut to a size of 5 × 2 cm and dried at room temperature and then at 105 °C. Subsequently, the samples were put in the stove up to 1050 °C.

For the geopolymers application, the foundry sand 17A corresponding to a higher ratio of aluminum to silicon, was chosen. As a matrix for geopolymer samples, metakaolin CSC1000-ARGICAL 1000 provided by Bal-Co (Sassuolo, MO, IT) was used. As activating solutions, sodium silicate provided by Ingessil (Montorio, Vr, IT) and NaOH 8 M provided by Merck-Sigma-Aldrich (Milano, IT) were used.

The fresh pastes were prepared by mixing metakaolin and waste sand with the activating solution (NaOH and $Na_2SiO_3$) into a container with a mechanical mixer for approx. 10 min. When the geopolymer paste was homogenous and achieved suitable workability, it was poured into a mold and covered with a plastic film. The samples were cured at room temperature for 28 days before further characterization.

A standard mixture of geopolymer was initially prepared using 100% metakaolin and reacted with 30 mL of NaOH 8M and 30 mL of sodium silicate. In the other formulations, part of metakaolin (30%, 50%, and 70% by weight) was replaced with waste sand 17A, and the collected results of chemical stability, durability in water, and mechanical resistance were compared.

### 2.5. *Ceramic Samples Characterization*

#### 2.5.1. Linear Shrinkage

Linear shrinkage has been calculated for samples according to the following expression:

$$LS\% = \frac{d_i - d_f}{d_i} \times 100$$

where:

- $d_i$ = initial diameter (40 mm),
- $d_f$ = final diameter.

The results are given as the mean of the 3 measurements taken for the same specimen.

#### 2.5.2. Water Absorption

For water absorption tests, samples with known weight were immersed in boiling water for two hours. Then they were extracted, dried on the surface, and weighed again. The water absorption was obtained as follows:

$$WA\% = \frac{W_B - W_S}{W_S} \times 100$$

where:

$W_B$ = wet sample weight

$W_S$ = dry sample weight.

#### 2.5.3. Colorimetry

The measures were performed using a PCE-CSM 6 instrument that operates in the CIE L*a*b* color space. The color space model allows for the accurate representation of colors using three coordinates. The L*a*b* mode covers the entire visible spectrum of the human eye and allows you to describe the set of colors perceived by sight.

to slightly less sintering, when the replacement of STD sand with foundry sand is higher than 50%, but spent foundry sand does not significantly affect linear shrinkage.

**Table 4.** Results of linear shrinkage (LS%), water absorption (AA%), and colorimetry (ΔE) with standard deviation (St.Dev) of porcelain stoneware samples.

| Porcelain Stoneware | LS% | St.Dev | WA% | St.Dev | ΔE |
|---|---|---|---|---|---|
| STD | 4.8 | 0.3 | 0.009 | 0.004 | 0 |
| SF10 | 4.7 | 0.1 | 0.0015 | 0.0002 | 0 |
| SF50 | 4.9 | 0.2 | 0.022 | 0.003 | 1 |
| SF100 | 4.3 | 0.1 | 0.02 | 0.01 | 1.73 |

For water absorption tests, it can be observed that the absorption obtained for each formulation is lower than 0.025% WA, despite a slight increase observed with the addition of sand in quantities exceeding 50%. From these results, it can be deduced that, despite the addition of foundry sand, the obtained stoneware remains a low porosity material, according to the norm UNI EN 176 ISO BI.

For colorimetry measures, the values of ΔE in relation to the reference sample STD show that the addition of foundry sand to the mixture does not lead to significant color variations. In fact, all the samples have almost the same color. The biggest difference is observed between the STD sample and the SF100% sample where ΔE = 1.73, so the color variation begins to be distinguishable, but the overall difference could be accepted.

The diffractograms obtained from the XRD analysis of SF100 and STD are both shown in Figure 3 for comparison. The main difference between them is the presence of the Microcline intermediate phase in sample SF100 (highlighted in the inset graph), which is not observed in the STD sample. It is also possible to observe a different intensity of the peaks at 28° 2θ corresponding to albite, higher for sample SF100 compared with the STD. Anyway, these differences can be considered relatively small, as both diffractograms show mainly the same crystalline phases (quartz, mullite, and albite).

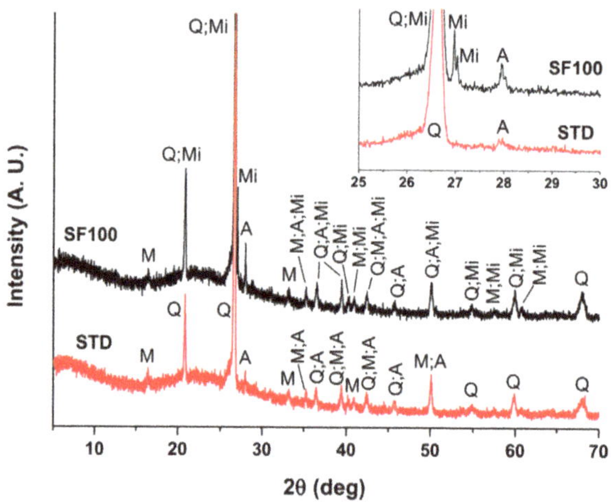


**Figure 3.** X-ray diffraction of samples SF100 and STD (Q = Quartz, Mi = Microcline intermediate, M = Mullite, A = Albite).

*3.4. Bricks*

The results of linear shrinkage, water absorption, and colorimetry of brick samples produced using 10%, 50%, and 100% of waste sand are reported in Table 5 and compared to the standard formulation STD. The post-firing linear shrinkage (LS%) shows a slight increase by adding the foundry sand. The presence of Fe in the foundry sand can influence this behavior.

**Table 5.** Results of linear shrinkage (LS%), water absorption (AA%), and colorimetry (ΔE) with standard deviation (St.Dev) of brick samples.

| Brick Sample | LS% | St.Dev | WA% | St.Dev | ΔE |
|---|---|---|---|---|---|
| F.STD | 5.2 | 0.9 | 8.6 | 0.4 | 0 |
| F.10 | 6.6 | 0.8 | 10.8 | 0.41 | 1.49 |
| F.50 | 6 | 1 | 12.1 | 0.5 | 1.15 |
| F.100 | 6 | 1 | 12.7 | 0.7 | 2.84 |

The water absorption values for brick samples increase with the amount of foundry sand added. The higher value of WA% was shown by sample F.100 with 12.7 WA%, compared to 8.65 WA% shown by the standard formulation with basalt. This means that, despite the addition of foundry sand, brick samples remain a highly porous material [16].

For colorimetry measures, the ΔE values of samples depend on the amount of sand added. The ΔE measures are in the range of 2–3 for all samples compared to the standard formulation, so it can be deduced that there is a distinguishable color variation.

In Figure 4, the diffractograms obtained for the brick formulations are reported. Even in this case, there are no particular differences between the diffractograms of F.100 and F.STD samples, which show peaks of the same crystalline phases (quartz, hematite, and albite), with a negligible variation in their intensity, due to the variation in the percentage of foundry sand added.

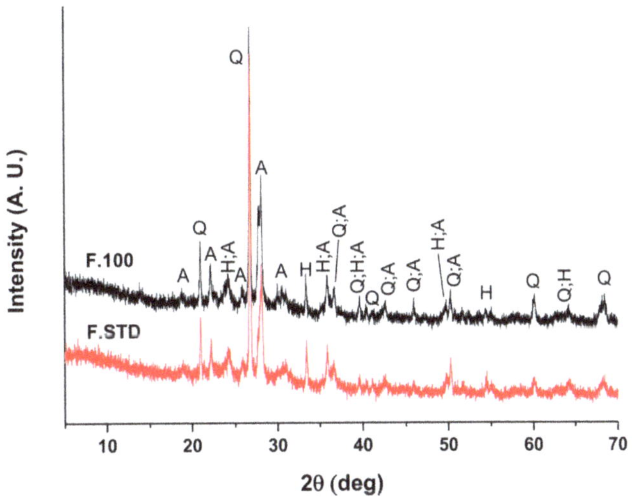


**Figure 4.** X-ray diffraction of F.100 and F.STD (Q = Quartz, H = Hematite, A = Albite).

From the analyses carried out, it is possible to conclude that the introduction of foundry sand as a secondary raw material for tiles and bricks in place of virgin sand modifies the physical properties of the standard samples in an acceptable way. The results obtained using waste foundry sands for ceramic materials may be considered acceptable, compared to similar studies with waste materials found in the literature for tiles and porcelain stoneware, e.g., in terms of water absorption and shrinkage [17,18] and for bricks [19], on the application which they are intended.

*3.5. Geopolymer*

The geopolymer formulations are reported in Table 6, with the respective Si/Al molar ratio, which increases by increasing the foundry sand content because the waste is poor in aluminum with respect to metakaolin.

**Table 6.** Sample compositions of geopolymer formulation.

| Name | MK (g) | WFS (g) | NaOH 8M (mL) | $Na_2SiO_3$ (mL) | Si/Al |
|---|---|---|---|---|---|
| 100% MK | 100 | 0 | 30 | 30 | 1.4 |
| 70% MK | 70 | 30 | 30 | 30 | 1.82 |
| 50% MK | 50 | 50 | 30 | 30 | 2.27 |
| 30% MK | 30 | 70 | 30 | 30 | 3 |

The integrity test of samples containing 30%, 50%, and 70% of foundry sand showed that the color of the immersion water became darker by increasing the amount of sand in the formulation, but it remained transparent. The change in color of the immersed water may be an indication of the release of elements from the sample into the water and their subsequent decrease in chemical stability.

When samples lose their structural consistency, and the water of immersion becomes cloudy, the formulation is not considered acceptable. This is a qualitative indication of the stability of formulations that will then be quantified by pH and ionic conductivity tests to indirectly evaluate the efficacy of reticulation reactions. Even when the alkaline solution does not react completely with aluminosilicate powders, it may have still leached out during immersion of the sample in water. In this case, almost all samples remain intact after 24 h, so the integrity test is considered passed.

In Figure 5 (left), the pH measurements of 30% MK, 50% MK, and 70% MK samples after 24 h are compared with 100% MK. It was noted that the pH value increased over time at short immersion times while remaining constant or even decreasing at longer immersion times. The pH of the eluate after 24 h is in a range between 10 and 11 for all samples, confirming the typical alkalinity of these samples. The standard 100% MK shows pH values between 11.19 at 30 min and 10.73 at 1440 min (24 h) [20].

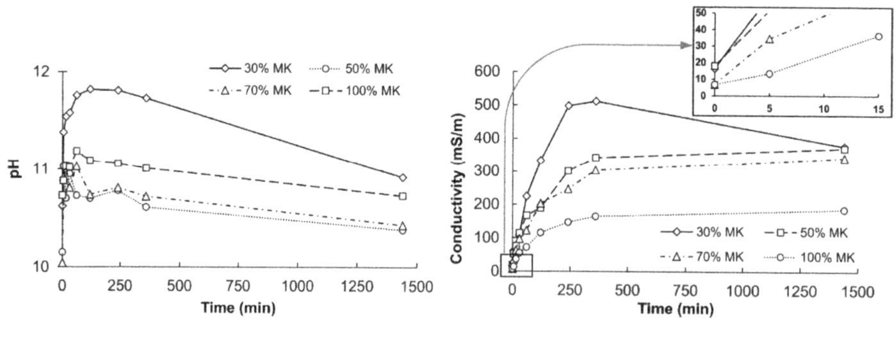


**Figure 5.** Results of pH (**left**) and ionic conductivity (**right**) of geopolymer samples cured at room temperature. Zoom of ionic conductivity: sharp rise of conductivity during the first minutes.

Figure 5 (right) shows the conductivity measurements carried out on the same samples. The highest ionic conductivity is recorded by the sample with the lowest amount of metakaolin (30% MK), which after 8 h reaches a value of about 500 mS/m but then stabilizes within 24 h at 378 mS/m, similar to the value of 50% MK sample with 372 mS/m. The sample having the highest amount of metakaolin (70% MK) exhibited an overall lower value, approaching the standard 100% MK, with a conductivity of 300 mS/m after 8 h of testing.

This behavior follows the typical trend of alkali-activated materials found in literature, and it is related to the lower stability of samples with a very high addition of sand [21].

Figure 6 shows the results of the compression strength tests on samples after the curing time of 28 days. There is a strong dependence of the values on the percentage of MK present in the formulations. In fact, the compressive strength decreases by decreasing the amount of MK. In Figure 6, the values of the Si/Al ratio for each sample are shown as orange points. Mechanical properties improve as the Si/Al ratio decreases to around 1.4–1.8. Acceptable results were, however, achieved for sample 70% MK showing a value around 21 MPa, compared to the standard 100% MK [22], so the addition of a 30% foundry sand is considered feasible.

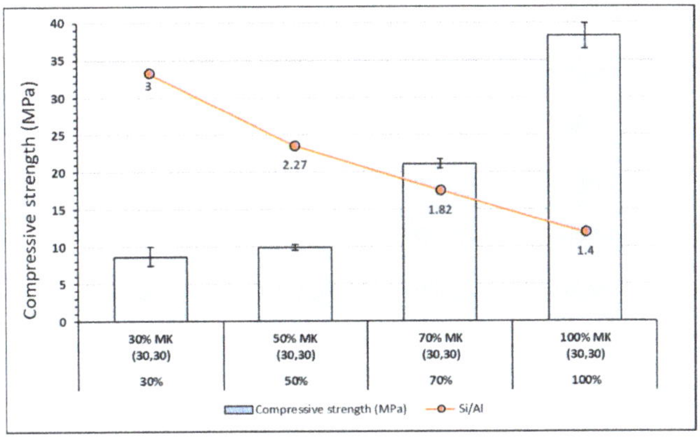


**Figure 6.** Results of compressive strength for geopolymer samples.

XRD patterns of 70% MK sample and 17A foundry sand are shown in Figure 7. In the sample with 70% MK, the presence of the broadband between 15–35° 2θ can be observed, which is characteristic of the amorphous phase that for pure metakaolin-based geopolymers is positioned between 25–32° 2θ [23]. The presence of this broadband (blue circled in Figure 7) confirms that the presence of foundry sand does not hinder the geopolymerization process, but some degree of crystallinity is maintained due to the nature of sand: this behavior is already observed in literature for geopolymers where wastes with a complex semicrystalline nature were used in place of metakaolin [24].

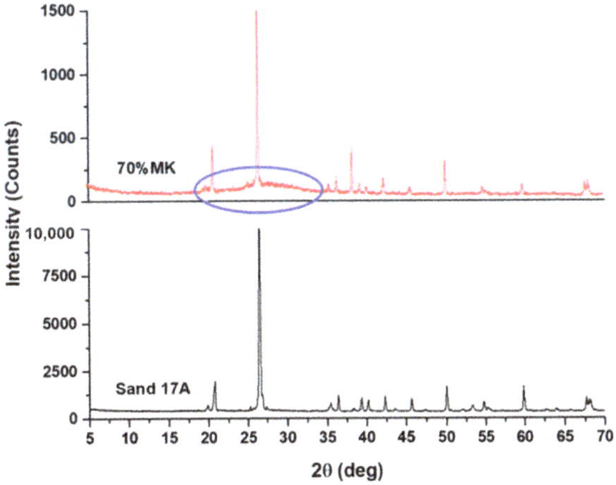


**Figure 7.** XRD patterns of 70% MK sample and 17A foundry sand.

## 4. Conclusions

In this work, the potential application of WFSs obtained from different Italian metal casting industries in geopolymers and ceramic materials was investigated.

For porcelain stoneware, the value of linear shrinkage was found to be in the range of 4.3–4.8% for all the formulations (10%, 50%, and 100%), and the values of water absorption (<0.02%) indicate that the obtained material, despite the addition of foundry sand, remains a low porosity material. The XRD diffractograms confirm that there are no relevant differences between them.

Colorimetry measures revealed that the biggest difference is observed between the STD sample and the SF100% sample with $\Delta E = 1.73$.

For bricks, the addition of foundry sand only slightly affects the post-firing linear shrinkage, showing an increase of the same, as well as for the water absorption. Even with the addition of foundry sand, brick samples remain a highly porous material, and no differences in the crystalline phases present are detected. Even the $\Delta E$ values of samples depend on the amount of sand added, with a distinguishable color variation.

For geopolymer formulations, the results of chemical stability, pH and ionic conductivity, and compression strength tests show that there is a strong dependence of the values on the percentage of MK present in the formulations, and the properties of samples worsen with the decrease of MK.

In summary, foundry sand can be used in partial or total substitution to the sand that is typically used as standard, at least when the colorimetric aspect is not of primary importance. So, it can potentially be used for formulations that are not associated with large tile formats where aesthetic needs are relevant.

It is thus possible to obtain acceptable results on the properties of the final materials by replacing virgin sand with foundry sand in geopolymer formulations (although in percentages not exceeding 30%). Further studies could be carried out to optimize the obtained results, thus improving the performance of geopolymer samples containing foundry sand.

**Author Contributions:** Conceptualization, C.S. (Cristina Siligardi) and I.L.; methodology, C.S. (Cristina Siligardi); investigation, E.P. and M.C.A.-T.; data curation, M.C.A.-T.; writing—original draft preparation, C.S. (Caterina Sgarlata); writing—review and editing, C.S. (Caterina Sgarlata) and I.L.; supervision, I.L. and C.S. (Cristina Siligardi); project administration, I.L.; funding acquisition, I.L. All authors have read and agreed to the published version of the manuscript.

**Funding:** This research was funded by INSTM and Fondazione Cariplo-2020_1216, project title: New recycling process for the foundry sands for Action D- Evaluation and assessment of novel routes for waste sands recovery.

**Institutional Review Board Statement:** Not applicable.

**Informed Consent Statement:** Not applicable.

**Data Availability Statement:** Not applicable.

**Acknowledgments:** The authors recognize Utilizzo Di Sabbie Di Fonderia Nella Produzione Di Ceramici E Geopolimeri Per L'edilizia Sostenibile within Project Cariplo-2020_1216 for Action D-Evaluation and the assessment of novel routes for waste sands recovery.

**Conflicts of Interest:** The authors declare no conflict of interest.

## References

1. Ahmad, J.; Zhou, Z.; Martínez-García, R.; Vatin, N.I.; De-Prado-Gil, J.; El-Shorbagy, M.A. Waste Foundry Sand in Concrete Production Instead of Natural River Sand: A Review. *Materials* **2022**, *15*, 2365. [CrossRef] [PubMed]
2. Tittarelli, F. Waste foundry sand. In *Waste and Supplementary Cementitious Materials in Concrete*; Elsevier: Amsterdam, The Netherlands, 2018; pp. 121–147.
3. Babu, G.M.; Dadapeer, A.B.S. An Experimental Investigation on Effects of Foundry sand on Concrete Interms of Characteristics of Strength. *J. Emerg. Technol. Innov. Res.* **2018**, *5*, 154–161.

4. Siddique, R.; Noumowe, A. Utilization of spent foundry sand in controlled low-strength materials and concrete. *Resour. Conserv. Recycl.* **2008**, *53*, 27–35. [CrossRef]
5. Son, B.T.; Long, N.V.; Hang, N.T.N. Fly ash-, foundry sand-, clay-, and pumice-based metal oxide nanocomposites as green photocatalysts. *RSC Adv.* **2021**, *11*, 30805–30826. [CrossRef]
6. Iqbal, M.F.; Liu, Q.-F.; Azim, I.; Zhu, X.; Yang, J.; Javed, M.F.; Rauf, M. Prediction of mechanical properties of green concrete incorporating waste foundry sand based on gene expression programming. *J. Hazard. Mater.* **2020**, *384*, 121322. [CrossRef]
7. Siddique, R.; Singh, G.; Singh, M. Recycle option for metallurgical by-product (Spent Foundry Sand) in green concrete for sustainable construction. *J. Clean. Prod.* **2018**, *172*, 1111–1120. [CrossRef]
8. Gambalonga, B.; Nicolini, J.L.; Inocente, J.M.; Pich, C.T.; Angioletto, E.; Pereira, F.R.; Montedo, O.P.K.; Arcaro, S. Valorization of waste foundry sand aggregates in hot-mix asphalt. *Process Saf. Environ. Prot.* **2023**, *173*, 277–288. [CrossRef]
9. Abdulhameed, H.; Mansi, A.; Mohammed, A.; Abdulhameed, A.; Hanoon, A. Study the use of Nano-limestone and Egg-shell Ash in Eco-friendly SCC: An Experimental and Statistical Evaluation Based on Computer Programming. In Proceedings of the 14th International Conference on Developments in eSystems Engineering (DeSE), Sharjah, United Arab Emirates, 7–10 December 2021; pp. 509–514. [CrossRef]
10. Murmu, A.L.; Patel, A. Towards sustainable bricks production: An overview. *Constr. Build. Mater.* **2018**, *165*, 112–125. [CrossRef]
11. Luqman, A.L.I.; Khan, M.N.A.; Rasheed, Y. Effect of Waste Foundry Sand (WFS) on Strength and Durability of Pressed Fired Clay Bricks. In Proceedings of the ND Conference on Sustainability in Civil Engineering (CSCE'20), Department of Civil Engineering, Capital University of Science and Technology, Islamabad, Pakistan, 12 August 2022.
12. Dyer, P.P.O.L.; Klinsky, L.M.G.; Silva, S.A.; Silva, R.A.; Lima, M.G. Macro and microstructural characterization of waste foundry sand reused as aggregate Road. *Mater. Pavement Des.* **2021**, *22*, 464–477. [CrossRef]
13. Patil, A.R.; Sathe, S.B. Feasibility of sustainable construction materials for concrete paving blocks: A review on waste foundry sand and other materials. *Mater. Today Proc.* **2020**, *43*, 1552–1561. [CrossRef]
14. Piccolo, F.; Andreola, F.; Barbieri, L.; Lancellotti, I. Synthesis and Characterization of Biochar-Based Geopolymer Materials. *Appl. Sci.* **2021**, *11*, 10945. [CrossRef]
15. Lancellotti, I.; Ponzoni, C.; Barbieri, L.; Leonelli, C. Alkali activation processes for incinerator residues management. *Waste Manag.* **2013**, *33*, 1740–1749. [CrossRef] [PubMed]
16. Hossiney, N.; Das, P.; Mohan, M.K.; George, J. In-plant production of bricks containing waste foundry sand—A study with Belgaum foundry industry. *Case Stud. Constr. Mater.* **2018**, *9*, e00170. [CrossRef]
17. Yuan, Q.; Robert, D.; Mohajerani, A.; Tran, P.; Pramanik, B.K. Utilisation of waste-to-energy fly ash in ceramic tiles. *Constr. Build. Mater.* **2022**, *347*, 128475. [CrossRef]
18. Dondi, M.; Raimondo, M.; Zanelli, C. Clays and bodies for ceramic tiles: Reappraisal and technological classification. *Appl. Clay Sci.* **2014**, *96*, 91–109. [CrossRef]
19. Arias-Trujillo, J.; Matías-Sanchez, A.; Cantero, B.; López-Querol, S. Effect of polymer emulsion on the bearing capacity of aeolian sand under extreme confinement conditions. *Constr. Build. Mater.* **2019**, *236*, 117473. [CrossRef]
20. Sun, Z.; Vollpracht, A.; van der Sloot, H.A. pH dependent leaching characterization of major and trace elements from fly ash and metakaolin geopolymers. *Cem. Concr. Res.* **2019**, *125*, 105889. [CrossRef]
21. Galiano, Y.L.; Pereira, C.F.; Vale, J. Stabilization/solidification of a municipal solid waste incineration residue using fly ash-based geopolymers. *J. Hazard. Mater.* **2011**, *185*, 373–381. [CrossRef]
22. Rajamma, R.; Labrincha, J.A.; Ferreira, V.M. Alkali activation of biomass fly ash–metakaolin blends. *Fuel* **2012**, *98*, 265–271. [CrossRef]
23. Pinheiro, D.D.R.; Gonçalves, L.R.; Sena, R.L.P.D.; Martelli, M.C.; Neves, R.D.F.; Ribeiro, N.F.D.P. Industrial Kaolin Waste as Raw Material in the Synthesis of the SAPO-34 Molecular Sieve. *Mater. Res.* **2020**, *23*. [CrossRef]
24. Sarkar, M.; Dana, K. Partial replacement of metakaolin with red ceramic waste in geopolymer. *Ceram. Int.* **2020**, *47*, 3473–3483. [CrossRef]

applied sciences

MDPI

*Article*

# Interface Behavior Analysis of Different Geomembrane Liner Systems Based on PIV Techniques

**Junli Gao * and Jiajun Wang**

School of Mechanics and Engineering Science, Shanghai University, Shanghai 200072, China; shuwjj@shu.edu.cn
* Correspondence: susan_jl@staff.shu.edu.cn

**Abstract:** The interfacial friction performance of the liner system has an important influence on the overall stability of modern landfills, but the interface of the liner system using conventional geomembranes may have problems such as shear failure and slip along the liner system. Accordingly, the concept of a ribbed geomembrane was proposed. Based on the laboratory model tests, the variation laws of *p–s* curves and the additional stress of sand were studied under different values of shape, rib spacing, and rib height. A series of particle image velocimetry (PIV) analyses of sand particles were performed to provide insight into the reinforcement mechanism of the ribbed geomembrane liner system. The results indicated that the ribbed geomembrane could obviously improve the stability of the liner system compared with the smooth geomembrane. The ribbed geomembrane formed an indirect influence area with sand in a specific range. The ribbed geomembrane with optimal values of rib spacing and rib height was beneficial to reduce the settlement of the upper sand and the stress redistribution of the sand particles.

**Keywords:** ribbed geomembrane; laboratory model test; particle image velocimetry (PIV); liner system; influence area

**Citation:** Gao, J.; Wang, J. Interface Behavior Analysis of Different Geomembrane Liner Systems Based on PIV Techniques. *Appl. Sci.* **2023**, *13*, 6614. https://doi.org/10.3390/app13116614

Academic Editor: José Manuel Moreno-Maroto

Received: 20 April 2023
Revised: 22 May 2023
Accepted: 27 May 2023
Published: 29 May 2023



## 1. Introduction

Landfill liner systems are generally composed of cohesive soil and geosynthetic materials (e.g., geomembranes, geotextiles, etc.). As the shear strength of the interface between geosynthetic materials is often less than the shear strength of the interface between the overlying refuse and geosynthetic materials, instability often occurs between geosynthetic materials in landfill liner systems. Therefore, the concept of the ribbed geomembrane was proposed by Gao et al. [1].

Direct shear tests and theoretical analyses of the interface between ribbed geomembranes and geotextiles have shown that ribbed geomembranes can, to a certain extent, effectively improve the frictional properties of the interface and mitigate the destabilization of the liner system. Similar structural materials such as ribbed geomembranes have also been investigated by domestic and international scholars. The theory of the indirect influence zone was proposed by Bao [2], who pointed out that the reinforcing effect of the reinforcing material would cause the soil around it to form a "reinforced soil body". Irsyam et al. [3] conducted a direct shear test on the geogrid using hot wax and obtained the shear surface and displacement vector distribution of loose sand and dense sand at different cross-rib spacing. The shear zone formation of H–V reinforced sand specimens was numerically simulated in detail by Zhang et al. [4] to reveal the progressive damage law of shear zone generation and expansion in H–V reinforced soils. Zhou et al. [5] and Yang et al. [6] investigated the reinforcement length and height of the reinforced soil foundation by indoor model tests and pointed out that the optimum reinforcement length was three times the foundation width, and the optimum number of reinforcement layers was two to three. Oda et al. [7] analyzed the generation and development of shear zones in the soil by finite element simulation of triaxial tests.

In order to study ribbed geomembranes in more depth, this paper investigates the effects of rib shape, rib spacing, and rib height on the load settlement pattern of the ribbed geomembrane–geotextile liner system and the additional stresses in the landfill sand based on preliminary direct shear tests and reinforcement theory [8,9], with the aim of providing a more comprehensive representation of the properties of the ribbed geomembrane–geotextile interface through model tests and PIV analysis techniques.

## 2. Laboratory Tests

### 2.1. Devices

As shown in Figure 1, the test devices used in this research included a model test system, a PIV system, and data acquisition systems. The model container (600 mm × 290 mm × 400 mm) was composed of a 25 mm thick steel plate welding skeleton and transparent high-strength plexiglass plates. The loading device was a dynamic triaxial apparatus (GCTS). A 200 mm × 290 mm hardwood loading board was placed in the middle of the top surface of the sand. The load was controlled at a uniform speed of 40 N/s with a controlled stress method. Due to the limitations of the loading device, the loading was terminated at 150 kPa. The test data were collected by a displacement gauge (YHD-100; range 0–50 mm) and three sand pressure cells (range 0–500 kPa). The test data were processed by the TK-ST-DA 16-channel wireless receiver, TK-ST wireless transceiver, and supporting acquisition processing software. The PIV system consisted of a charge-coupled device (CCD) camera (B5M16; 5 megapixels; image acquisition rate 11.3 frames per second) and particle image velocimetry software PIVlab 2.40.

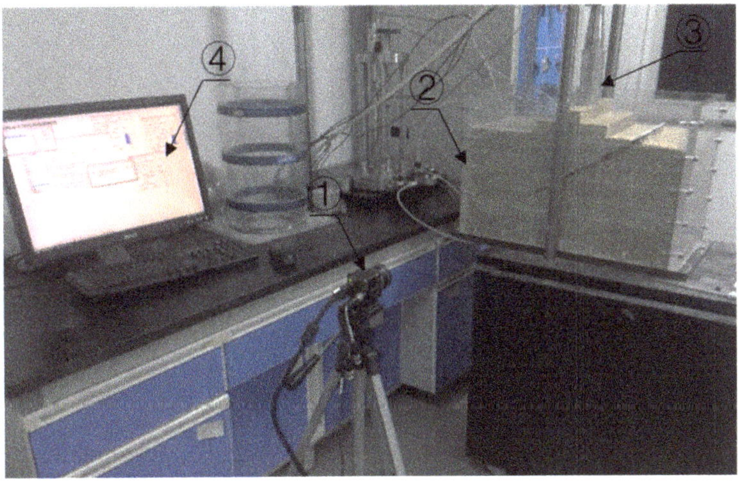


**Figure 1.** Model test equipment for PIV investigation. ① CCD camera; ② Model container; ③ Loading device; ④ PIV analysis software.

### 2.2. Materials

The specific gravity of the sand was 2.712, the water content was 1.51, the coefficient of uniformity was 2.07, and the coefficient of curvature was 0.87. The smooth geomembrane samples were made of high-density polyethylene (HDPE) with a minimum density of 0.939 g/cm$^3$. The size of the smooth geomembrane sample was 360 mm × 290 mm × 1.5 mm, the yield strength was 22 N/mm, and the breaking strength was 40 N/mm. The ribbed geomembrane samples were made by sticking ribs to the surface of the smooth geomembrane. In order to prevent the ribs from falling off, they were nailed to the smooth geomembrane. The size of the strip rib was 30 mm × 290 mm, and the block rib size was 30 mm × 30 mm. The layout plan is shown in Figure 2.

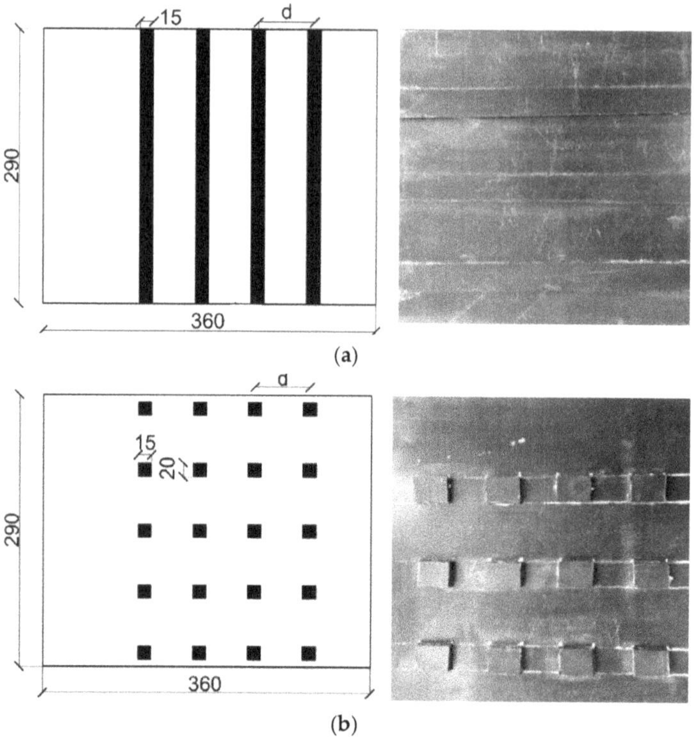


**Figure 2.** Layouts of ribbed geomembrane samples (unit: mm). (**a**) Strip rib. (**b**) Block rib.

A non-woven geotextile is a kind of permeable geosynthetic made of synthetic fibers by acupuncture or knitting. During the test, the non-woven geotextile was tightly attached to the geomembrane under the extrusion of the upper sand. The physical properties of the non-woven geotextile sample are shown in Table 1.

**Table 1.** The physical properties of the non-woven geotextile sample.

| Thickness (mm) | Minimum Density (g/m$^2$) | Breaking Intensity (kN/m) | Puncture Resistance of CBR (kN) | Tearing Strength (kN) |
|---|---|---|---|---|
| 3 | 400 | 12.5 | 2.1 | 0.33 |

## 2.3. Test Plan

This paper used a ribbed geomembrane–geotextile liner system to study the settlement and additional stress of landfill sand under different working conditions. The slope ratio used in the laboratory tests was 1:1.5, and the landfill height was 300 mm, as shown in Figure 3. As shown in Table 2, 19 test conditions were designed considering different rib shapes, rib spacings, and rib heights. Three parallel tests were carried out for each condition with the smooth geomembrane as a comparison.

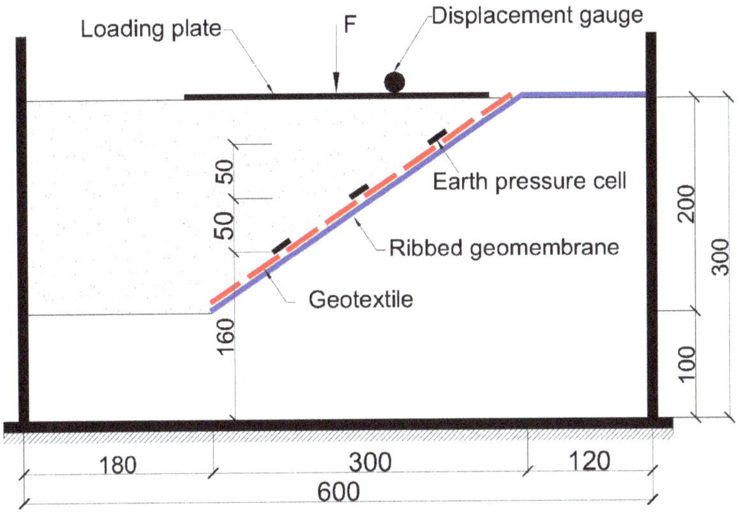


**Figure 3.** Model test arrangement (unit: mm).

**Table 2.** Test conditions.

| Rib Shape | Rib Spacing *d* (mm) | Rib Height *h* (mm) |
|---|---|---|
| Strip rib, block rib, smooth | 50, 60, 80 | 3, 4.5, 6 |

### 2.4. Test Procedure

The scale line and slope shape were marked on the model container before the test. The compactness of the sand was ensured to 80% during the sand filling process, and the foam board model was used to build the slope. In the process of filling sand, three sand pressure cells were embedded in the specified positions. During the filling process, the sand interface was kept flat with the scale line, and the slope surface was corrected after filling to the specified height. After that, the geomembrane and geotextile were laid on the slope foundation. Then the displacement gauge and the loading board were installed vertically, and the loading device was aligned with the center of the loading board. After connecting the sand pressure cells, the displacement gauge and wireless receiver were opened on the computer. After the model preloading was completed, the equipment was checked for normal operating. Each group of tests was completed when loading to the final load.

## 3. Results and Discussion

### 3.1. Model Settlement Analysis

The relative reduction in the settlement of the ribbed and smooth samples was calculated at the final load. The maximum load of this model test was 150 kPa. When the settlement value increased suddenly and the sand around the loading board rose laterally, the model was regarded as reaching the ultimate load. As shown in Table 3, it can be seen that the ribbed geomembrane reduced the settlement of the upper sand by 13.35–52.32% compared with the smooth geomembrane system. It was proved that the ribbed geomembrane reduced the settlement of upper sand significantly.

**Table 3.** Analysis of settlement under different conditions.

| Test Conditions | | | Settlement | Reduction |
|---|---|---|---|---|
| **Rib Shape** | **Rib Spacing** *d* **(mm)** | **Rib Height** *h* **(mm)** | **Settlement** *S* **(mm)** | **Reduction** $\Delta$ **(%)** |
| Smooth | / | / | 5.39 | / |
| Strip rib | 50 | 3 | 3.63 | 35.25 |
| | | 4.5 | 3.43 | 36.36 |
| | | 6 | 3.98 | 26.16 |
| | 60 | 3 | 3.45 | 36.01 |
| | | 4.5 | 3.11 | 42.30 |
| | | 6 | 3.41 | 36.73 |
| | 80 | 3 | 4.25 | 21.15 |
| | | 4.5 | 4.3 | 20.22 |
| | | 6 | 4.45 | 17.44 |
| Block rib | 50 | 3 | 3.49 | 35.25 |
| | | 4.5 | 3.41 | 36.73 |
| | | 6 | 4.01 | 25.60 |
| | 60 | 3 | 3.18 | 41.00 |
| | | 4.5 | 2.57 | 52.32 |
| | | 6 | 3.66 | 32.09 |
| | 80 | 3 | 4.32 | 19.85 |
| | | 4.5 | 4.05 | 24.86 |
| | | 6 | 4.67 | 13.35 |

Note: The reduction in the model settlement of the ribbed geomembrane relative to that of the smooth geomembrane was calculated as $\Delta = |S_{Rib} - S_{Smooth}| / S_{Smooth} \times 100\%$.

### 3.2. Model *p–s* Curve Analysis

The *p–s* curves under different rib shapes, rib spacings *d*, and rib heights *h* were analyzed in this section. The *p–s* curves of the strip and block ribbed geomembrane were similar to the variation of rib spacings and rib heights. Therefore, the following section focuses on analyzing the *p–s* curves of block ribbed geomembranes under different rib spacings *d* and rib heights *h*.

Figure 4 shows the effect of different rib shapes on the *p–s* curve. From Figure 4a–c, it can be seen that the ribbed geomembrane effectively reduced the settlement of the liner system model compared with the smooth geomembrane. In addition, the settlement of the strip ribbed geomembrane liner system was smaller than that of the block ribbed geomembrane liner system in each working condition. The settlement difference was maximum at $d$ = 60 mm. On one hand, compared with block rib, strip rib had a larger friction force when sliding with the soil, as well as a more robust extrusion and occlusal effect, thus providing a more significant passive impedance force. On the other hand, the end-bearing resistance of ribs increased the overall stiffness of the surrounding soil to a greater extent and then reduced the settlement.

As shown in Figure 5, when the rib spacing increased from 50 to 60 mm, the settlement of corresponding rib height $h$ = 3 mm, 4.5 mm, and 6 mm decreased by 8.88%, 24.63%, and 8.73%, respectively. It could be seen that when the rib height was too high, the reinforcement mechanism would be weakened. The reason was the rib height made the geomembrane bond too closely with the soil above it, resulting in a shear plane at the lower edge of the geomembrane, thus increasing the settlement. When the rib spacing was further increased to 80 mm, the corresponding settlement increased by 35.8, 57.57, and 27.04%. It could be seen that when the rib spacing was too high, it also had the opposite effect, which was directly caused by weakening the stiffening mechanism.

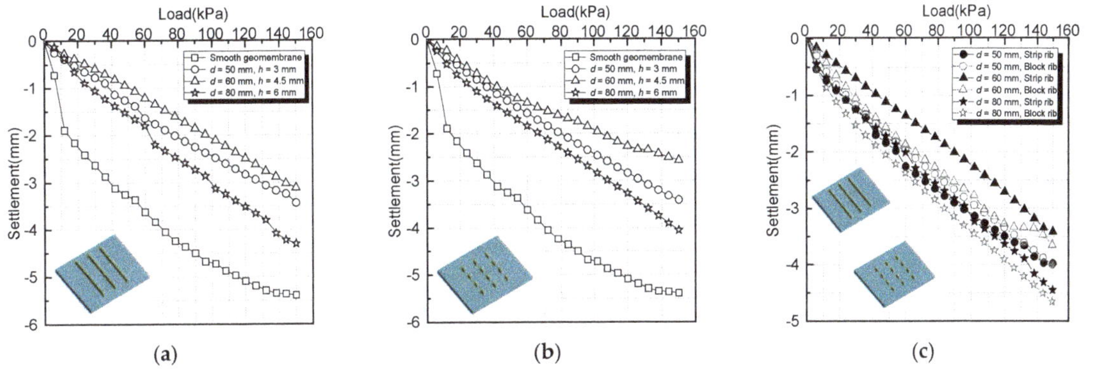


**Figure 4.** Effect of rib shapes on *p–s* curves. (**a**) Smooth and strip rib. (**b**) Smooth and block rib. (**c**) Strip and block rib.

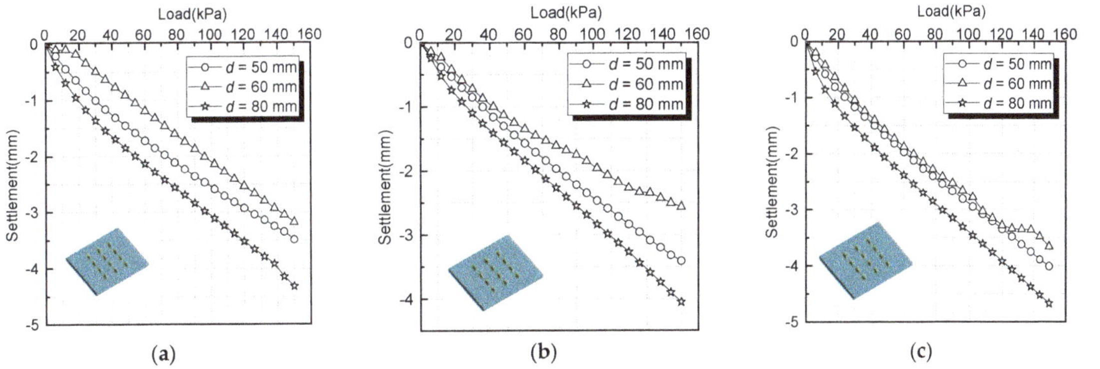


**Figure 5.** Influence of rib spacing on *p–s* curves. (**a**) $h$ = 3 mm. (**b**) $h$ = 4.5 mm. (**c**) $h$ = 6 mm.

The above test results show that the rib shapes, rib spacings $d$, and rib heights $h$ directly affected the reinforcing effect of the ribbed geomembrane on the sand. Within a specific range, the influence of the rib shapes was inconspicuous when the rib spacing was small, and the settlement of the strip ribbed geomembrane liner system decreased with the increase in the rib spacing. The friction and interlocking forces of the ribs on the sand increased with the increase in the rib spacing $d$ and rib height $h$. However, the complex formed by the ribs and the surrounding sand exceeded the limit value for the reinforcing effect of the ribs. After exceeding the limit value, the reinforcing effect of bearing resistance on the sand was weakened. The formation range of the composite was reduced, resulting in decreased overall stability and bearing capacity.

### *3.3. Additional Stress Analysis of Sand*

The additional stress is the incremental stress caused by the load in the foundation. It is the main cause of deformation due to the loss of stability of the foundation.

The additional stress distribution of the smooth geomembrane model is shown in Figure 6. From Figure 6, it can be seen that the additional stress of the sand in the smooth geomembrane liner system gradually decreased from the top to the bottom, indicating that the upper load was transmitted downward. The attenuation of the upper load was consistent with the law that the additional stress in the sand decreases with the increase in the buried depth [10].

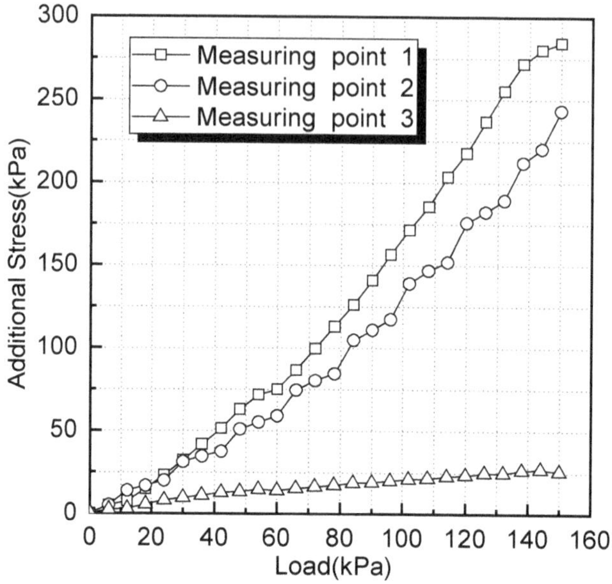


**Figure 6.** Additional stress distributions of the smooth geomembrane model.

The additional stress distribution of the ribbed geomembrane model is shown in Figure 7. From Figure 7, it can be seen that the additional stress of the sand in the ribbed geomembrane liner system decreased uniformly from top to bottom, indicating that the ribs affected the distribution of additional stress in the sand during the downward transfer of upper load. It can be seen that the additional stress of the model decreased from top to bottom, which accorded with the law of load transfer in the model. Compared with the smooth geomembrane condition, the additional stress of the ribbed geomembrane model was uniform. This was because the ribs produced a vertical component force under the upper load, which could offset the upper load to reduce the load. It made the internal stress distribution of the model change so that the upper load was partly transferred to the bottom. When the ribbed geomembrane liner system was subjected to the upper load, part of the upper load was offset by the vertical upward force generated by the ribs. This changed the additional stress distribution of the sand and avoided the overall instability or reduction of the bearing capacity of the model.

Taking the rib height $h$ = 6 mm as an example to analyze the effect of the rib spacing $d$ on the additional stress distribution of the block ribbed geomembrane model (Figure 8), compared with rib spacings of $d$ = 50 mm and 80 mm, the additional stress of the sand at measuring points 1 and 2 was significantly lower when $d$ = 60 mm. The additional stress of the sand at measuring point 3 and the additional stress of the sand with rib spacing $d$ = 60 mm were significantly higher than those of other conditions. This showed that when the rib spacing $d$ = 60 mm, the ribs blocked the downward movement of sand, reduced the tensile stress on the geomembrane when sand slipped, and improved the overall stability model. When the rib spacing $d$ = 80 mm, the additional stress of measuring point 3 and measuring point 2 at the final load were almost the same. This showed that when the rib spacing exceeded a specific value, the overall stability of the model may be destroyed. It was also indicated that there was an optimal value for the rib spacing of the ribbed geomembrane.

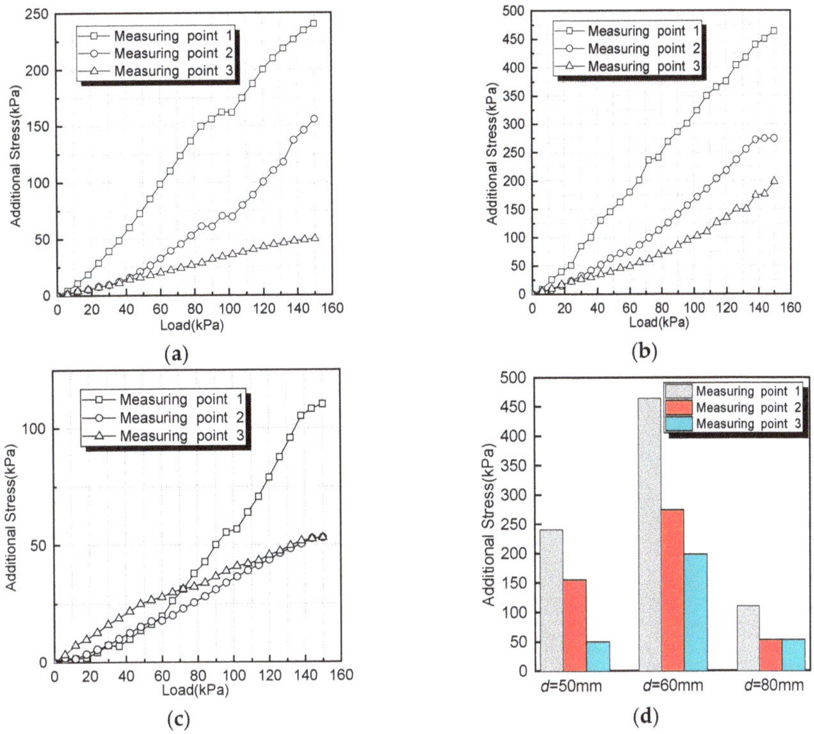


**Figure 7.** Distribution of additional stress in the block rib geomembrane model. (**a**) $d$ = 50 mm, $h$ = 4.5 mm. (**b**) $d$ = 60 mm, $h$ = 4.5 mm. (**c**) $d$ = 80 mm, $h$ = 4.5 mm. (**d**) $p$ = 150 kPa, $h$ = 4.5 mm.

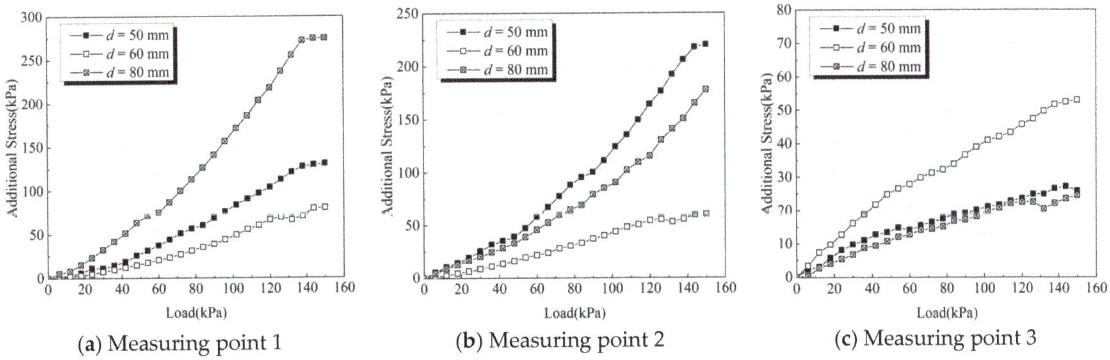


**Figure 8.** Influence of rib spacing on the additional stress of sand.

Due to the similar effects of different rib heights on the additional stress of the sand, 80 mm rib spacing conditions were used as an example to analyze the additional stress distribution of rib height on the ribbed geomembrane model. As shown in Figure 9, it can be seen from the figure that when the rib height $h$ = 3 mm, the values at measuring points 1 to 3 varied reached maximums. The additional stress of the sand at different measurement points decreased with the increase in the landfill depth. When the rib height $h$ = 4.5 mm, from measuring points 1 to 3, there was additional stress of the sand with the increase in the landfill depth. The overall change trends were relatively stable, which showed that the blocks played an active role in bearing the blockage of sand movement. When the rib height $h$ = 6 mm, the additional stress of the sand was about 270 kPa at measuring point 1 and about 30 kPa at measuring point 3. The effect was slightly better than with the rib height $h$ = 3 mm but was still inferior to that with $h$ = 4.5 mm. This showed that the inappropriate values of $h$ may weaken the reinforcing effect of the ribs. The optimal rib height could strengthen the reinforcing effect, increase the end-resistance of the ribs, and enhance the interaction between the ribbed geomembrane and the sand. It can be concluded that the optimal rib height was beneficial to the redistribution of the additional stress of the sand.

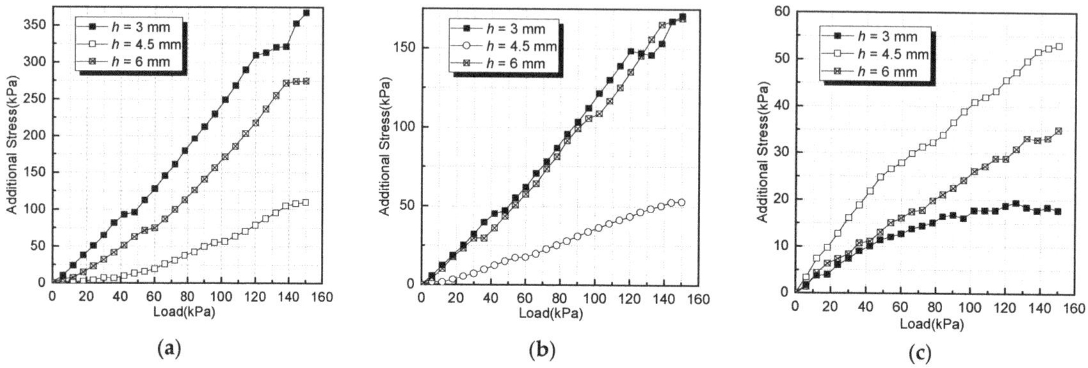


**Figure 9.** Influence of rib height on the additional stress of sand. (**a**) Measuring point 1. (**b**) Measuring point 2. (**c**) Measuring point 3.

### 3.4. PIV Image Analysis

The whole process of the sand particle displacement recorded by this research provided suitable conditions for analyzing the liner system model deformation. Since the blocks were not set along the edge, the displacement of sand particles can hardly be observed. Thus, the analysis focused on the conditions of the smooth geomembrane and the strip ribbed geomembrane.

Figures 10 and 11 show the sand movement under the conditions of the smooth geomembrane and the strip ribbed geomembrane. The sand particles indicated in the figure reflected the movement trends of some particles around. In the figure, 1–4 are sand particles located in four different positions. Under the conditions of smooth geomembrane, sand particles 1 and 2 moved vertically downward, sand particle 3 deflected down to the left, and sand particle 4 deflected down to the right. Under the conditions of ribbed geomembrane, sand particle 1 deflected down to the left, sand particles 2 and 3 slid downward along the interface, and sand particle 4 deflected down to the right. Due to the lower upper load, the sand particles under the loading board gradually experienced vertical downward displacements in the initial stage. When t = 2~4 s, the sand particles between the loading board and geomembrane moved down to the left, and the influence scope of the upper load increased. When t = 4~6 s, due to the limitation of the geomembrane interface, the displacements of the sand particles under the loading board gradually formed a circular slip surface. When t = 6~10 s, the displacements of the sand particles near the ribs were relatively low, but the displacements of the sand particles on the observation surface increased significantly. At this time, the range of the sliding surface expanded, and the sand particles overflowed to both sides of the loading board.

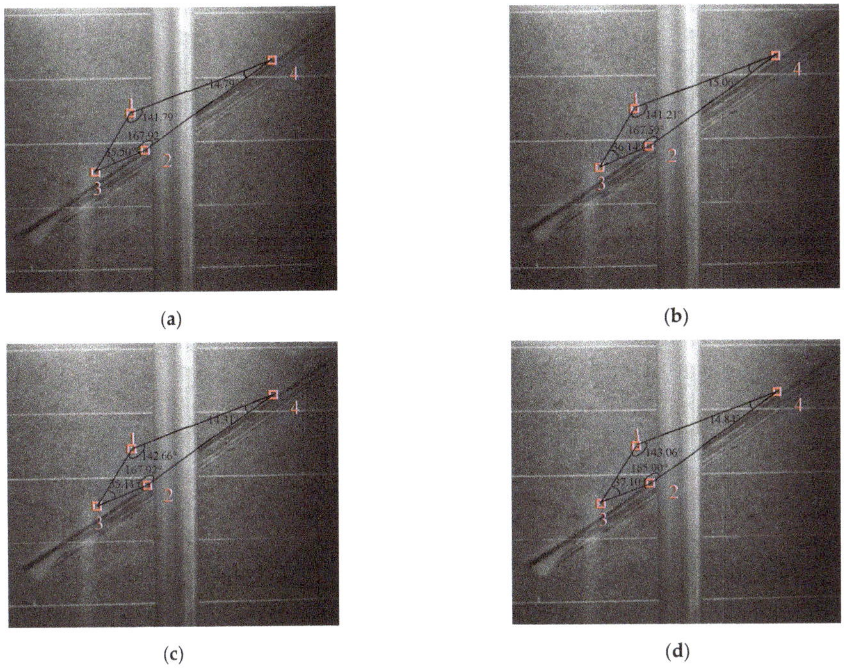

**Figure 10.** Schematic diagram of the sand movement under the conditions of the smooth geomembrane. (**a**) t = 2 s. (**b**) t = 4 s. (**c**) t = 6 s. (**d**) t = 10 s.

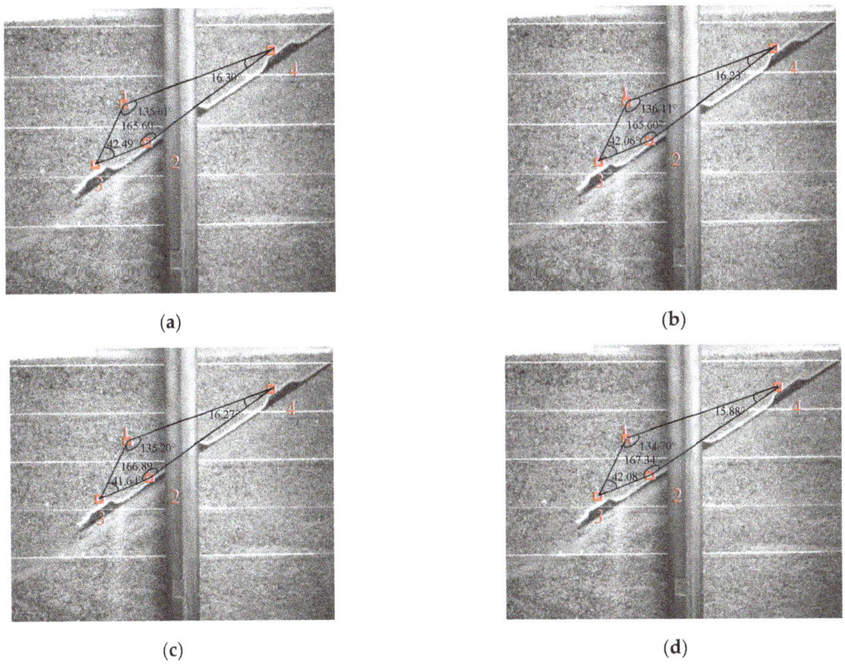

**Figure 11.** Schematic diagram of the sand movement under the conditions of the ribbed geomembrane. (**a**) t = 2 s. (**b**) t = 4 s. (**c**) t = 6 s. (**d**) t = 10 s.

### *3.5. Sand Displacements Analysis of Different Geomembranes*

The photos of the smooth geomembrane and the ribbed geomembrane during the loading process were intercepted and calibrated, respectively, and then the cross-correlation calculation was carried out with the initial time to obtain the displacement nephogram at the corresponding time. To study the effect of ribs on the sand displacements, the horizontal and vertical displacement nephograms of the smooth geomembrane conditions were compared with the strip rib geomembrane conditions, as shown in Figures 12 and 13. The shadows in the middle of the figures were caused by the test equipment blocking 3the model container in the shooting direction. Its influence on the calculation result was eliminated by drawing masks in the PIV image settings. Figure 12 shows that the absolute value of the horizontal displacement of sand above the smooth geomembrane interface was 0.0825 mm. The absolute value of the horizontal displacement of sand above the ribbed geomembrane interface was 0.0063. As shown in Figure 13, the absolute value of the horizontal displacement of sand above the smooth geomembrane interface was 0.2275 mm, and the absolute value of the horizontal displacement of sand above the ribbed geomembrane interface was 0.0588 mm. This showed that the ribbed geomembrane reduced the horizontal and vertical displacements effectively.

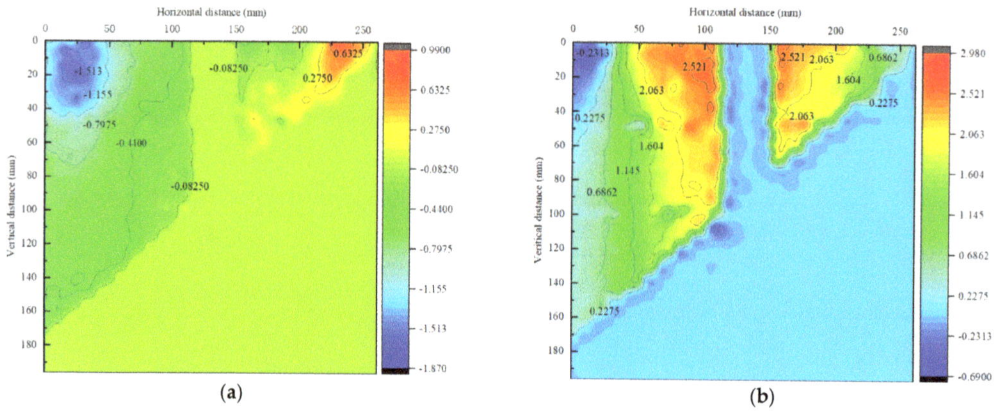


**Figure 12.** Displacement nephograms of the smooth geomembrane condition. (**a**) Horizontal displacement. (**b**) Vertical displacement.

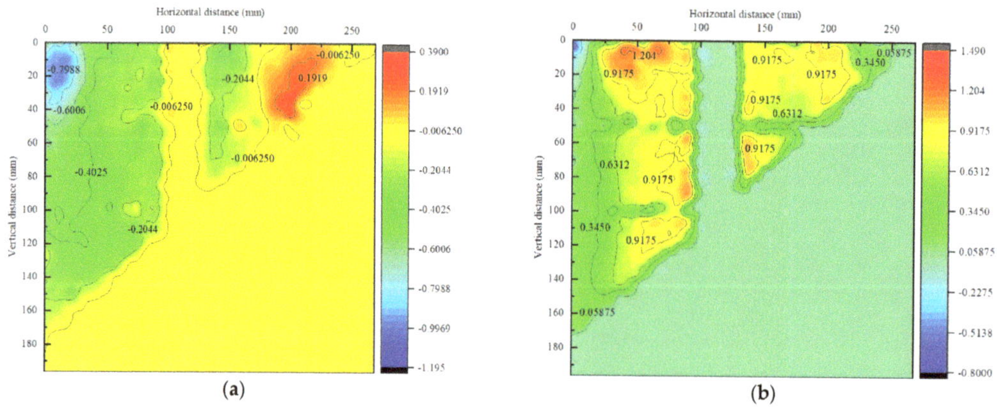


**Figure 13.** Displacement nephograms of the ribbed geomembrane condition ($d$ = 60 mm, $h$ = 4.5 mm). (**a**) Horizontal displacement. (**b**) Vertical displacement.

To analyze the displacements of the sand particles above the geomembrane interface, the data in the loading process were extracted and plotted, as shown in Figures 14 and 15. It was shown that the ribs directly resisted the sand particles, formed a more comprehensive range of indirectly affected areas, and combined the sand closely. This area prevented the interface of the geomembrane liner system from slipping and constrained the deformation of the liner system. This was a kind of quasi-cohesive force, which improved the strength of the slope surface. With the increasing normal stress of the ribs, the bearing resistance of the ribs enhanced the interface shear strength significantly. At the same time, the friction resistance of the ribs increasing with the incremental upper load. Therefore, compared with the smooth geomembrane conditions, the ribbed geomembrane conditions could effectively improve the stability of the liner system, which was in accordance with the interaction mechanism between the sand and the denti-strip [11].

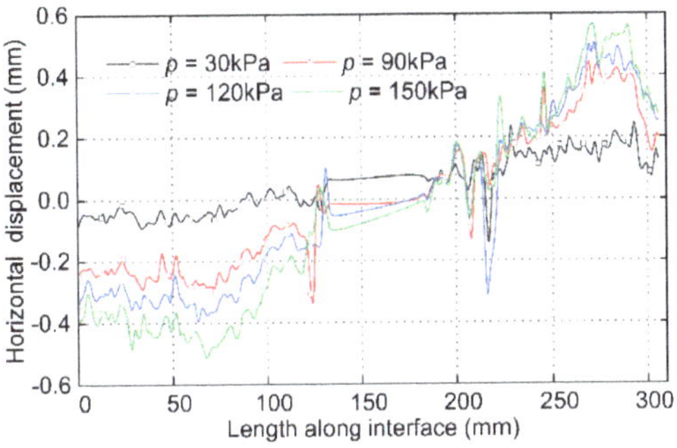


(a)

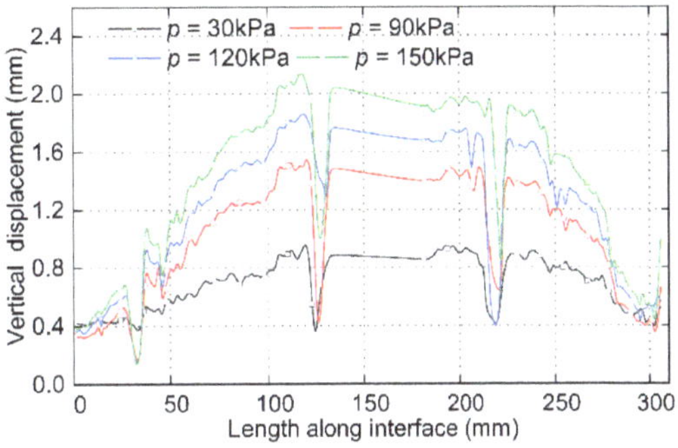


(b)

**Figure 14.** Displacement of sand above the smooth geomembrane interface. (**a**) Horizontal displacement. (**b**) Vertical displacement.

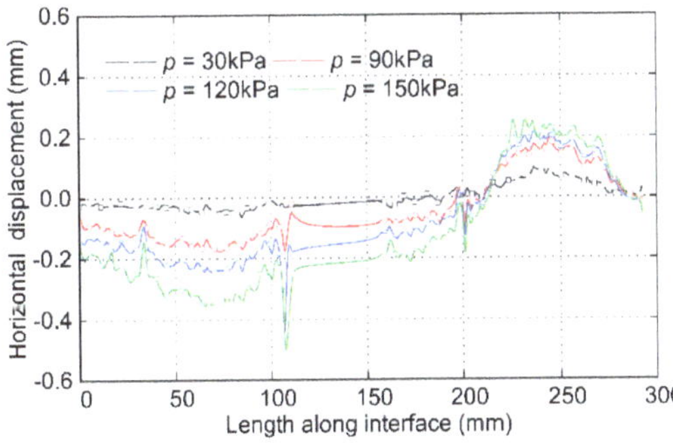


(a)

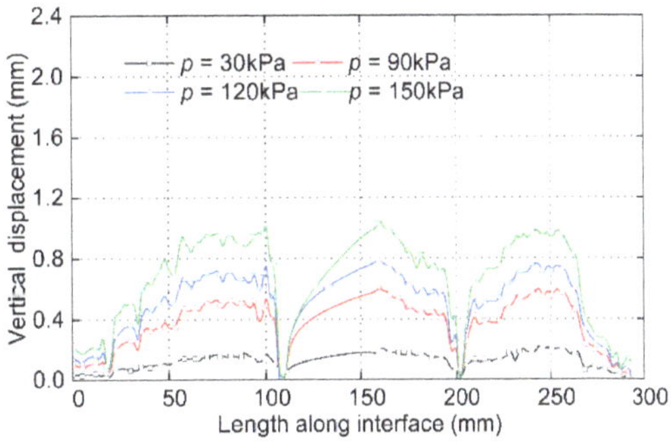


(b)

**Figure 15.** Displacement of sand above the ribbed geomembrane interface. (**a**) Horizontal displacement. (**b**) Vertical displacement.

### 3.6. Analysis of Rib Spacing on Sand Displacements

Figures 16 and 17 show the horizontal and vertical displacements of different rib spacings. Compared with Figure 13, it was found that the absolute values of the sand horizontal displacement above the geomembrane interface of three working conditions (rib spacing $d$ = 50 mm, $d$ = 60 mm, and $d$ = 80 mm) were 0.1625, 0.0063, and 0.1975 mm, respectively. At the same time, the absolute values of the sand vertical displacement above the geomembrane interface of the three working conditions (rib spacing $d$ = 50 mm, $d$ = 60 mm, and $d$ = 80 mm) were 0.4525, 0.0588, and 0.3650 mm, respectively. This was due to the better effect of the 60 mm rib spacing on the cohesion of sand particles around the ribs. The interaction between the ribs and the sand was not evident under the condition of 50 mm rib spacing. The reinforcing effect of the ribs on the surrounding sand was weakened under 80 mm rib spacing.

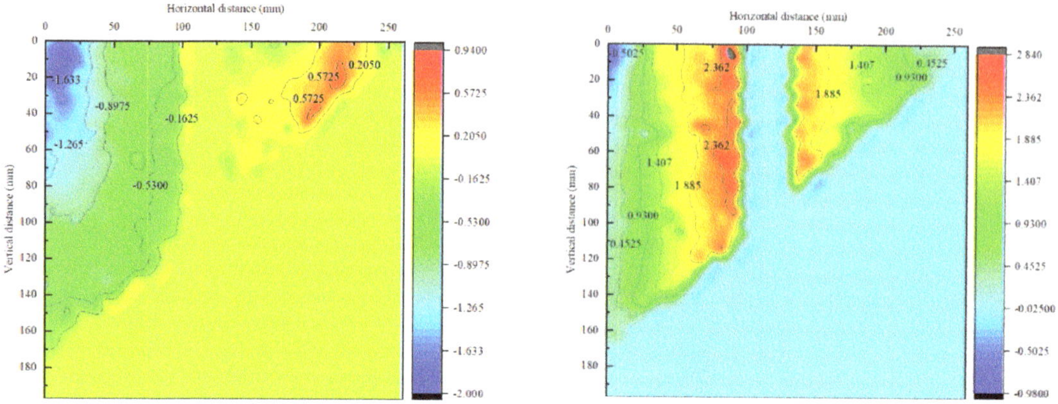


(a) (b)

**Figure 16.** Displacement nephograms of smooth geomembrane conditions ($d$ = 50 mm, $h$ = 4.5 mm). (**a**) Horizontal displacement. (**b**) Vertical displacement.

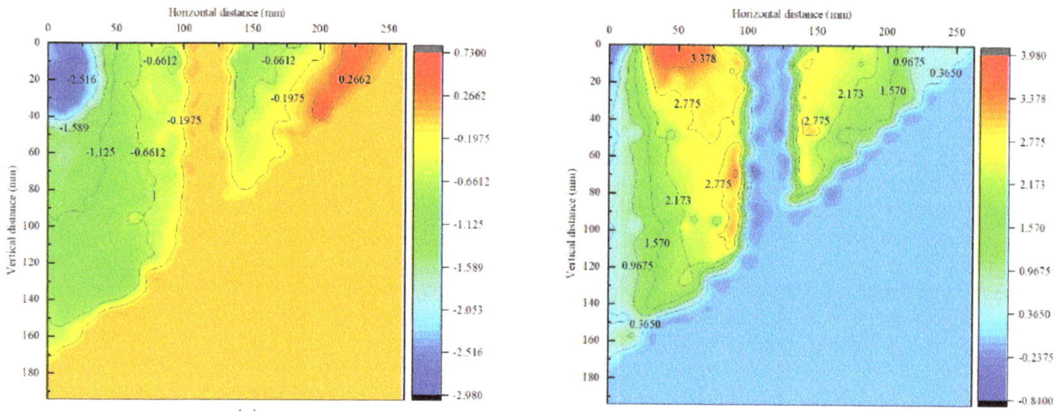


(a) (b)

**Figure 17.** Displacement nephograms of ribbed geomembrane conditions ($d$ = 80 mm, $h$ = 4.5 mm). (**a**) Horizontal displacement. (**b**) Vertical displacement.

As shown in Figures 18 and 19, the sand displacements of different rib spacings were analyzed. Combined with Figure 15, the results showed that the horizontal and vertical displacements of the sand particles above the geomembrane interface were smaller when the rib spacing $d$ = 60 mm. It could indicate that the 60 mm rib spacing blocked and interlocked the movements of the surrounding sand particles. It proved the optimal rib spacing of the ribbed geomembrane existed, and this was consistent with the previous experimental results.

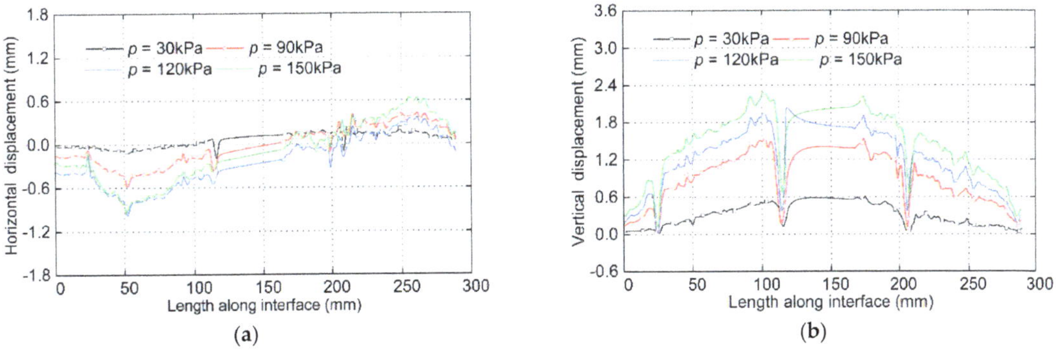


**Figure 18.** Displacement of sand above ribbed geomembrane conditions ($d$ = 50 mm, $h$ = 4.5 mm). (**a**) Horizontal displacement. (**b**) Vertical displacement.

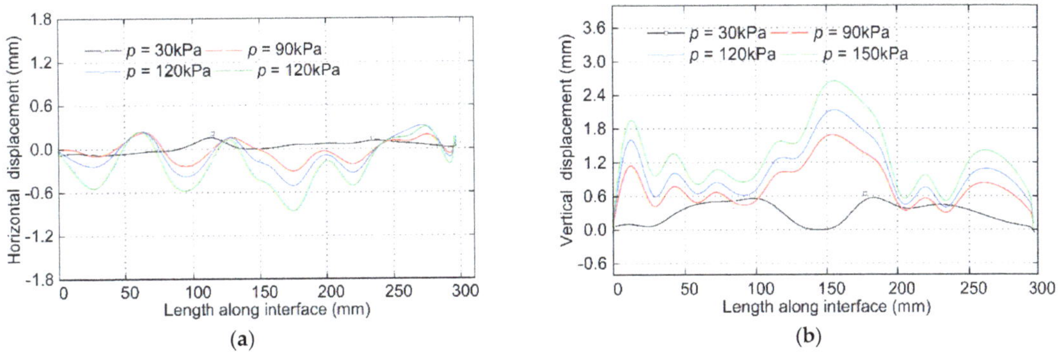


**Figure 19.** Displacement of sand above ribbed geomembrane conditions ($d$ = 80 mm, $h$ = 4.5 mm). (**a**) Horizontal displacement. (**b**) Vertical displacement.

### *3.7. Analysis of Rib Height on Sand Displacements*

Figures 20 and 21 show the horizontal and vertical displacements of different rib heights. Compared with Figure 13, it was found that the absolute values of the sand horizontal displacement above the geomembrane interface of three working conditions (rib height $h$ = 3 mm, $h$ = 4.5 mm, and $h$ = 6 mm) were 0.0500, 0.0063, and 0.0725 mm, respectively. At the same time, the absolute values of sand vertical displacement above the geomembrane interface of the three working conditions (rib height $h$ = 3 mm, $h$ = 4.5 mm, and $h$ = 6 mm) were 0.1200, 0.0588, and 0.0625 mm, respectively. This was because the elastic modulus of the sand increased more in the indirect influence area created by the 4.5 mm rib height, which inhibited the movement trend of the sand particles. The interaction between the ribs and the sand was small when the rib height was 3 mm, and the reinforcing effect of the ribs on surrounding sand declined when the rib height was 4.5 mm.

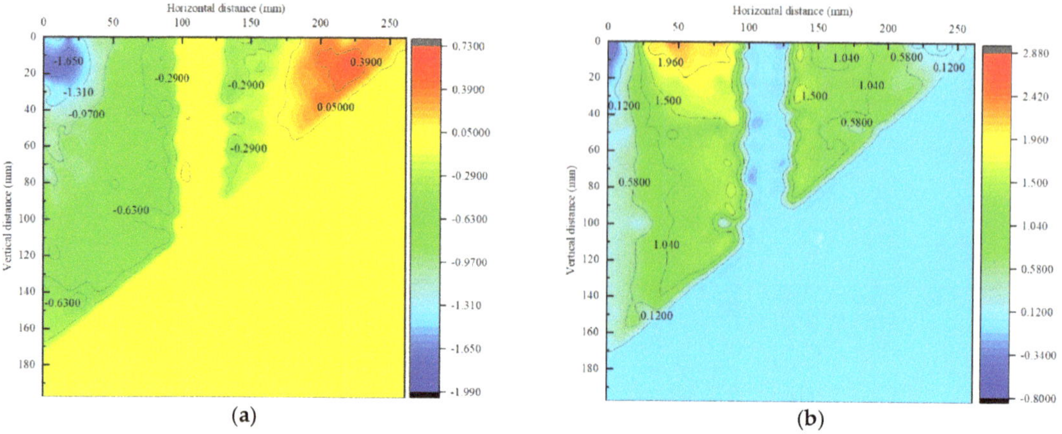


**Figure 20.** Displacement nephograms of smooth geomembrane conditions ($d$ = 60 mm, $h$ = 3 mm). (**a**) Horizontal displacement. (**b**) Vertical displacement.

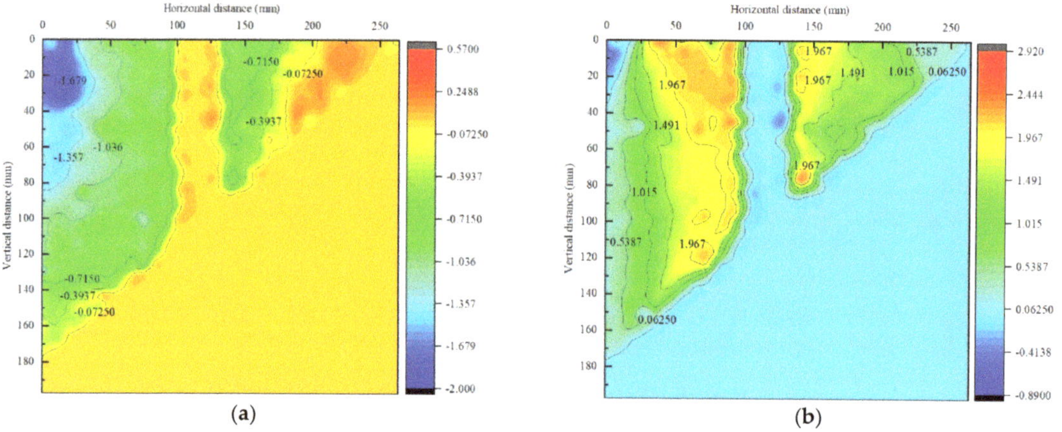


**Figure 21.** Displacement nephograms of ribbed geomembrane conditions ($d$ = 60 mm, $h$ = 6 mm). (**a**) Horizontal displacement. (**b**) Vertical displacement.

As shown in Figures 22 and 23, the sand displacements of different rib heights were analyzed. Compared with Figure 15, the results showed that the horizontal and vertical displacements of the sand particles above the geomembrane interface were smaller when the rib height $h$ = 4.5 mm. It can be concluded that the 4.5 mm rib height limited the surrounding sand particles and reduced the risk of liner system interface sliding. It verified the optimal rib height of the ribbed geomembrane, which was the same as the previous experimental results.

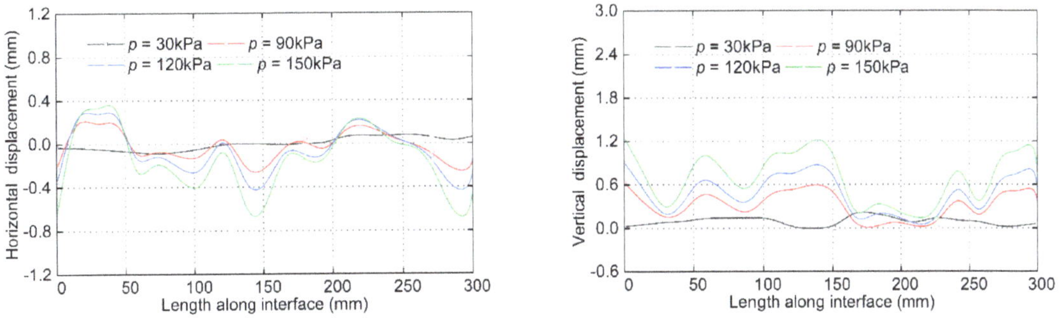


(a) (b)

**Figure 22.** Displacement of sand above ribbed geomembrane conditions ($d$ = 60 mm, $h$ = 3 mm). (**a**) Horizontal displacement. (**b**) Vertical displacement.

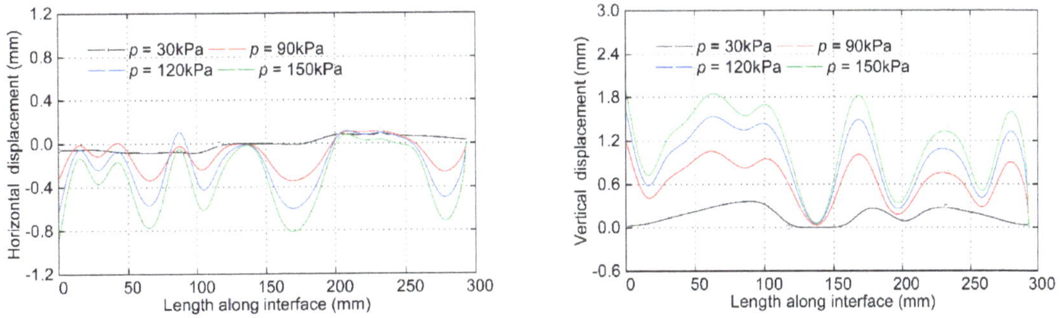


(a) (b)

**Figure 23.** Displacement of sand above ribbed geomembrane conditions ($d$ = 60 mm, $h$ = 6 mm). (**a**) Horizontal displacement. (**b**) Vertical displacement.

## 4. Conclusions

1. It can be seen from the comparative analysis of working conditions that the settlement of the ribbed geomembrane was significantly less than that of the smooth geomembrane. It was proved that the ribs enhanced the stability of the liner system.
2. According to the analysis of the test results, there was an optimum value of 60 mm for the rib spacing and 4.5 mm for the rib height. When the value was less than the optimal value, the ribbed geomembrane could not effectively improve the stability of the liner system. When the value was more than the optimal value, the reinforcing effect of the ribbed geomembrane on the sand gradually decreased.
3. By analyzing the additional stress of the sand, it can be concluded that the ribbed geomembrane was beneficial to the stress redistribution, thereby improving the overall stability of the liner system.
4. According to the PIV image analysis, the ribbed geomembrane can form an indirect influence area with the sand within a specific range to restrain the displacement of nearby sand particles. The indirect influence area was conducive to the stability of the liner system.

## 5. Prospect

This paper effectively demonstrated the advantages of a ribbed geomembrane and geotextile liner system using indoor model tests and PIV image velocimetry techniques. The results of this thesis can provide a theoretical basis for practical applications in projects such as landfills.

**Author Contributions:** Conceptualization, J.G.; Investigation, J.G.; Data curation, J.W.; Writing—original draft, J.G.; Writing—review & editing, J.W.; Project administration, J.G. All authors have read and agreed to the published version of the manuscript.

**Funding:** This research received no external funding.

**Institutional Review Board Statement:** Not applicable.

**Informed Consent Statement:** Not applicable.

**Data Availability Statement:** The data that support the findings of this study are available from the corresponding author, upon reasonable request.

**Conflicts of Interest:** The authors declare no conflict of interest.

## References

1. Gao, J.L.; Zhang, M.X.; Zhang, W.J. Interface property between sand and reinforced geomembrane. *Rock Soil Mech.* **2011**, *32*, 3220–3230.
2. Bao, C.G. Study on interface behavior of geosynthetics and soil. *Chin. J. Rock Mech. Eng.* **2006**, *25*, 1735–1744.
3. Irsyam, M.; Hryciw, R.D. Friction and passive Resistance in soil reinforced by plane ribbed Inclusions. *Géotechnique* **1991**, *41*, 485–498. [CrossRef]
4. Zhang, M.X.; Zhang, S.L. Behaviour of soil reinforced with H-V inclusions by PFC^2D. *Chin. J. Geotech. Eng.* **2008**, *30*, 625–631.
5. Zhou, J.; Kong, X.L.; Wang, X.C. Bearing capacity behaviours and failure modes of reinforced grounds. *Chin. J. Geotech. Eng.* **2008**, *30*, 1265–1269.
6. Yang, Q.; Zhang, K.; Lua, M.T. Study of model test on performance of soil foundation reinforced by geogrids. *J. Dalian Univ. Technol.* **2006**, *46*, 390–394.
7. Oda, M.; Nemat-Naaser, S.; Konish, J. Stress-induced anisotropy in granular masses. *Soil Found.* **1985**, *25*, 85–97. [CrossRef] [PubMed]
8. Gao, J.L.; Zhang, M.X.; Zhang, W.J. Interface Frictional Property Between Sand and Geomembrane. In *Advances in Environmental Geotechnics, Proceedings of the International Symposium on Geoenvironmental Engineering, Hangzhou, China, 8–10 September 2009*; Springer: Berlin/Heidelberg, Germany, 2010; pp. 822–827.
9. Gao, J.L.; Zhang, M.X.; Lin, Y.L.; Qiu, C.C. Analysis of interaction mechanism of reinforced geomembrane and sandy sand interface. *Rock Sand Mech.* **2012**, *33*, 2465–2471. [CrossRef]
10. Gao, J.J.; Xu, H.; Qian, J.W. Settlement Behavior of Soft Subgrade Reinforced by Geogrid-Encased Stone Column and Geocell-Embedded Sand Cushion: A Numerical Analysis. *Adv. Civ. Eng.* **2022**, *2022*, 8874520. [CrossRef]
11. Zhang, M.X.; Zhang, S.L.; Huang, J. Behavior of interface between denti-strip geosynthetic reinforcements and sand under low surcharge. *Chin. J. Geotech. Eng.* **2007**, *29*, 1623–1629.

*applied sciences*

MDPI

*Article*

# Development and Application of a Microsurfacing Mix Design Method to Assess the Influence of the Emulsion Type

**Caroline F. N. Moura [1], Joel R. M. Oliveira [1], Hugo M. R. D. Silva [1,*], Carlos A. O. F. Palha [2] and Cesare Sangiorgi [3]**

1 Department of Civil Engineering, Institute for Sustainability and Innovation in Structural Engineering, University of Minho, 4800-058 Guimaraes, Portugal; id8972@alunos.uminho.pt (C.F.N.M.); joliveira@civil.uminho.pt (J.R.M.O.)
2 Department of Civil Engineering, University of Minho, 4800-058 Guimaraes, Portugal; cpalha@civil.uminho.pt
3 Department of Civil, Chemical, Environmental, and Materials Engineering, University of Bologna, 33-40126 Bologna, Italy; cesare.sangiorgi4@unibo.it
* Correspondence: hugo@civil.uminho.pt; Tel.: +351-253-510200

**Abstract:** Microsurfacing asphalt mixtures are a preventive maintenance technology comprising the application of a slurry (produced with a modified asphalt emulsion), aggregate, filler, and water on top of an existing pavement at ambient temperature. Although it is a widely used technology, further studies on the mix design procedures are necessary to ensure an adequate composition. Thus, this study contributes to developing an improved mix design procedure for microsurfacing asphalt mixtures. Different mixtures were prepared, and the influence of the type and amount of asphalt emulsion and the amount of added water and filler (cement) on the characteristics of the mixture were evaluated. Two preliminary tests, referred to as the "pizza test" and the "ball test", were proposed to determine the initial proportions of added water and cement in the mixture, respectively. Then, consistency, cohesion, and shaking abrasion tests were performed to determine the optimum content of each component and evaluate their influence on the mixture characteristics. The results showed that these tests are essential to optimize the mix composition, even though it was found that the mix design of microsurfacings is a complex task because the mixture is a system with chemical interactions strongly influenced by its composition.

**Keywords:** microsurfacings; mix design; asphalt emulsion; consistency; cohesion; abrasion

**Citation:** Moura, C.F.N.; Oliveira, J.R.M.; Silva, H.M.R.D.; Palha, C.A.O.F.; Sangiorgi, C. Development and Application of a Microsurfacing Mix Design Method to Assess the Influence of the Emulsion Type. *Appl. Sci.* **2023**, *13*, 7925. https://doi.org/10.3390/app13137925

Academic Editor: José Manuel Moreno-Maroto

Received: 14 June 2023
Revised: 30 June 2023
Accepted: 2 July 2023
Published: 6 July 2023



## 1. Introduction

Cold mix asphalt (CMA) combines unheated aggregates, mineral fillers, and bitumen emulsion to produce flexible pavement materials. This technology allows the manufacture of mixtures at ambient temperatures without heating vast amounts of aggregates and bitumen, decreasing $CO_2$ emissions and saving energy and thereby playing a fundamental role in developing sustainable and ecological pavements [1–3]. Therefore, they are gaining popularity, including in recycling work [4], due to their low environmental impact and high sustainability compared with hot paving technologies.

Microsurfacing asphalt mixtures can be described as a cold mix technology for preventive paving maintenance, which involves the application of a combination of polymer-modified asphalt emulsion, well-graded fine aggregate, mineral filler (usual cement), water, and chemical additives (if any) at ambient temperature [5–8]. Extensive studies conducted by the Texas Transportation Institute (TTI) have confirmed the effectiveness of microsurfacing as a valuable method for preventative maintenance and pavement preservation, leading to a substantial increase in the service life of existing pavements. However, it is essential to note that while microsurfacing extends pavement life successfully by three to four years, its ability to reduce long-term cracking is somewhat limited. Nevertheless, microsurfacing guarantees the provision of a functional and long-lasting pavement surface.

The same transportation institute highlighted the crucial role of careful project selection in successfully implementing microsurfacing. Although microsurfacing has proven effective in addressing issues such as rutting, bleeding, loss of surface friction, oxidation, and raveling, its effectiveness may be constrained when it is applied to pavement with structural deficiencies [9,10].

They are a usual solution for pavement maintenance, acting as a protective surface course that can address issues related to skid resistance, raveling, or oxidation on structurally sound pavements, improving the road surface characteristics [5,11,12]. Microsurfacing is also employed to correct the shape or profile rutted wheel paths and obtain a faster traffic opening [13], with a 40% reduction in the number of original rutting and substantial increases in the friction characteristics of the pavement being reported in previous works [14]. In addition, since this technology requires fewer material resources, is cold-produced, and laid at a low thickness, it has a low cost, short construction period, and rapid traffic opening [15,16].

Lonbar and Nazirizad [17] stated that microsurfacing is generally classified as a preventative instead of a corrective maintenance treatment due to the significant ability of microsurfacing to seal and restore pavement surfaces. According to Broughton, Lee [14], the first step to effectively using microsurfacing when treating existing flexible pavements is to ensure the pavement is structurally sound and appropriately prepared to accept the treatment. However, Gransberg [18] alleged that microsurfacing is a technique for preventive, routine, and corrective conservation; this flexibility was justified because it is a thin coating that can be placed in a thickness of two to three times the size of the mixture's largest aggregates.

According to Dash and Panda [3], cold mix technologies present several advantages, including energy-saving, the inexistence of harmful gas emissions, and facilitated application in varying climatic conditions. However, it lags far behind in research and applications. The authors stated that, unlike the hot mix design, a universally accepted cold mix design procedure is still not established. The laboratory procedures followed by different researchers/agencies vary widely regarding compaction methods, curing conditions, and other mixing procedures for evaluating relevant engineering properties.

The same situation occurs with microsurfacing. In contrast to the conventional methods used for hot mix asphalt, the mix design of microsurfacing mixtures is complicated because it is a chemical system with several variables, including the type and content of asphalt emulsion, types and gradation of aggregates, water, and filler content, which affect the final properties of the mixtures [5,19,20]. Emulsion characteristics play a pivotal role in ensuring the workability and durability of microsurfacing, contributing to its final behavior. The emulsion's properties mainly depend on the design of the components selected for its production, like emulsifier type and dosage, acid content, binder grade and content, and polymer modifier type, and dosage [21]. In addition, the consistency and adhesion between aggregates and bitumen depend on the mineralogical composition of the aggregates. Consequently, each specific composition of aggregates and asphalt emulsion has unique chemical properties that affect the bond between the aggregate and bitumen emulsion [6,12].

Another essential parameter affecting the microsurfacing mix design is the filler type and amount, which contribute significantly to the breaking and curing process due to the large surface area [21]. The mixtures typically contain 0–3% mineral filler, usually Portland cement. This component significantly influences the microsurfacing's performance, as the filler–mastic interaction critically affects the mixture cohesion [5,22]. Regarding water, its addition improves the workability of the mixture, and its primary purpose is to wet, dissolve and adhere to other components, and moderate the chemical reaction [5,12]. The optimum amount of water must be determined because a lack of water leads to the weak coating of aggregates and hardening of the mixture, and excessive water content can lead to the separation of aggregates or lengthen the mixture's curing time [6].

As each component of the microsurfacing mixture interacts with the other components to form a chemically complex compound, the amount and type of asphalt emulsion, cement, and water have a significant influence on the test results used to design the microsurfacing mixture [19,21,23]. Therefore, Gransberg [18] stated that the primary mix design objectives of microsurfacing are ensuring good compatibility among the mixture ingredients and satisfying the strength and durability parameters.

Thus, several mix design guidelines have been developed for an optimum job mix formula. The International Slurry Surfacing Association (ISSA), the Texas Transportation Institute (TTI), the American Society for Testing and Materials (ASTM), and the California Department of Transportation (Caltrans) developed microsurfacing mix design procedures, and the ISSA's guidelines are the most accepted and used worldwide. Similar standards have also been developed in the European Union and South Africa to support the design of slurry and microsurfacing materials [5,19,20,23,24].

However, various researchers have reported difficulties when using the existing mix design procedures, namely the repeatability and reproducibility of test results and poor correlation between field and laboratory results [5]. Some authors, like Robati and Carter [19] and Kumar and Ryntathiang [25], have evaluated the ISSA method for microsurfacing mix design. They have concluded that modification to the current mix design procedure might be necessary to enhance reproducibility and provide successful mix designs on the basis of performance-related tests in the design method. Thus, the microsurfacing mix design methodologies described by the ASTM and ISSA should be used only as a guide [25].

Therefore, this work aimed to significantly contribute to developing an improved mix design method for microsurfacing mixtures that also included test methods from the European standards [26–28]. The method was also applied to study the influence of the type and amount of asphalt emulsion, filler (cement), and added water content on the composition and characteristics of a specific microsurfacing mixture. This paper is the result of an interlaboratory study of the working group TG2 (Cold bitumen emulsion mixtures) of RILEM's Technical Committee 280-CBE: multiphase characterization of cold bitumen emulsion materials.

## 2. Materials and Methods

### 2.1. Materials

Microsurfacings are mixtures comprising an asphalt emulsion, aggregates, filler, and water. According to Bhargava et al. [5], as microsurfacing performance is significantly influenced by the materials' quality, understanding the behavior and parameters that affect the properties of each component is fundamental to ensuring the mixture's durability.

The materials used in this project were a 0/6 mm basalt aggregate, Portland cement as filler, two types of modified asphalt emulsion from different origins (emulsion C65BP4, named Emulsion 1, and emulsion C60BP4, named Emulsion 2), and water to increase the workability of the mixture.

### 2.2. Structure of the Microsurfacing Mix Design Method Proposed in This Work

The present work proposes a new procedure for a microsurfacing mix design based on tests developed according to a set of European Standards for slurry surfacing. Figure 1 presents the flowchart of the mix design procedure proposed in this work.

In the first step, two preliminary tests, developed by the RILEM TC 280-CBE, were carried out to determine the cement and added water content. Subsequently, the consistency test (EN 12274-3) was performed to confirm the amount of added water in the mixture. Finally, the cohesion test (EN 12274-4) was performed to identify the ideal relationship between the amount of added water, cement, and emulsion. In the end, the shaking abrasion test (EN 12274-7) was performed to determine the water sensitivity of the mixtures.

The present work produced two mixtures, one with Emulsion 1 and one with Emulsion 2, for each test. Thus, the microsurfacing mixtures were always produced using the same type of aggregate and cement, varying the amounts of the asphalt emulsion, added water, and cement. For mixture preparation, the dry aggregate amount was considered a reference for the amount of asphalt emulsion (typically between 10% and 15%), added water (typically around 10%), and cement (typically between 2% and 3%), i.e., the percentage of these materials in each mixture was always defined considering the mass of aggregate as 100%.

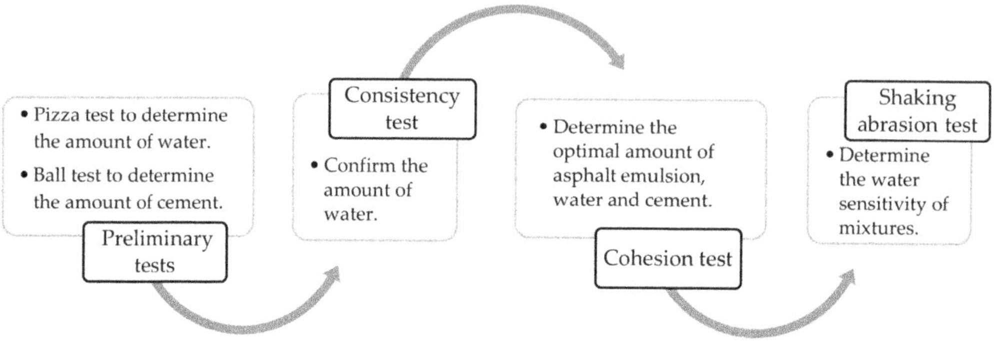


**Figure 1.** Flowchart of the microsurfacing mix design procedure proposed in this work.

### 2.3. Characterization of Aggregates Used in Microsurfacing Mixtures

The aggregates used in pavement surface courses must comply with the European Standard EN 13043 requirements, which specify the properties and test methods suitable for their characterization. Tests were carried out for particle size distribution, wear resistance, density, water absorption, and aggregate-binder affinity to better understand the aggregates' properties and their behavior in the manufacturing of the mixture.

The particle size distribution of the selected aggregates was determined following the EN 933-1 standard to verify whether they fulfilled the Portuguese Specifications of Infraestruturas de Portugal [29]. The test was carried out with samples of 0/6 mm aggregate collected using the quartering method. Initially, the samples were dried in an oven at 110 °C for 24 h. After drying, the samples were weighed and placed in the sieve column with a larger opening of 10 mm and a smaller opening of 0.063 mm.

The aggregate particles' resistance to wear was tested by abrasion in a wet environment, as specified by the European Standard EN 1097-1, and was expressed by the micro-Deval (MDE) coefficient. The MDE coefficient was determined as a function of the portion of the original 500 g sample reduced to a size of less than 1.6 mm after a wear period of 120 min. The test measured the aggregate's wear produced by an abrasive load inside a rotating drum with water.

The affinity of the specific basaltic aggregate used in this work to both asphalt emulsions in this study was tested following EN 12697-11. The amount of emulsion added was adjusted to use the same residual binder amount mentioned in the standard, and after mixing the emulsion with the 8/11 aggregate fraction, the mixture was spread and maintained on a watch glass for 24 h curing. Using the same procedure for both emulsions made it possible to know the potential bonding efficiency between the basaltic aggregate and each emulsion used in the microsurfacing mixture. Furthermore, the density and water absorption of the aggregates were determined according to the EN 1097-6 standard.

### 2.4. Characterization of Asphalt Emulsions Used in Microsurfacing Mixtures

This project used two types of polymer-modified asphalt emulsions (C60BP4 and C65BP4) produced in Italy with a class 4 breaking value, the first with 60% ± 2% and the latter with 65% ± 2% residual binder. The Portuguese Specifications [29] consider these emulsions adequate for microsurfacings.

The breaking value of the emulsion was determined following the EN 13075-1 standard. A reference filler was added to a specified amount of cationic asphalt emulsion at a uniform rate with constant agitation in a specific apparatus. When the emulsion broke down completely, the amount of filler added was determined by weighing, and the break index value corresponds to the filler mass (in grams) multiplied by 100 and divided by the amount of emulsion (in grams).

The emulsion adhesion test was based on the European Standard EN 13614. Initially, 200 g of aggregate was weighed, washed, and left in an oven at 110 °C for 2 h. Subsequently, the aggregates were added to the emulsion, which contained 10 g of residual binder, and mixed with a spatula. The mixture was then spread on a tray lined with parchment paper and put in an oven at 60 °C for 24 h. Then, the mixture was transferred to a 300 mL water glass beaker at 60 °C and returned to the oven at 60 °C for 24 h. Finally, the surface of the emulsion-coated aggregate was evaluated.

The procedure specified in the EN 13074-1 standard was used to recover the binder of each emulsion. According to this standard, recovery involves evaporating all the water in the emulsion, leaving only the binder. The procedure consisted of spreading a known mass of emulsion on a tray lined with parchment paper and exposing the tray with the emulsion at room temperature under standard laboratory conditions for 24 h. After this, the tray was placed in an oven at 50 °C for another 24 h. At the end of this period, the tray was removed from the oven and allowed to cool until it reached room temperature, and then all residual binder was collected from the tray.

After the binder recovery, to evaluate which properties could influence the microsurfacing mixtures, wide-ranging binder characterization tests were carried out, such as penetration at 25 °C (EN 1426), softening point temperature determined by the ring and ball method ($T_{R\&B}$) (EN 1427), dynamic viscosity (EN 13302), cohesion energy by force ductility at 5 °C (EN 13589), elastic recovery at 25 °C (EN 13398), and determination of the complex shear modulus on the dynamic shear rheometer (EN 14770). The rheology tests were performed in the linear viscoelastic regime using an 8 mm plate with a 2 mm gap at temperatures below 30 °C and a 25 mm plate with a 1 mm gap at higher temperatures.

### 2.5. Preparation of Microsurfacing Mixtures

According to Destrée et al. [16], when comparing different test methods of the EN 12274 standards, the mixing procedure description is not always the same, and it is not sufficiently detailed. Regarding the standards used in this work, only standard EN 12274-7 describes the preparation sequence of the mixture. Thus, this study proposed a mixing procedure that determined the addition sequence of each material to be used in all tests where the standard does not determine the procedure for the mixture preparation.

The mixture preparation started with the sieving of the aggregates in a 2 mm sieve to separate the coarse material (over 2 mm) from the fine material passing through the sieve (Figure 2a). These materials were added to the mixture at different times. The experimental procedure showed that adding separate fractions of aggregates and the total amount of water in two phases of the mixture preparation made it more consistent without excess water. Regarding the asphalt emulsion, its storage container was inverted for at least one day before its use and shaken until its contents were well mixed to ensure adequate homogeneity.

The mixtures were always prepared using the same procedure. First, 52% of the aggregates (the fine fraction) were placed in a bowl, and half the amount of water defined was added (Figure 2b), mixing until all the aggregates were homogeneously wet. Then, the defined amount of cement was added and mixed homogeneously (Figure 2c).

Subsequently, the other 48% of the aggregates (coarse part) were added (Figure 3a), together with the remaining water that would complete the predefined water content, mixing continuously until the mixture was homogeneous. Finally, the amount of emulsion defined for each mixture was added (Figure 3b), and the resulting microsurfacing material was obtained, as shown in Figure 3c. The added percentage of fine and coarse aggregates was defined according to the particle size distribution of the aggregates specified for microsurfacing mixtures.

(a) (b) (c)

**Figure 2.** Initial phase of microsurfacing production: (**a**) aggregate fraction separation, (**b**) mixture of fine aggregates with half the total water, and (**c**) addition of cement.

(a) (b) (c)

**Figure 3.** Final phase of microsurfacing production: (**a**) incorporation of coarse aggregates, (**b**) addition of asphalt emulsion, and (**c**) microsurfacing mixture immediately after production.

## 2.6. Microsurfacing Mix Design Method

### 2.6.1. Preliminary Mix Design Tests

In the study's first step, two preliminary tests, developed by the RILEM TC 280-CBE, were carried out to estimate the initial amount of added water and cement content necessary to produce the microsurfacing mixtures. The amount of added water was determined using the "pizza" test, which consists of molding a sample of the mixture into a pizza shape. After 30 min, the pizza's surface should be black, and when broken in half, its interior should be brownish. After 60 min, the entire pizza should be black. The added water must be reduced if the desired color is not achieved within that time.

The "ball" test was carried out to determine the initial amount of cement by compressing the mixture into a ball and constantly squeezing between the hands to remove all excess water. After 60 min, the ball must be dropped to the ground from a height of 1.50 m and should not break. If the ball breaks, the amount of cement must be increased.

### 2.6.2. Consistency Test

The consistency test aims to optimize the amount of added water content after the preliminary tests, assessing the workability and segregation potential of the microsurfacing material. This test aimed to determine whether the added water amount determined by the pizza test reflected the ideal water content for each mixture.

The test followed the European Standard EN 12274-3 and was conducted at room temperature using a frustum of a cone with a height of 75 mm, a diameter of 40 mm at the top and a diameter of 90 mm at the bottom; a spatula; and a metal plate graduated in eight concentric circles, each increasing the radius by 10 mm. The test consisted of preparing 400 g of the mixture and pouring it into the cone in the center of the metal plate.

Immediately, the cone was removed, and the sample could flow freely. The outflow of the mixture was then measured from the inner circle at four points 90° apart, and the average of the four flow values was taken as the consistency test result. In this test, a microsurfacing mixture with 25 mm to 35 mm flow values has adequate workability [30].

#### 2.6.3. Cohesion Test

The cohesion test (EN 12274-4) is used to determine the development of the curing process of a microsurfacing mixture over time when subjected to torque. Thus, this test determines the time required for the microsurfacing to be open to traffic. Also, this test was used in this work to determine the ideal relationship between the percentage of emulsion, added water, and cement in microsurfacing mix designs.

The test was carried out with specific equipment and measured the torque applied to microsurfacing samples at different curing times, i.e., 5, 10, 15, 30, 60, 90, and 120 min. The first step of carrying out this test was to prepare the mixture samples and place them in metallic molds soon removed to inhibit the mixture from adhering to them while waiting for the pre-set curing time of each sample. After the predetermined curing period, the sample was placed centrally under the equipment piston, which was gently lowered, applying an air pressure of 200 kPa. Soon after, the torque meter was placed on the upper end of the cylinder rod and twisted in a smooth and firm horizontal movement through an arc of 90° to 120°, and the applied torque was recorded using dedicated software.

#### 2.6.4. Shaking Abrasion Test

This test was carried out following the EN 12274-7 standard; it determines the loss of adhesiveness of microsurfacing mixtures due to their sensitivity to water, as it measures the loss of material from compacted standard specimens when they are placed in cylinders filled with water that is turned from end to end in specific equipment. Four cylindrical samples, each with a height of 25 mm and a diameter of 30 mm, were tested.

The specimens were obtained by adding approximately 45 g of the mixture (prepared following the EN 12274-7 standard) in a mold, compacted by applying a 10 KN load at a speed of 20 mm/min. Subsequently, the water absorption of each specimen is calculated according to the procedure presented in the EN 12274-7 standard.

After this process, each sample was placed inside a different stirring cylinder previously filled with 750 mL of fresh drinking water at 25 °C. The mechanical stirrer operated at 20 rpm at room temperature until reaching 3600 rotations, i.e., after 3 h of testing. Finally, the samples were removed from the cylinders, washed to remove all loose material, allowed to dry, and reweighed. The mass loss was then calculated to determine each specimen's adhesiveness loss.

## 3. Results

### *3.1. Aggregate Characteristics*

The grading curve of the mixture of aggregates studied was adjusted to the grading envelope defined by Infraestruturas de Portugal [29]. Only the most commonly used microsurfacing asphalt mixture among the three alternatives presented in Infraestruturas de Portugal [29] was studied. Thus, the grading envelope of the second layer in a double microsurfacing was selected. The best fit for the recommended grading envelope is presented in Figure 4. The percentage of aggregates passing through the 2 mm sieve dividing the coarse and fine fractions previously mentioned was 52%.

The value of the MDE coefficient of the basaltic aggregate (between the dimensions of 4 mm and 6.3 mm) obtained in this study was 9.7%. This value is significantly lower than the threshold value of 25% established in the specifications from Infraestruturas de Portugal [29]. The aggregates showed a 2.78 $Mg/m^3$ density and a water absorption value of 1.85%, typical of a basalt aggregate.

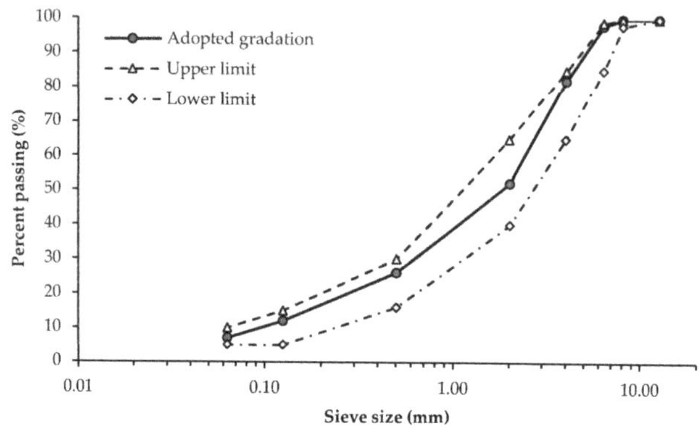


**Figure 4.** Aggregate gradation adopted for each microsurfacing asphalt mixture.

The results of the aggregate's affinity to both emulsions studied (the percentage of the aggregate surface covered with binder) are presented in Table 1. Figure 5 shows the visual appearance of some aggregate particles covered with Emulsion 1 after 6 h, 24 h, and 48 h of testing to demonstrate the reduction in the area of the aggregate surface covered with binder over time.

**Table 1.** Result of the affinity test between the basalt aggregate and the studied emulsions.

| Time | Emulsion 1 | Emulsion 2 |
|---|---|---|
| After 6 h | 95% | 90% |
| After 24 h | 70% | 50% |
| After 48 h | 50% | 30% |

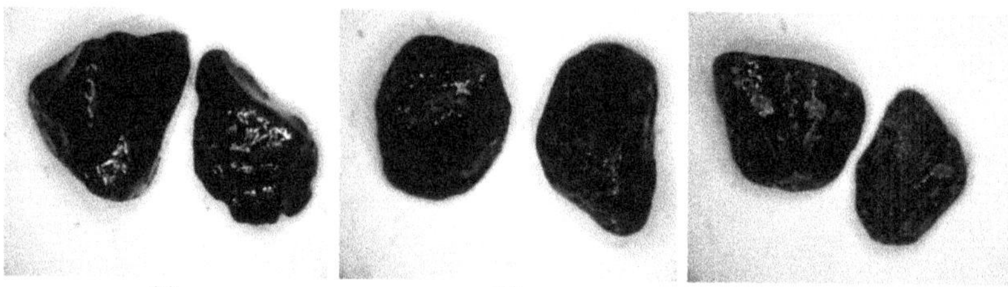

(a) (b) (c)

**Figure 5.** Example of basalt aggregates surface covered with bitumen when using Emulsion 1 after the following testing periods: (**a**) 6 h, (**b**) 24 h, and (**c**) 48 h.

This test showed a better affinity of the basalt aggregates with Emulsion 1 compared with Emulsion 2, which could result in a better performance of the corresponding microsurfacing mixture. The visual assessment of the aggregate surface covered with binder over time was more difficult using basalt due to its color, but the brightness of the binder in water facilitates this task.

### 3.2. *Asphalt Emulsions' Characteristics*

The breaking values obtained were 121.9 for Emulsion 1 and 158.3 for Emulsion 2. According to the EN 13808 standard, the emulsions can be classified as class 4 for the breaking value parameter (values between 110 and 195). The asphalt content and the emulsifier's amount and type influence the emulsion's breaking value. Therefore, emulsions with the same classification but different suppliers may have different breaking values.

The adhesion value obtained for Emulsion 1 was 90%, and for Emulsion 2, it was 75%. Concerning the EN 13808 standard for this parameter, Emulsion 2 can be classified as class 2 (≥75%), while Emulsion 1 belongs to class 3 (≥90%). These results align with those previously obtained when studying the aggregates' affinity to both emulsions (EN 12697-11).

After the binder recovery by evaporation, characterization tests were carried out. The results are shown in Figure 6. Regarding the EN 13808 standard for cationic asphalt emulsions, the penetration values of the bitumen recovered from the two studied emulsions can be classified as class 3 (≤100 × 0.1 mm). Regarding the softening temperature result, the Emulsion 1 bitumen belongs to class 3 ($T_{R\&B}$ > 55 °C), and the Emulsion 2 bitumen belongs to class 2 ($T_{R\&B}$ > 60 °C).

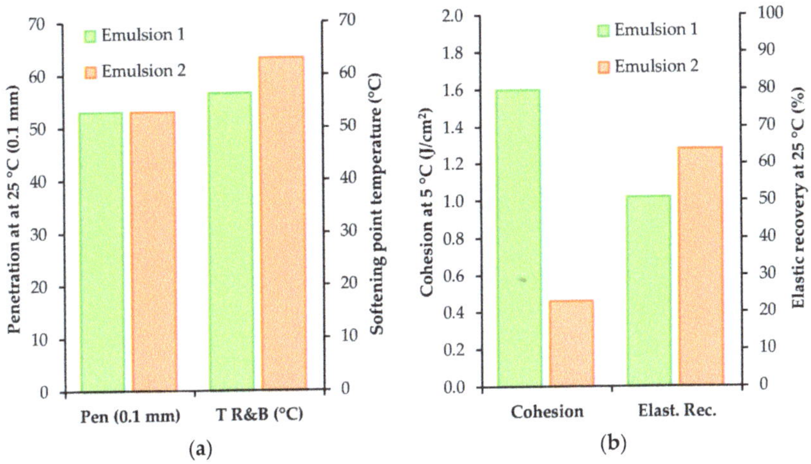


**Figure 6.** Characterization test results of binders recovered from asphalt emulsions: (**a**) penetration and softening point and (**b**) cohesion and elastic recovery.

According to the cohesion results, the bitumen recovered from Emulsion 1 had a value greater than 1 J/cm$^2$ at a temperature of 5 °C, corresponding to class 4 of the EN 13808 standard. The bitumen recovered from Emulsion 2 did not reach the minimum cohesion value established for this test temperature, corresponding to a lower performance class. The results of the elastic recovery test showed that the bitumen binders recovered from the two studied emulsions belonged to class 5, with values greater than 50%.

Figure 7 shows the dynamic viscosity results obtained for both emulsions using the dynamic shear rheometer (DSR) for temperatures up to 88 °C (parallel plates configuration) and the rotational viscometer for temperatures above 90 °C (cup and bob configuration).

According to the results, it is possible to conclude that at lower temperatures (below 40 °C), the Emulsion 1 binder presents higher viscosity values. A different behavior was observed above the temperature of 40 °C: there was a faster decrease in the viscosity with increasing temperature in the Emulsion 1 binder, and the Emulsion 2 binder presented viscosity values that were less susceptible to temperature variations.

The master curves of the stiffness modulus and phase angle values obtained in the DSR after carrying out frequency sweep tests at different temperatures under controlled strain levels are presented in Figure 8 for a reference temperature of 40 °C. The equipment software carried out the time–temperature superposition using the WLF equation because the tests were performed in the linear viscoelastic regime.

The stiffness modulus results showed that the binder recovered from Emulsion 1 was stiffer than that recovered from Emulsion 2 at higher frequencies (i.e., above 1 Hz), demonstrating that this binder had a higher strength at temperatures lower than 40 °C. At higher temperatures, the stiffness of the binder recovered from Emulsion 2 became higher, confirming the viscosity and softening point test results. These results suggest that the base bitumen used to produce Emulsion 1 was more viscous than that of Emulsion 2, but the latter should have a higher polymer content that increases the strength at higher temperatures.

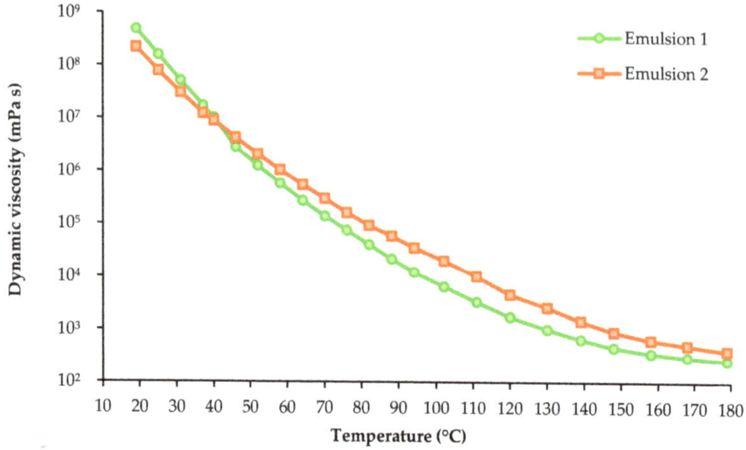


**Figure 7.** Dynamic viscosity of the binders recovered from the studied emulsions.

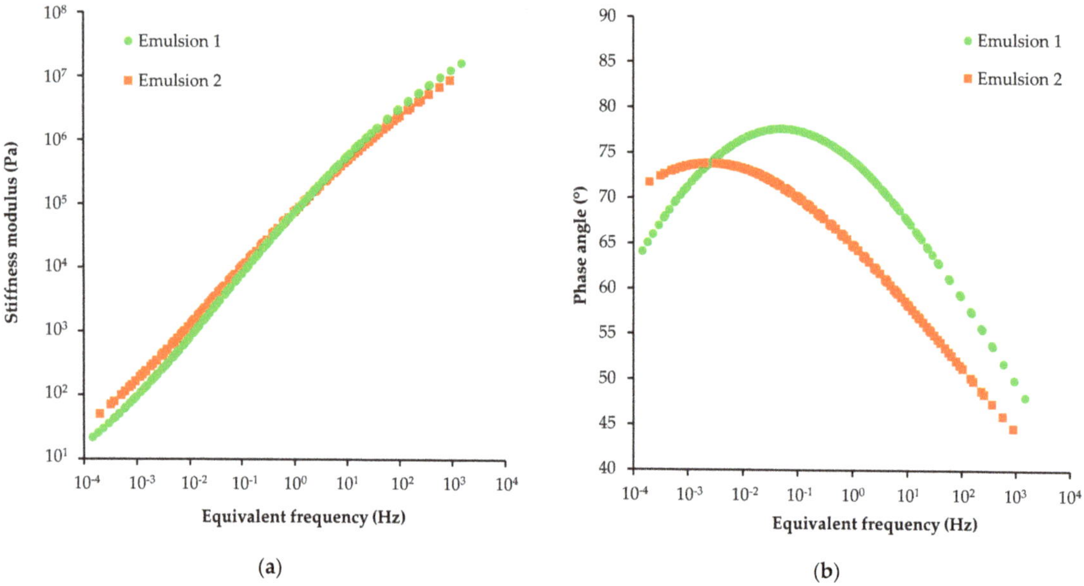


**Figure 8.** Rheological characterization of binders recovered from asphalt emulsions (master curves obtained for a reference temperature of 40 °C): (**a**) stiffness modulus and (**b**) phase angle.

The phase angle results also suggested a lower modification level of the binder recovered from Emulsion 1 compared with Emulsion 2 since it achieved a higher peak value of around 80° at a higher equivalent frequency, which, according to Peralta et al. [31], corresponds to a lower relaxation time (i.e., the inverse of frequency) and reduced resistance to permanent deformation at high temperatures.

Additionally, the rheological characterization of the binders recovered from both emulsions provided their high-temperature PG grades, which are typically used to compare the performance of asphalt binders. Thus, the asphalt binder recovered from Emulsion 1 was a PG 70 binder, while that recovered from Emulsion 2 was a PG 76 binder.

### 3.3. *Microsurfacing Test Results*

#### 3.3.1. Preliminary Design of Microsurfacing Mixtures

Several preliminary tests were performed for each microsurfacing mixture with two test repetitions per composition to determine the initial amounts of added water and cement. According to a previous study by RILEM TC 280-CBE, asphalt emulsion was fixed at 12% for all mixtures to estimate the starting percentages of added water and cement for performance testing. The results obtained at this stage for the mixtures produced with both asphalt emulsions are presented in the following paragraphs.

The first microsurfacing mixture produced with asphalt Emulsion 1 used 10% water and 2.0% cement. This mixture showed poor workability, and molding the pizza and the ball was impossible. Then, the amount of cement was increased to 2.5%, allowing the pizza and ball samples to be shaped. However, the mixture still presented a dry aspect with a low consistency. Thus, a new mixture was prepared with 2.5% cement, increasing the amount of water to 11%. This composition presented improved consistency, fulfilling the expected results in both preliminary tests. The pizza cured after one hour, and the ball did not break after falling to the ground. Figure 9 shows the test samples of this final microsurfacing mixture with Emulsion 1.

(a) (b) (c)

**Figure 9.** Samples used for preliminary design of microsurfacing asphalt mixtures: (**a**) pizza sample just after molding, (**b**) pizza sample after 60 min, (**c**) sample for ball test.

The first microsurfacing mixture produced with Emulsion 2 was based on the last composition studied for Emulsion 1, comprising 11% water and 2.5% cement. However, the water content was too high, delaying the pizza curing process. Thus, a new mixture was designed for Emulsion 2 with a lower water content (10%) and the same cement content (2.5%). This final mixture fulfilled the expected results in both preliminary tests: the pizza cured after one hour, and the ball did not break after falling from a 1.5 m height.

#### 3.3.2. Consistency Test Results

The consistency test results for the microsurfacing mixtures produced with both asphalt emulsions studied in this work are presented below. The initial compositions studied at this stage were based on the preliminary test results for Emulsions 1 and 2.

Firstly, 400 g of microsurfacing mixture was produced with emulsion 1, using 52% fine aggregates and 48% coarse aggregates, 2.5% cement, 12% emulsion, and 11% water. The average flow value of the mixture measured at four points in perpendicular directions was 25 mm, confirming that 11% water assures adequate workability for mixtures produced with Emulsion 1. Figure 10 shows the test preparation and its final result.

The initial mixture produced with Emulsion 2 used 52% fine aggregates and 48% coarse aggregates, 2.5% cement, 12% emulsion, and 10% water. However, the average flow of the mixture was higher than 35 mm, showing that 10% water was excessive for mixtures produced with Emulsion 2. Therefore, a new microsurfacing mixture was designed with less water (9%), which presented an average flow of 30 mm measured at four points in perpendicular directions, thus improving the workability of the mixture produced with Emulsion 2.

(a) (b)

**Figure 10.** Example of the different phases of the consistency test for the microsurfacing asphalt mixture prepared with one of the emulsions: (**a**) test preparation and (**b**) final result.

The results of this test demonstrated that microsurfacing mixtures can use varying amounts of added water to guarantee adequate workability depending on the type of emulsion used. The results of this work suggest that emulsions with lower residual binder contents, such as Emulsion 2, may result in microsurfacing mixtures demanding lower amounts of added water.

3.3.3. Cohesion Test Results

This test was carried out following the EN 12274-4 standard. Seven samples were produced for each composition and tested after 5, 10, 15, 30, 60, 90, and 120 min of curing. Figure 11 shows the molds used to produce the test samples and the samples in the curing process before the test. According to the ISSA [32], the most critical parameters assessed in the cohesion test are the torque values obtained after 30 and 60 min of microsurfacing curing. For this test, the ISSA defines the set time as the point at which the microsurfacing system reaches a minimum of 12 kgf cm torque, and the straight rolling traffic time corresponds to a minimum torque of 20 kgf cm. Therefore, a quick-set system should reach 12 kgf cm of torque within 30 min, while a quick-traffic system should reach 20 kgf cm of torque within 60 min.

(a) (b)

**Figure 11.** Preparation of microsurfacing asphalt samples for cohesion test: (**a**) molds used in the test and (**b**) samples in the curing process.

The initial composition studied in this test for the microsurfacing mixture produced with Emulsion 1 was based on the mix design resulting from the consistency test, i.e., 52% fine aggregates and 48% coarse aggregates, 12% emulsion, 2.5% cement, and 11% water. However, the mixture did not reach the cohesion values defined by ISSA for a quick-set and quick-traffic system after 30 and 60 min of curing. Thus, new compositions were designed to improve the microsurfacing cohesion values.

The second microsurfacing mixture studied increased the emulsion content to 13% while maintaining all the other composition parameters. This mixture presented a cohesion value of 17.12 kgf cm within 30 min of curing and 21.11 kgf cm within 60 min, fulfilling the target results defined by the ISSA for a quick-set and quick-traffic system and confirming that this composition can be assumed to have adequate performance. Nevertheless, new mix compositions were studied to assess the possibility of producing more sustainable solutions with equivalent cohesion values.

The third mixture kept the added water at 11%, reducing the asphalt emulsion to 12% and the cement to 2.0%. This composition slightly increased the cohesion value after 30 min of curing, but decreased it after 60 min. Finally, two additional compositions were studied to evaluate the effect of increasing the emulsion content to 13% and 14%, keeping the added water at 11% and the cement at 2%. Table 2 summarizes all the results obtained for Emulsion 1.

**Table 2.** Mixture compositions produced with Emulsion 1 for the cohesion test.

| Mixture | Design of the Microsurfacing Mixture | | | Cohesion (kgf cm) | |
|---|---|---|---|---|---|
| | Water (%) | Cement (%) | Emulsion (%) | After 30 min | After 60 min |
| 1 | 11% | 2.5% | 12% | 16.40 | 19.06 |
| 2 | 11% | 2.5% | 13% | 17.12 | 21.11 |
| 3 | 11% | 2.0% | 12% | 17.70 | 19.89 |
| 4 | 11% | 2.0% | 13% | 17.36 | 20.74 |
| 5 | 11% | 2.0% | 14% | 19.12 | 18.22 |

Since the amount of water was fixed at 11%, the results showed the mutual influence of the emulsion and cement contents on the cohesion test performance of these microsurfacing mixtures. The amount of Emulsion 1 showing the best compromise between the 30 and 60 min curing times was 13%, although the quick-set performance may be improved for higher emulsion contents. Increasing the amount of cement improved the cohesion values only if the emulsion content was also increased.

Figure 12 shows the variation of cohesion values according to the amount of material added to the mixture produced with Emulsion 1. The second composition, with 11% water, 2.5% cement, and 13% emulsion, presented the best overall cohesion test performance and was selected for the shaking abrasion test.

Figure 13 presents the final appearance of the samples after the cohesion test for the selected microsurfacing mixture. According to the visual assessment criteria defined by EN 12274-4, the samples with 5 to 30 min of curing time were considered disintegrated, those with 60 and 90 min of curing time were classified as cracked, and the sample with 120 min of curing time was considered solid. These observations demonstrate the rapid evolution of the cohesion with the curing time for this mixture composition.

Regarding the microsurfacing mixtures produced with Emulsion 2, the first composition studied was also based on the final mix design obtained in the consistency test, i.e., 52% fine aggregates and 48% coarse aggregates, 12% emulsion, 2.5% cement, and 9% water. However, the mixture with this configuration did not achieve the target cohesion values mentioned by the ISSA after 30 and 60 min of curing. Thus, new compositions were also designed to improve the cohesion results.

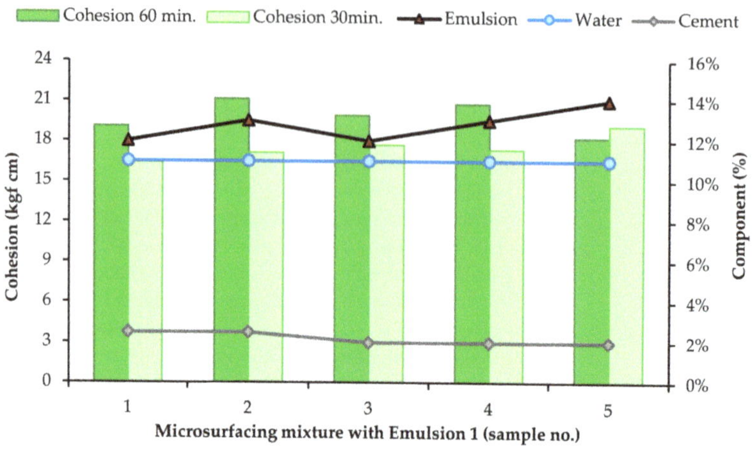


**Figure 12.** Variation of cohesion values according to the material added to the mixture produced with Emulsion 1.

**Figure 13.** Samples' final appearance after the cohesion test for the selected mixture produced with Emulsion 1.

The second mixture was only increased in the emulsion content by 1%, but that change had little effect on the microsurfacing cohesion values, which were still typical of a slow-set and slow-traffic system. Therefore, a third mixture was produced with 13% emulsion and 3% cement, improving the cohesion results and turning it into a quick-set system, confirming that the active filler (cement) is vital in the mixture cohesion. However, this solution could not be considered a quick-traffic system due to insufficient cohesion after 60 min of curing. Thus, a new composition was assessed by increasing the water content by 1%, which slightly lowered the cohesion value after 30 min but increased the cohesion value after 60 min (although not reaching a quick-traffic class).

Intending to look for a better combination of the amounts of emulsion, added water, and cement to improve cohesion, new mix designs were defined that varied the amount of added water and emulsion but kept the amount of cement at 3%. Table 3 summarizes all the results obtained for Emulsion 2.

**Table 3.** Mixture compositions produced with Emulsion 2 for the cohesion test.

| Mixture | Design of the Microsurfacing Mixture | | | Cohesion (kgf cm) | |
|---|---|---|---|---|---|
| | Water (%) | Cement (%) | Emulsion (%) | After 30 min | After 60 min |
| 1 | 9 | 2.5 | 12 | 8.09 | 11.34 |
| 2 | 9 | 2.5 | 13 | 8.74 | 13.99 |
| 3 | 9 | 3.0 | 13 | 13.49 | 16.04 |
| 4 | 10 | 3.0 | 13 | 12.68 | 16.71 |
| 5 | 11 | 3.0 | 13 | 10.62 | 13.63 |
| 6 | 10 | 3.0 | 14 | 10.35 | 14.88 |
| 7 | 10 | 3.0 | 12 | 10.02 | 13.50 |

After fixing the amount of cement at 3%, the results of mixtures 3 to 7 showed the mutual influence of the emulsion and added water contents on the cohesion of these microsurfacing mixtures. After 30 and 60 min, the best cohesion values were obtained for 10% water and 13% emulsion. Increasing the water or emulsion contents reduced the 30-min curing time cohesion values by increasing the amount of water needing evaporation. Reducing the emulsion content also had a negative effect due to a reduction in the amount of residual binder available in the microsurfacing mixture.

Figure 14 shows the variation of cohesion values according to the amount of material added to the mixture produced with Emulsion 2. As can be observed, the composition showing the best cohesion values was that with 10% water, 3% cement, and 13% emulsion, which, according to ISSA, can be classified as a quick-set and slow-traffic system.

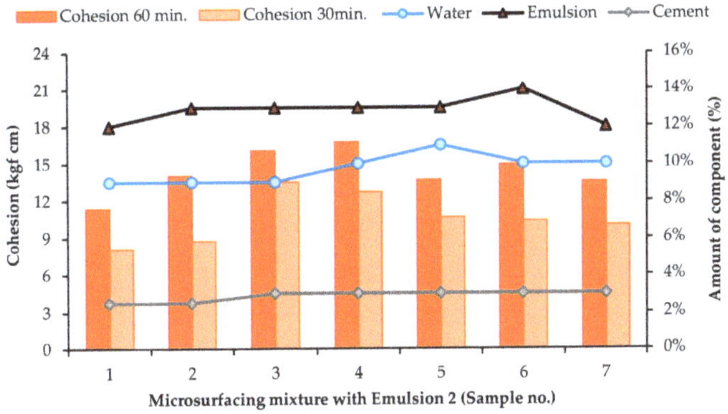


**Figure 14.** Variation of cohesion values according to the material added to the mixture produced with the Emulsion 2.

Figure 15 presents the samples' final appearance after the cohesion test for the selected mixture produced with Emulsion 2. According to the visual assessment defined by EN 12274-4, the 5 and 10 min samples were considered disintegrated, the 15 and 30 min samples were classified as cracked, and the 60 to 120 min samples were considered solid. These results confirmed the quick-set class of this microsurfacing mixture.

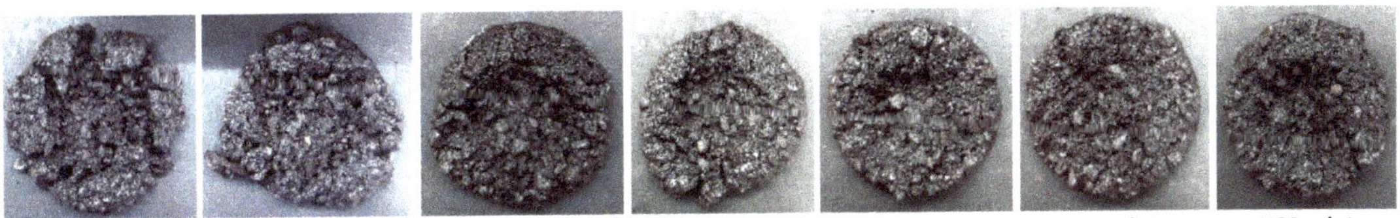


**Figure 15.** Samples' final appearance after the cohesion test for the selected mixture produced with Emulsion 2.

After analyzing each microsurfacing mixture separately, Figure 16 shows the evolution of the cohesion values with the curing time for the two selected compositions.

This graph demonstrates how the mixtures developed strength with time. It can be noted that both mixtures showed similar patterns, even though cohesion increased more rapidly and for higher values in the case of Emulsion 1. According to the ASTM [33] classification, the pattern of cohesion-time curves was a quick-set, quick-traffic system for Emulsion 1 and a quick-set, slow-traffic system for Emulsion 2. These results could be related to the binder cohesion values presented previously (1.00 J/cm$^2$ and 0.46 J/cm$^2$ for bitumen recovered from Emulsion 1 and 2, respectively). Thus, the bitumen cohesion could be used as an indicator of the shear strength of the microsurfacing mixture. Nevertheless, the ultimate cohesion values observed after 120 min were similar for both emulsions and higher than 20 kgf cm, showing that the mix design method proposed in the present work could adequately select the compositions for different types of emulsions.

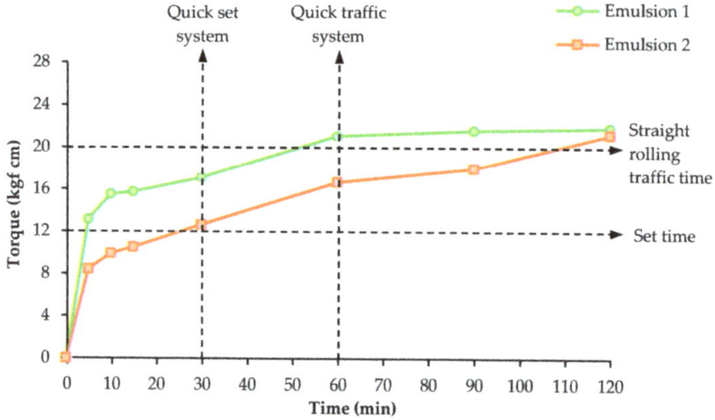


**Figure 16.** Cohesion values for all curing times for better mix designs for both mixtures.

### 3.3.4. Shaking Abrasion Test

The shaking abrasion test was carried out to determine the loss of adhesiveness of mixtures studied due to their water sensitivity. This test was performed for the compositions with higher cohesion values previously obtained for both emulsions.

Figure 17 presents the appearance of a sample before and after the test; it can be seen that the edges of the specimens are polished after the test.

**Figure 17.** Samples before (**left**) and after (**right**) carrying out the shaking abrasion test.

Table 4 presents the results of the tests carried out with the microsurfacing mixtures produced with both emulsions. The abrasion results are expressed as the mass loss percentage and were obtained by averaging the results of the four test specimens. The mixture with the best performance (lowest mass loss) was prepared with Emulsion 2. Table 4 also presents the water absorption of the mixtures and the volume of the specimens (Vv) after vacuum application, showing that the samples produced with Emulsion 1 had slightly higher volumes than those produced with Emulsion 2. Even though there were no significant differences between the two mixtures, the mixture with higher mass loss also presented a higher volume and water absorption.

The abrasion test rated the microsurfacing mixtures produced with Emulsions 1 and 2 differently from the cohesion test. This demonstrates that the properties of the emulsions used in microsurfacing mixtures significantly influence the final performance of the material. The lower abrasion of the mixture with Emulsion 2 may be related to the potential higher polymer modification of this emulsion, as inferred from its residual binder's higher elastic recovery and softening point temperature. The higher amount of cement in that mixture may have also contributed to the lower abrasion and air void content values measured, which indirectly reduced water absorption.

**Table 4.** Abrasion results for the microsurfacing asphalt mixtures produced with both emulsions studied in this work.

| Emulsion | Sample | Vv (cm$^3$) | Water Absorption (%) | Abrasion (%) |
|---|---|---|---|---|
| Emulsion 1 | 1 | 18.2 | 13.59 | 1.95 |
| | 2 | 18.0 | 13.26 | 1.97 |
| | 3 | 18.0 | 12.71 | 2.21 |
| | 4 | 17.8 | 13.48 | 2.49 |
| | Mean | 18.0 | 13.26 | 2.15 |
| Emulsion 2 | 1 | 17.9 | 12.22 | 0.74 |
| | 2 | 17.3 | 12.07 | 0.51 |
| | 3 | 17.9 | 11.11 | 0.49 |
| | 4 | 17.9 | 11.73 | 0.74 |
| | Mean | 17.8 | 11.78 | 0.62 |

## 4. Conclusions

This work proposed a new mix design procedure for microsurfacing mixtures, which was used to assess the influence of the type and amount of asphalt emulsion and the added water and cement contents on the characteristics of the mixture.

The study started with materials characterization, which provided valuable insights to help elucidate the mixtures' behavior in the following tests. The basalt aggregates demonstrated better compatibility with Emulsion 1 than Emulsion 2, suggesting that Emulsion 1 may be more suitable for optimal performance. Emulsion 1 also exhibited higher adhesion and cohesion results than Emulsion 2, meeting the relevant standards. Regarding rheological properties, Emulsion 1 displayed higher viscosity at lower temperatures, while Emulsion 2 showed less susceptibility to temperature changes. Additionally, Emulsion 1 had higher stiffness at lower temperatures, whereas Emulsion 2 showed higher stiffness at higher temperatures, likely due to different polymer contents. The phase angle results indicated that Emulsion 1 had lower modification levels than Emulsion 2, affecting its resistance to permanent deformation at high temperatures.

After the materials characterization, two mixtures were produced with emulsions C65BP4 (Emulsion 1) and C60BP4 (Emulsion 2). A sequence for adding each material was proposed to standardize the procedure of mixture production since these indications are missing in the European standards and significantly influence the mixture's workability. Furthermore, two preliminary tests were suggested to determine the mixture's initial amount of cement and added water: the pizza and ball tests. This work showed that the pizza and ball tests had good capability in evaluating the effects of water and cement on the material's behavior. These tests save time for the subsequent mix design by quickly approximating the values of water and cement needed for adequate performance.

The consistency test can be used to confirm the amount of added water and the workability of the mixture. In this work, the amount of water determined in the pizza and ball tests was confirmed for the microsurfacing mixture with Emulsion 1, while the mixture with Emulsion 2 required a reduction of 1% in the water content to obtain the expected workability.

The cohesion test was the most time-consuming phase of the mix design procedure, as it is necessary to identify the best relationship between the amount of asphalt emulsion, added water, and cement until a higher cohesion value is found. The cohesion results of this test were generally higher for Emulsion 1, resulting in a quick-set quick-traffic system for that emulsion and a quick-set slow-traffic system for Emulsion 2. Moreover, the ultimate cohesion values observed after 120 min were higher than 20 kgf cm and similar for both emulsions, showing that the mix design method proposed in the present work could adequately select the mix compositions for different types of emulsions.

The mixture produced with Emulsion 2 presented better results than those obtained with Emulsion 1 in the shaking abrasion test, which may be related to this emulsion's potential higher polymer modification, as inferred from its residual binder's higher elastic recovery and softening point temperature.

Summing up, microsurfacing systems are strongly influenced by asphalt emulsions (usually modified), aggregate types, and the water and cement contents. Therefore, their performance depends on several physical interactions and chemical reactions that are difficult to isolate. Nevertheless, the mix design process presented in this paper was able to select the best microsurfacing mix composition for two different emulsions.

**Author Contributions:** Conceptualization, C.F.N.M., J.R.M.O., H.M.R.D.S., C.A.O.F.P. and C.S.; methodology, C.F.N.M., J.R.M.O., H.M.R.D.S., C.A.O.F.P. and C.S.; validation, J.R.M.O., H.M.R.D.S. and C.A.O.F.P.; formal analysis, C.F.N.M., J.R.M.O. and H.M.R.D.S.; investigation, C.F.N.M., J.R.M.O. and H.M.R.D.S.; writing—original draft preparation, C.F.N.M., J.R.M.O., H.M.R.D.S. and C.A.O.F.P.; writing—review and editing, C.F.N.M., J.R.M.O., H.M.R.D.S. and C.S.; supervision, J.R.M.O. and H.M.R.D.S. All authors have read and agreed to the published version of the manuscript.

**Funding:** This research was funded by Fundação para a Ciência e a Tecnologia through a Ph.D. grant (number 2021.08004.BD). This work was also partly financed by FCT/MCTES through national funds (PIDDAC) under the R&D Unit Institute for Sustainability and Innovation in Structural Engineering (ISISE), reference UIDB/04029/2020, and under the Associate Laboratory Advanced Production and Intelligent Systems ARISE, reference LA/P/0112/2020.

**Institutional Review Board Statement:** Not applicable.

**Informed Consent Statement:** Not applicable.

**Data Availability Statement:** Data sharing does not apply to this article.

**Acknowledgments:** The authors would like to acknowledge the support from the University of Minho staff involved in this work and RILEM's Technical Committee 280-CBE for providing the materials.

**Conflicts of Interest:** The authors declare no conflict of interest. The funders had no role in the study's design; in the collection, analyses, or interpretation of data; in the writing of the manuscript; or in the decision to publish the results.

## References

1. Shanbara, H.K.; Dulaimi, A.; Al-Mansoori, T.; Al-Busaltan, S.; Herez, M.; Sadique, M.; Abdel-Wahed, T. The future of eco-friendly cold mix asphalt. *Renew. Sustain. Energy Rev.* **2021**, *149*, 111318. [CrossRef]
2. Wulandari, P.S.; Tjandra, D. The effect of crumb rubber in dense graded and open graded cold mixture asphalt. In Proceedings of the IOP Conference Series: Earth and Environmental Science, Surabaya, Indonesia, 20 August 2021.
3. Dash, S.S.; Panda, M. Influence of mix parameters on design of cold bituminous mix. *Constr. Build. Mater.* **2018**, *191*, 376–385. [CrossRef]
4. Xiao, F.P.; Yao, S.L.; Wang, J.G.; Li, X.H.; Amirkhanian, S. A literature review on cold recycling technology of asphalt pavement. *Constr. Build. Mater.* **2018**, *180*, 579–604. [CrossRef]
5. Bhargava, N.; Siddagangaiah, A.K.; Ryntathiang, T.L. State of the art review on design and performance of microsurfacing. *Road Mater. Pavement Des.* **2020**, *21*, 2091–2125. [CrossRef]
6. Zalnezhad, M.; Hesami, E. Effect of steel slag aggregate and bitumen emulsion types on the performance of microsurfacing mixture. *J. Traffic Transp. Eng. (Engl. Ed.)* **2020**, *7*, 215–226. [CrossRef]
7. Bae, A.; Stoffels, S.M. Economic effects of microsurfacing on thermally-cracked pavements. *KSCE J. Civ. Eng.* **2008**, *12*, 177–185. [CrossRef]
8. Madane, A.; Taye, S.; Mamidwar, S.; Chavan, S. A laboratory investigation on type III micro-surfacing with different fillers. *Int. Res. J. Eng. Technol. (IRJET)* **2019**, *6*, 3321–3324.
9. Broughton, B.; Lee, S.-J. *Microsurfacing in Texas*; Texas State University: San Marcos, TX, USA, 2012.
10. Freeman, T.; Pinchett, D.; Haobo, R.; Spiegelman, C. *Analysis and Treatment Recommendations from the Supplemental Maintenance Effectiveness Research Program (SMERP)*; Texas Transportation Institute: College Station, TX, USA, 2002.
11. Callai, S.C.; Sangiorgi, C. A review on acoustic and skid resistance solutions for road pavements. *Infrastructures* **2021**, *6*, 41. [CrossRef]
12. Keymanesh, M.R.; Ziari, H.; Zalnezhad, H.; Zalnezhad, M. Mix design and performance evaluation of microsurfacing containing electric arc furnace (EAF) steel slag filler. *Constr. Build. Mater.* **2021**, *269*, 121336. [CrossRef]
13. Grilli, A.; Graziani, A.; Carter, A.; Sangiorgi, C.; Pivoto Specht, L.; Copetti Callai, S. Slurry surfacing: A review of definitions, descriptions and current practices. *RILEM Tech. Lett.* **2019**, *4*, 103–109. [CrossRef]
14. Broughton, B.; Lee, S.-J.; Kim, Y.-J. 30 Years of Microsurfacing: A Review. *ISRN Civ. Eng.* **2012**, *2012*, 1–7. [CrossRef]
15. Xia, Y.F.; Jia, J.; Chen, Q. Road Performance Comprehensive Evaluation of Polymer Modified Emulsified Asphalt Fiber Microsurfacing. *Adv. Mater. Sci. Eng.* **2022**, *2022*, 8179137. [CrossRef]

16. Destrée, A.; Vansteenkiste, S.; Tanghe, T.; De Visscher, J.; Grilli, A. Reliable Laboratory Tests: A Prerequisite for the Design of High-Quality Slurry Surfacing Mixtures. *Adv. Mater. Sci. Eng.* **2022**, *2022*, 1–20. [CrossRef]
17. Lonbar, M.S.; Nazirizad, M. Laboratory Investigation of Materials Type Effects on the Microsurfacing Mixture. *Civ. Eng. J.* **2016**, *2*, 86–94. [CrossRef]
18. Gransberg, D.D. *Microsurfacing: A Synthesis of Highway Practices. NCHRP Synthesis 411*; Transportation Research Board: Washington, DC, USA, 2010.
19. Robati, M.; Carter, A.; Perraton, D. Evaluation of a modification of current microsurfacing mix design procedures. *Can. J. Civ. Eng.* **2015**, *42*, 319–328. [CrossRef]
20. Robati, M.; Carter, A.; Perraton, D. Repeatability and reproducibility of micro-surfacing mixture design tests and effect of aggregates surface areas on test results. *Aust. J. Civ. Eng.* **2013**, *11*, 41–55. [CrossRef]
21. Bhargava, N.; Siddagangaiah, A.K.; Ryntathiang, T.L. Systematic approach to address challenges in microsurfacing mix design. *Constr. Build. Mater.* **2021**, *270*, 121759. [CrossRef]
22. Poursoltani, M.; Hesami, S. Performance evaluation of microsurfacing mixture containing reclaimed asphalt pavement. *Int. J. Pavement Eng.* **2020**, *21*, 1491–1504. [CrossRef]
23. Robati, M. Evaluation and Improvement of Micro-Surfacing Mix Design Method and Modelling of Asphalt Emulsion Mastic in Terms of Filler-Emulsion Interaction. Ph.D. Thesis, École de Technologie Supérieure, Université du Québec, Montreal, QC, Canada, 2014.
24. Robati, M.; Carter, A.; Perraton, D. Evaluation of test methods and selection of aggregate grading for type III application of micro-surfacing. *Int. J. Pavemnet Eng. Asph. Technol.* **2013**, *14*, 11–66. [CrossRef]
25. Kumar, R.; Ryntathiang, T.L. New Laboratory Mix Methodology of Microsurfacing and Mix Design. *Transp. Res. Procedia* **2016**, *17*, 488–497. [CrossRef]
26. *European Standard EN 12274-3*; CEN. Slurry Surfacing—Test Methods—Part 3: Consistency. European Committee for Standardization: Brussels, Belgium, 2018.
27. *European Standard EN 12274-4*; CEN. Slurry Surfacing—Test Methods—Part 4: Determination of Cohesion of the Mix. European Committee for Standardization: Brussels, Belgium, 2018.
28. *European Standard EN 12274-7*; CEN. Slurry Surfacing—Test Methods—Part 7: Shaking Abrasion Test. European Committee for Standardization: Brussels, Belgium, 2005.
29. Infraestruturas de Portugal. *Caderno de Encargos Tipo Obra. 14.03—Pavimentação. Características dos Materiais*; Infraestruturas de Portugal: Almada, Portugal, 2014.
30. *Austroads Test Method AGPT-T270-18*; Determination of Optimum Amount of Added Water for Bituminous Slurry (Consistency Test). Austroads: Sydney, Australia, 2018.
31. Peralta, J.; Hilliou, L.; Silva, H.M.R.D.; Machado, A.V.; Pais, J.C.; Oliveira, J.R.M. Rheological Quantification of Bitumen Aging: Definition of a New Sensitive Parameter. *Appl. Rheol.* **2010**, *20*, 63293. [CrossRef]
32. *Technical Bulletin No. 139*; Test Method to Determine Set and Cure Development of Slurry Surfacing Systems by Cohesion Tester. International Slurry Surfacing Association: Glen Ellyn, IL, USA, 2017.
33. *ASTM D6372-15*; Standard Practice for Design, Testing, and Construction of Micro-Surfacing. American Society for Testing and Materials: West Conshohocken, PA, USA, 2016.

applied sciences

MDPI

Article

# Analysis of Failure Mode of Reinforced Embankments Overlying Voids Based on Discrete Method

Qi Zhang and Yongliang Lin *

School of Mechanics and Engineering Science, Shanghai University, Shanghai 200444, China; zq86184265@shu.edu.cn
* Correspondence: lin_yliang@163.com

**Abstract:** The mode of the reinforced embankment overlying voids was generated based on discrete element software. By changing the vertical distance H and the horizontal distance L, the influence of the void position on the bearing capacity characteristics, displacement field, stress field and its reinforcement deformation law of the reinforced embankment was analyzed when a local overload was applied. Numerical simulation results show that the vertical displacement of the geogrid is symmetrical around the center of the loading plate, and the transverse displacement of the geogrid shows a centrosymmetric trend around the center of the loading plate at different void locations. In addition, the failure mode of the embankment at different void positions is proposed. Four different failure modes exist for the reinforced embankments overlying voids under local overloading: perforation failure, collapse perforation failure, void side failure and no impact failure. When L = 0, as H increases, the embankment failure mode changes from perforation failure to collapse perforation failure; and when L > 0, as L increases, the failure mode changes from void side failure to no impact failure.

**Keywords:** failure modes; reinforced embankments; voids; discrete elements

**Citation:** Zhang, Q.; Lin, Y. Analysis of Failure Mode of Reinforced Embankments Overlying Voids Based on Discrete Method. *Appl. Sci.* **2023**, *13*, 9270. https://doi.org/10.3390/app13169270

Academic Editors: Raffaele Zinno and José Manuel Moreno-Maroto

Received: 10 July 2023
Revised: 12 August 2023
Accepted: 14 August 2023
Published: 15 August 2023



## 1. Introduction

With the gradual laying of the traffic road network in recent years, the environment of road construction has become more and more complex. Some voids are difficult to detect during the survey process and are gradually enlarged by the self-weight of the embankment or the traffic load during road operation, thus causing road collapse accidents [1–4]. Currently, the horizontal reinforcement method is widely used in engineering to prevent the sudden settlement and instability of embankments caused by overlying voids [5,6]. Based on the demand of engineering applications, many experts have performed a lot of research in the field of horizontal reinforcement method to prevent voids. The reasonableness of geosynthetic materials to prevent voids was verified via tests [7,8]. Kinney [9] analyzed the deformation characteristics of reinforced bodies. The results demonstrated that a geosynthetic material deflection curve did not completely fit a circular nor a parabolic shape. Huang [10] revealed that the friction angle and particle size within the soil had obvious influence on the deformation and tension of the reinforcement material with the help of a trapdoor test model. Lu [11] analyzed the stress–strain characteristics of the reinforced body and found that the peak tensile strain in the bottom layer of the geogrid was at the edge of the void. Huckert [12] analyzed the effect of void span on soil displacement and reinforced body deformation during traffic loading. Chevalier [13] analyzed the load transfer mechanism and the slip surface shape of the upper part of the void. Benmebarek [14] used PLAXIS to study the effect of reinforcement length and stiffness on the settlement of reinforced embankments overlying voids. By means of finite element combined with discrete element methods, the mechanism of force transfer between the soil and reinforcement was investigated and revealed from a microscopic viewpoint [15,16].

Li [17] compared and analyzed the deformation of the reinforced body under three types of load distribution, and concluded that it was economical and reasonable to use the inverted triangular distribution. Feng [18] considered the vertical deformation of the soil in the anchorage area and proposed a method for calculating the deformation of the reinforced body. Chen [19] considered the effect of soil cohesion in the design of the calculation method. A more systematic study of reinforced embankments where the void has collapsed below the reinforced body has been presented in the above literature.

In practical engineering, a portion of soil may exist between the void and the reinforcement material in the embankment. For such conditions, Das [20] investigated the relationship between void depth and embankment bearing capacity. Wu [21] studied the effect of eccentric load and load eccentricity on the bearing capacity characteristics of the void embankment. Zhou [22] took a square void embankment as the object of study to investigate the effect of the property of the fill and the relative position of the void on the failure mechanism of the embankment. Lai [23] proposed a failure model for a reinforced embankment overlying a void with a full load on the embankment surface. For rigid pavements, this treatment is feasible.

In the case of flexible pavements, the vehicle traffic load is equivalent to a local overload acting on the embankment under road operational conditions. The effects of void depth, reinforced body stiffness and reinforced body burial depth on the bearing capacity of embankments under strip loading have been studied [24,25], but the failure mode of reinforced embankments overlying voids under local overload has not yet been investigated. The discrete element simulation technique based on the mechanics of particles can consider the discrete characteristics of soil particles, which can effectively reveal the failure mode and microscopic mechanism of soil structure [26,27].

In this paper, we take the reinforced embankment overlying voids under local overload as the research object, and establish a discrete element model based on a model test. We analyze the influence of void location on the bearing capacity characteristics of reinforced embankment, soil displacement field, contact force chain and the deformation law of the reinforced body, and reveal the failure mode of reinforced embankment overlying voids from a microscopic perspective.

## 2. Model Building

### 2.1. Test Profiles

Figure 1 shows a schematic diagram of the test model with dimensions of 600 mm × 300 mm. The width of the top surface of the embankment is 240 mm, the slope ratio of the embankment slope is 1:1.5 and the length of the geogrid is 255 mm (solid red line). The width of the embankment simulated in the test was 9.6 m and the height of the embankment was 0.8 m. Considering that the embankment is an axisymmetric structure, 1/2 of the embankment was taken for the test. The geometric ratio of 1:20 is used to reduce the size of the embankment and the glass fiber geogrid. Due to conditions, this test was not scaled down strictly based on similarity ratios, but the reinforced embankment overlying voids test is still of some research value.

In order to better represent the relative position of the void in the embankment, the vertical and horizontal positions of the void are expressed by H and L in the diagram, respectively, where the vertical distance H represents the distance from the top of the void to the bottom of the loading plate, and the horizontal distance L represents the distance from the core of the void to the center of the loading plate. The width of the loading plate B is 60 mm, the distance from the center of the loading plate to the wall at the edge of the model box is 120 mm, and the depth of the reinforcement body h is 30 mm.

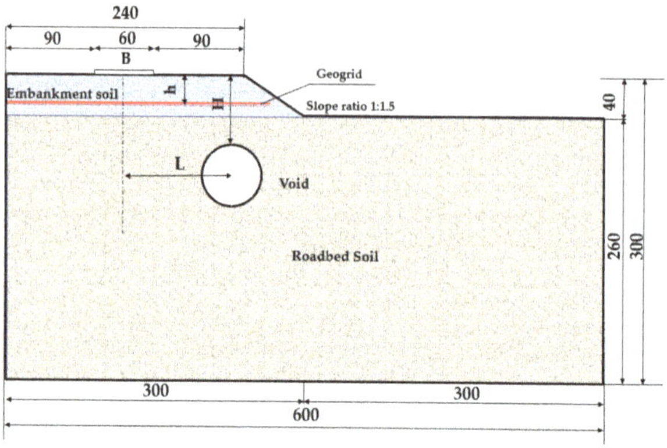


**Figure 1.** Test model (in mm).

The testing devices and materials mainly include the model box, loading and measuring devices, the geogrid and water pumping equipment (Figure 2). The size of the model box is 600 mm × 400 mm × 300 mm (length × width × height, respectively), and the dynamic and static triaxial loading device developed by the company GCTS, U.S.A., is used to load the soil body (the starting load is 0 kPa, and the loading speed is 3 N/s), which can achieve the real-time collection of load and top displacement data. The size of the loading plate is 288 mm × 60 mm × 30 mm (length × width × height, respectively), and the material of the loading plate is high-strength and low-density epoxy resin. The dynamic and static triaxial loading device is loaded in the center area of the loading plate, and geogrids are made of glass fiber material. The breaking strength of geogrids in the transverse and longitudinal directions is 60 kN/m, and the elongation at the break of the geogrid in the longitudinal and transverse directions is 6%. In addition, a soft plastic hollow cylinder with a diameter of 60 mm is used to simulate the void. A plastic membrane is wrapped around the inner and outer sides of the soft plastic cylinder, so that on the one hand, water is pumped inside the inner membrane by means of water pumping equipment to form a void, and on the other hand, the outer film prevents water from leaking into the soil.

(**a**)

(**b**)

**Figure 2.** *Cont.*

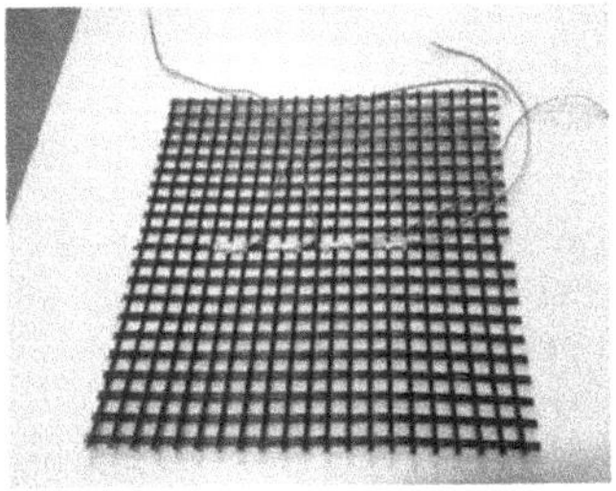

(c)

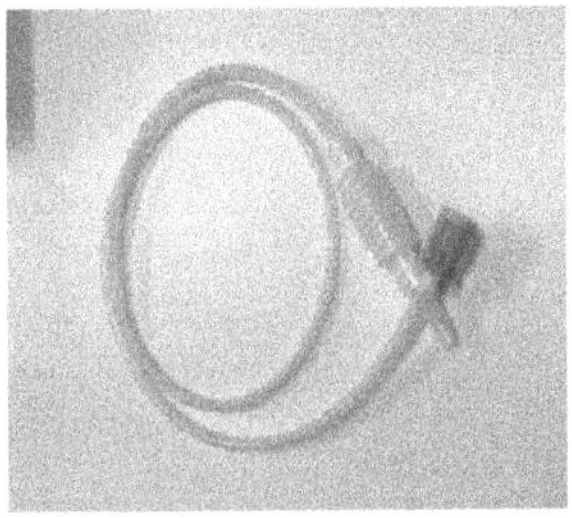

(d)

**Figure 2.** Schematic diagram of testing devices and materials: (**a**) model box; (**b**) loading and measuring devices; (**c**) geogrid; (**d**) water pumping equipment.

This laboratory test uses Fujian standard sand as a filler. After the sieve ratio test, it is determined that the particle size of the soil is mainly concentrated on the range of 0.5 to 2 mm. The gradation curve is shown in Figure 3 below, and the physical properties of the sand soil are shown in Table 1.

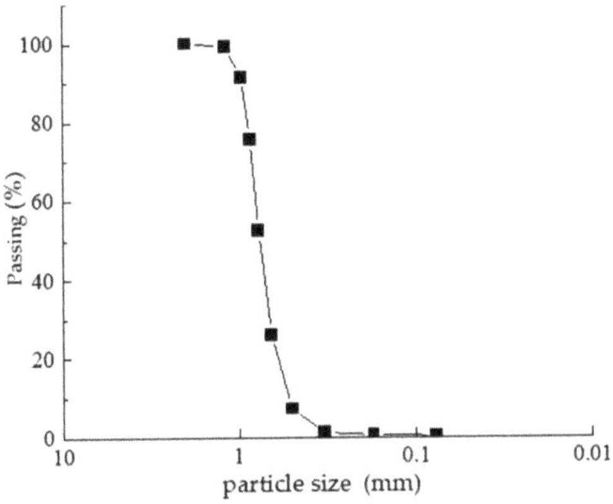


**Figure 3.** Grain size distribution curves of sandy soils.

**Table 1.** Physical properties of sandy soils.

| Property | Value |
|---|---|
| Gravity, γ (kN/m$^3$) | 17.03 |
| Specific gravity of particles, Gs | 2.65 |
| Porosity ratio, e | 0.585 |
| Water content, w (%) | 0.15 |
| Uniformity factor, Cu | 1.39 |
| Coefficient of curvature, Cc | 0.94 |

## 2.2. *Construction of Discrete Element Model*

The PFC particle flow method decomposes the bulk system into discrete units, which are used as the basic units to visually express the interaction states and motion properties between the units at each moment in time. The software replaces the selection of the constitutive model of the material itself by setting the contact between particles. Since the discrete elements do not need to satisfy the deformation coordination relationship, it has a greater advantage in simulating the large deformation of discontinuous media such as soil. At present, the PFC software has become a recognized numerical simulation software and has penetrated into almost all directions in the field of geotechnical engineering.

Based on the above laboratory experiment dimensions, PFC2D software was used to establish a computational model of the reinforced embankment overlying a void, by generating spherical particles to simulate the sandy soil and the geogrid, and by creating wall units to simulate the model boxes and loading plates. Figure 4 shows a schematic of the numerical simulation model. In the model, blue particles represent soil particles, red particles represent the geogrid and blue lines represent walls. The surrounding boundary walls is fixed. The wall is a rigid body and does not produce deformation. The bottom surface of the loading plate is set as the horizontal observation surface M. The position of the observation surface M decreases as the loading plate decreases. The load P in the load–settlement curve is obtained by dividing the loading plate contact force F monitored on profile M by the loading plate width B. The settlement ratio S/B is obtained by dividing the wall displacement S directly monitored by the loading plate width B in the Fish language that comes with the PFC.

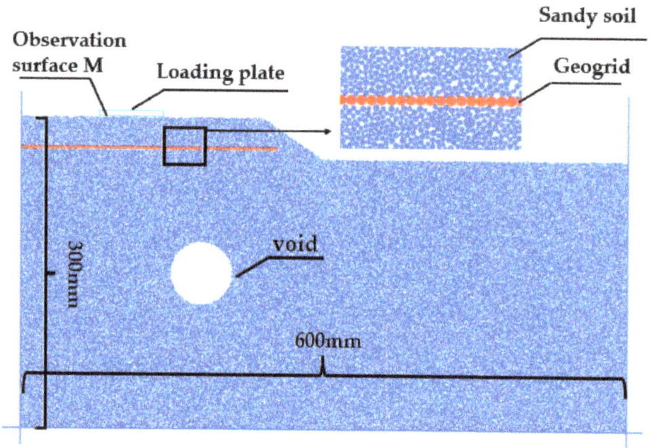


**Figure 4.** Numerical simulation model.

Based on the computational resources of the hardware, the generation of particles in discrete element software according to laboratory-tested soil particle grading curves is difficult to realize; therefore, it is necessary to extend the radius of the soil particles. When the radius multiplier is too small, the number of soil particles will be extensive while improving the accuracy of the simulation, and thus, the calculation cannot be converged; when the radius multiplier is too large, the simulation results of the discrete element will have a large deviation from the actual situation. Based on the requirements of calculation efficiency and simulation accuracy, this paper has made several attempts to enlarge the particle size of the soil, so as to obtain reasonable enlargement results. After attempts, we finally obtained a radius multiplier of 2, and the particle size of sandy soils in the simulation ranged from 1 to 4 mm. The porosity of the sandy soil was 0.17 upon converting from 3D to 2D [28]. The contact model of sandy soil was chosen as a linear contact model, the contact between particles of the geogrid was the parallel bonding model and the contact between sandy soil and geogrid was the linear contact model. The particle size of the geogrid was 1.5 mm. Based on the idea of inverse simulation, this paper has changed the microscopic parameters of the model to carry out a large number of simulations on sandy embankment, reinforced body tensile test and reinforced embankment load test, respectively, and compared the simulation results with the laboratory test results through the trial-and-error method, so as to finally obtain more reasonable microscopic parameters of the discrete element model. The microscopic parameters of each part of the material for the numerical simulation are shown in Table 2.

**Table 2.** Microscopic parameters of each material for the PFC model.

| Parameters | Soils | Geogrid | Model Box |
|---|---|---|---|
| Normal contact stiffness of particles (N/m) | $5.0 \times 10^8$ | $5.0 \times 10^9$ | $1.0 \times 10^{11}$ |
| Shear contact stiffness of particles (N/m) | $5.0 \times 10^8$ | $5.0 \times 10^9$ | $1.0 \times 10^{11}$ |
| Normal stiffness of cementation (N/m$^3$) | — | $6.5 \times 10^8$ | — |
| Shear stiffness of cementation (N/m$^3$) | — | $6.5 \times 10^8$ | — |
| Normal strength of cementation (N/m$^2$) | — | $1.0 \times 10^{11}$ | — |
| Shear strength of cementation (N/m$^2$) | — | $1.0 \times 10^{11}$ | — |
| Friction coefficient | 0.8 | 0.6 | 0.2 |

Based on the microscopic parameters in Table 2, sandy embankments, reinforced bodies and reinforced embankments were built, and the discrete element simulation results were compared to the laboratory tests by changing the location of the voids. In order to better verify the accuracy of the model parameters, the following working conditions were established to compare the simulation and test results, respectively, to finally verify the rationality of the parameters; the specific working conditions are shown in Table 3.

**Table 3.** Comparison conditions.

| Work Conditions | Void Diameter (mm) | Reinforced Layers | H/B | L/B |
|---|---|---|---|---|
| ZX1 | 0 | 0 | — | — |
| ZX2 | 60 | 0 | 2 | 0 |
| ZX3 | 60 | 1 | 2 | 0 |
| ZX4 | 60 | 1 | 2 | 1 |

Figure 5 shows the load–settlement curves of the embankment under the test and simulation, and Figure 6 is the stress–strain relationship curve for the uniaxial tensile test of the geogrid. From Figure 5, it can be seen that there is a certain gap between the discrete element test and simulated load–settlement curves, but the general trend is consistent, and the similarity between the two is high, and the peak bearing capacity remains at the same level. Therefore, it can be concluded that the sandy embankment and the reinforced embankment can be better simulated using the microscopic parameters in Table 2.

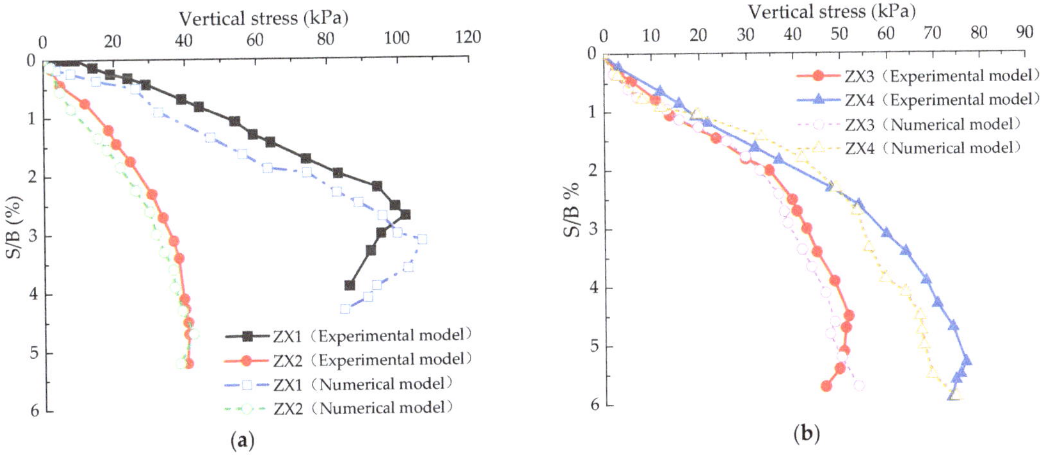


**Figure 5.** Load–settlement curve for the embankment: (**a**) sandy embankments; (**b**) reinforced embankments.

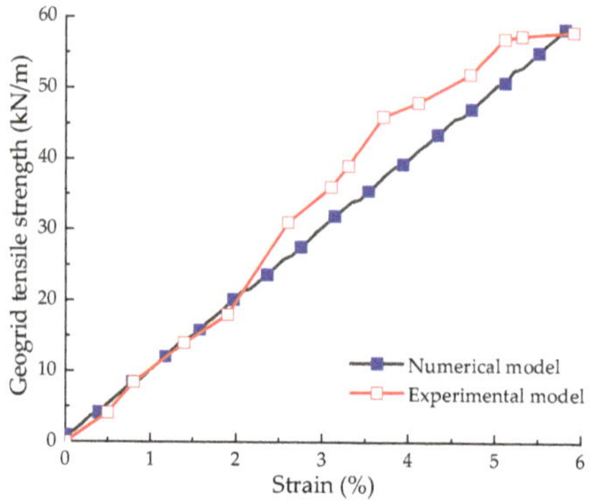


**Figure 6.** Geogrid stress–strain curve.

It can be seen from Figure 6 that there is a gap between the discrete element test and the simulated uniaxial tensile curves after loading for a certain period of time, which is due to the fact that the tensile strength of the reinforcing body in the indoor test changes with the increase in the strain of the reinforcing body, which is not taken into account in the discrete element simulation, but the general trend between the two is consistent. In both laboratory tests and simulations, the geogrid achieved a tensile strength of 60 kN/m at a strain of 6%. Therefore, the above selection of microparameters for the geogrid is reasonable.

This paper mainly analyzed the effects of different void locations on reinforced embankments under local overload conditions. To provide some basis for the design of reinforced bodies in the embankments overlying voids in practical engineering, the effects of vertical distance H or horizontal distance L on reinforced embankments overlying voids were analyzed, and the failure modes of embankment was proposed. The same loading speed was used for the models of different working conditions, and the specific simulation test scheme is shown in Table 4. The failure criteria in this paper are based on the provisions of the Engineering Geology Manual for foundation failure under shallow plate tests [29].

**Table 4.** Simulated operating conditions.

| Work Conditions | H/B | L/B |
|---|---|---|
| A1 | 1 | 0 |
| A2 | 2 | 0 |
| A3 | 3 | 0 |
| B2 | 2 | 1 |
| C2 | 2 | 2 |
| D2 | 2 | 3 |
| E2 | 2 | 4 |

## 3. Results

### *3.1. Characterization of Load Bearing Capacity*

Figure 7 shows the embankment load–settlement curves at different void locations. Figure 7a shows the load–settlement curves at different H/B. The load–settlement curves of each working condition basically overlap under the first 5 kPa load case. After 5 kPa, load differences start to appear. When H increases from B to 3B, the bearing capacity increases with the increase in the vertical distance H under the same foundation settlement. A3 and A2 have an increase of 2.5 times and 1.5 times, respectively, in the ultimate bearing capacity compared to A1. It can be seen from the above that the ultimate bearing capacity of the embankment increases non-linearly with the increase in H.

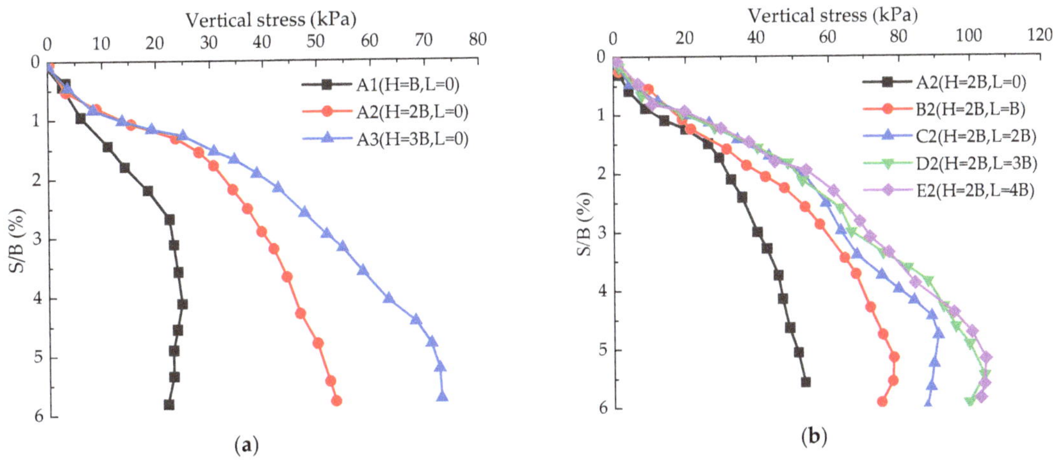


**Figure 7.** Load–settlement curves for different void locations: (**a**) at different H/B; (**b**) at different L/B.

From Figure 7b, it can be obtained that before the first 20 kPa load, the settlement curves of each working condition basically overlap, after 20 kPa load, there is a difference. When the horizontal distance L changes from 0 to 3B, the ultimate bearing capacity of the embankment increases with the increase in L. When $L \geq 3B$, the load–settlement curves of D2 and E2 basically coincide, the change in L has a small effect on the ultimate bearing capacity of the embankment at this point.

### *3.2. Analysis of Displacement Field*

In order to analyze the effect of H/B on embankment soil displacement, three representative regions, 1, 2 and 3, were intercepted in the embankment. Region 1 and 3 are at the lower soil of the reinforced body, region 3 is further away from the loading plate compared to region 1, and region 2 is at the upper part of the reinforced body further away from the loading plate. Figure 8 shows a diagram of the relative positions of the regions (the solid red line represents the reinforced body).

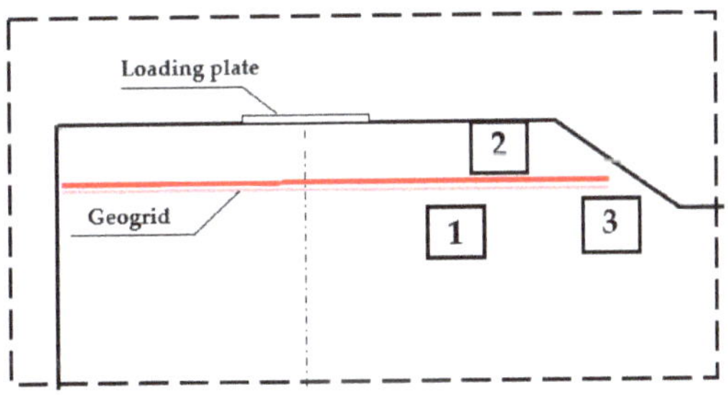


**Figure 8.** Schematic diagram of the relative positions of the regions.

Figure 9 shows the displacement vector diagram of the soil inside each region of the embankment at different H/B. From the diagram, it can be obtained that when H = B (Figure 9a), the soil displacements in regions 1 and 3 are generally in a downward direction. The left part of region 2 is in a disorderly state, and other parts of region 2 is moving downward as a whole because of gravity. When H = 2B (Figure 9b), soil displacements are generally to the right in region 1 and generally downward in region 3. When H = 3B (Figure 9c), soil displacements are generally to the right in regions 1 and 3. As H/B increases, the soil in regions 1 and 3 is transformed from an overall downward vertical displacement to a horizontal displacement to the right. This is because as H/B increases, the soils in regions 1 and 3 are subjected to more loads transferred from the embankment surface. The transferred loads cause a tendency for the displacements to move horizontally to the right. Comparing the soil displacement of region 2 at different H/B, it can be found that when H/B is transformed from 1 to 2, there is a significant increase in the width of the local slip surface in the upper part of the reinforced body, and when H/B is transformed from 2 to 3, the increase in the width of the slip surface is not significant at this time.

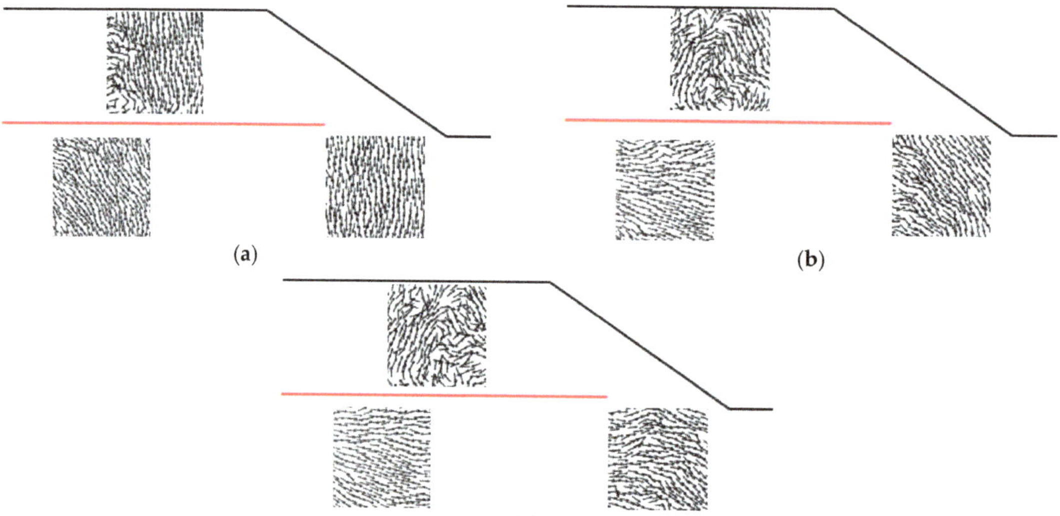

**Figure 9.** Soil displacement vector diagram of embankment at different H/B: (**a**) H = B, L = 0; (**b**) H = 2B, L = 0; (**c**) H = 3B, L = 0.

In order to analyze the effect of L/B on embankment soil displacement, the four regions, 5, 6, 7 and 8, were intercepted in the embankment. The depth of the four regions was the same, and the top of the region was equal to the road base surface. The distance between the center of each region and the center of the loading plate was B, 2B, 3B and 4B, respectively. Figure 10 shows a schematic diagram of the relative positions of the regions (the solid red line represents the reinforced body).

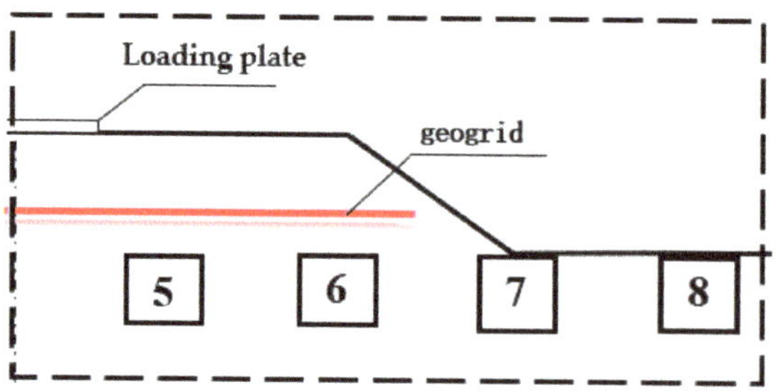


**Figure 10.** Schematic diagram of the relative positions of the regions.

Figure 11 shows the soil displacement vectors in each region of the embankment at different L/B. When L/B is 1, 2, 3 and 4, respectively, the void is located directly below regions 5, 6, 7 and 8. When comparing the different L/B conditions, it can be found that the width of the soil region dominated by horizontal displacements in the upper part of the void gradually increases as L increases. When L/B < 4, the soil on the upper left side of the void is displaced horizontally to the right, while the soil on the upper part of the void and the upper right side of the void is moved downward. This is due to the fact that the load transmitted from the embankment surface acts on the upper left area of the void, which is much greater than the self-weight of the soil, which causes the soil to have a tendency to move horizontally to the right. When L changes, the soil displacement in region 8 is always predominantly downward, which indicates that the change in L has less of an impact on the soil in region 8.

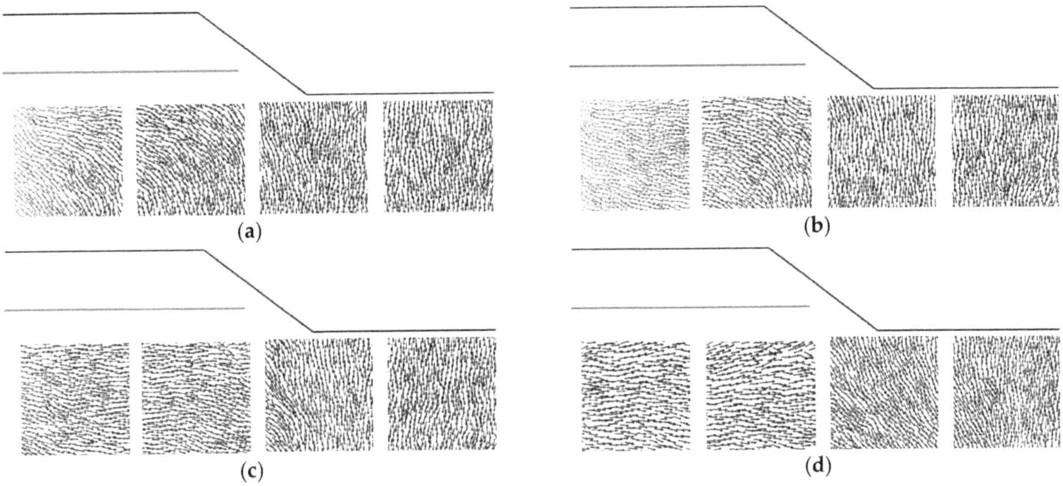

**Figure 11.** Soil displacement vectors of embankment at different L/B: (**a**) H = 2B, L = B; (**b**) H = 2B, L = 2B; (**c**) H = 2B, L = 3B; (**d**) H = 2B, L = 4B.

### *3.3. Analysis of Contact Force Chain*

Figure 12 shows the contact force chain diagram of the embankment at different H/B, where the circular dashed line represents the position of the void and the red continuous particles represent the position of the reinforcement body. From the diagram, it can be seen that the contact force chain is columnar in the upper part of the reinforced body, and the contact force chain has a triangular blank area at the position above the void. The contact force chain extends from the embankment surface to both sides of the void. The contact force chain in the area below the void is thinner than in other areas at the same depth, due to the void preventing the transfer of load to the lower area of the void. When H/B = 1 (Figure 12a), the void is close to collapsing into the lower part of the reinforced body, and the contact force chain at the bottom of the embankment is thicker at this time, which is due to the fact that the embankment is subjected to a smaller load, making the original contact force of the embankment soil relatively larger. As the vertical distance H increases, the contact force chain at the top of the void becomes denser and the number of thick chains increases, the embankment soil can withstand more load at this time.

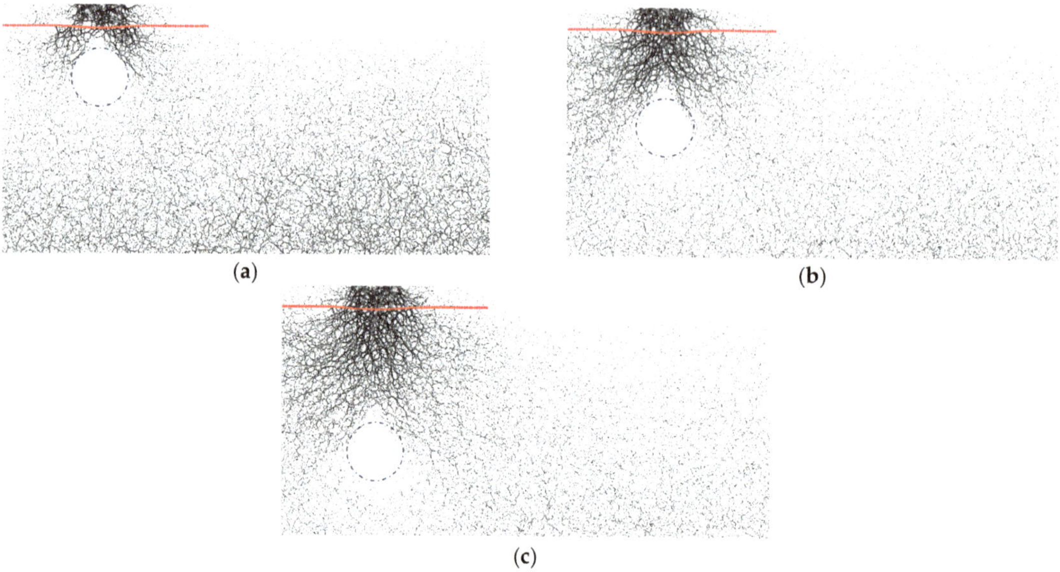

**Figure 12.** Diagram of the contact force chain at different H/B: (**a**) H = B, L = 0; (**b**) H = 2B, L = 0; (**c**) H = 3B, L = 0.

Figure 13 shows the contact force chain diagram of the embankment at different L/B. From the diagram, it can be obtained that the contact force chain on the left side of the void is significantly larger than that in the right side of the void in terms of coarseness and density of distribution, which indicates that the existence of the void prevents the transfer of the embankment surface load to the soil area on the right side of the void. By comparing the contact force chains at different L/B, it can be seen that the range of action of the coarse force chain gradually increases as L increases.

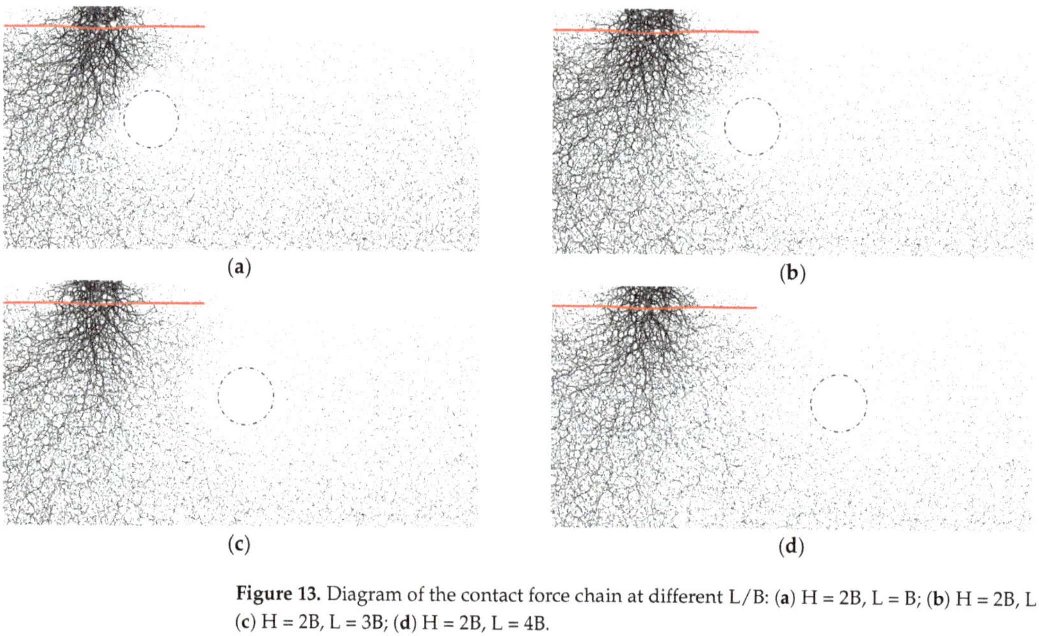

**Figure 13.** Diagram of the contact force chain at different L/B: (**a**) H = 2B, L = B; (**b**) H = 2B, L = 2B; (**c**) H = 2B, L = 3B; (**d**) H = 2B, L = 4B.

### 3.4. Deformation Analysis of Reinforced Bodies

Figure 14 shows the geogrid displacement curves of the embankment of different H/B at the time of failure. Figure 14a shows the vertical displacement curves of the geogrid at different H/B. From Figure 14a, the maximum vertical settlement of the geogrid shows a trend of increasing and then decreasing as H increases. The vertical displacement curve is symmetrical around the center of the loading plate, and the vertical displacement of the geogrid within a distance of 60 mm from the center of the loading plate is fitted with a quadratic parabola. The vertical displacement of the reinforced body gradually decreases with increasing distance from the center of the loading plate, and the vertical displacement approaches zero at 60 mm from the center of the loading plate. The deformation of the reinforcement body around 60 mm from the center of the loading plate shows a circular segment, and the geogrid in this area has a certain negative deformation, caused by the extrusion of the surrounding soil. Figure 14b shows the transverse displacement curves of the geogrid at different H/B. From the figure, the transverse displacement of the geogrid is centrosymmetric around the center of the loading plate (the direction of displacement of the geogrid is positive to the right), and the transverse displacement at the center of the loading plate increases and then decreases as H increases. The transverse displacement of the geogrid 60 mm outside the center of the loading plate remains almost constant; the transverse strain of the geogrid in this area is small.

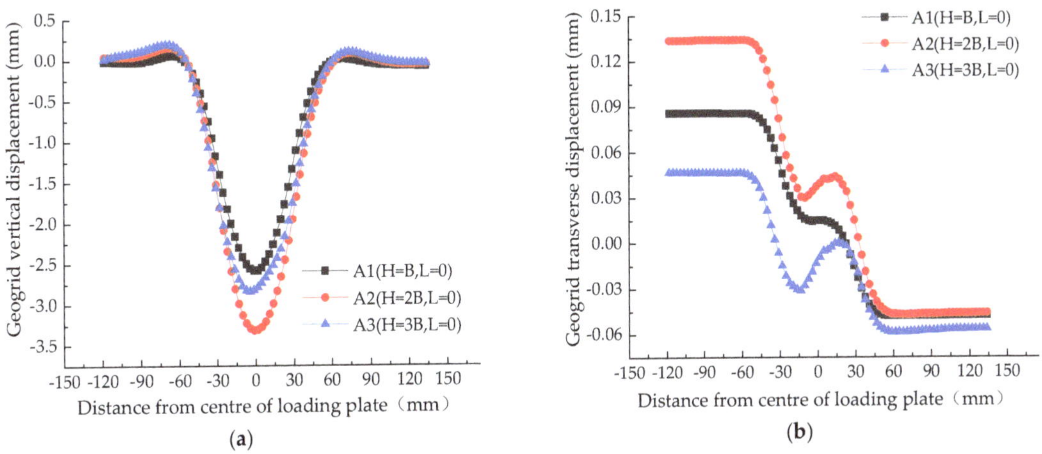


**Figure 14.** Geogrid displacement curves at different H/B: (**a**) vertical displacement; (**b**) transverse displacement.

Figure 15 shows the geogrid displacement curves of the embankment at different L/B. From Figure 15a, when L $\leq$ 3B, the maximum vertical displacement of the geogrid decreases as L increases, when L $\geq$ 3B, the maximum vertical displacement of the geogrid roughly overlaps. The vertical displacement curves of the geogrid at different L/B are symmetrical around the center of the loading plate and decrease with increasing distance from the center. This indicates that the deformation of the geogrid under ultimate load is mainly influenced by the downward displacement of the loading plate and is less influenced by the horizontal distance L.

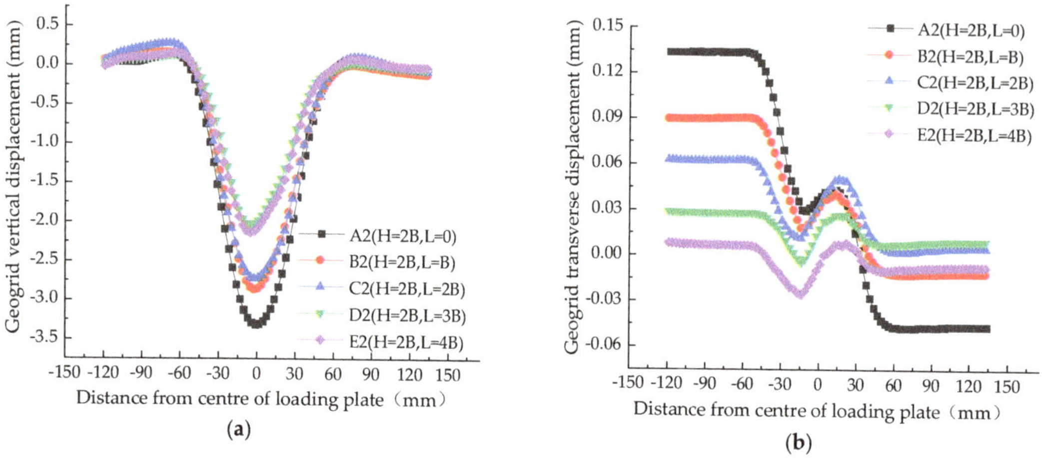


**Figure 15.** Geogrid displacement curves at different L/B: (**a**) vertical displacement; (**b**) transverse displacement.

From Figure 15b, it can be seen that the displacement direction of the geogrid at the center of the loading plate is to the right. The relative displacement of the reinforced soil interface on the left side of the geogrid is greater than that on the right side, probably due to the tendency of the soil at the lower right side of the geogrid to move toward the slope during the loading process, which drives the overall movement of the geogrid to the right. As L increases from 0 to 3B, the transverse strain rate of the geogrid at the edge of the loading plate (located 30mm from the center of the loading plate) decreases significantly. When L increases from 3B to 4B, the transverse strain rate of the geogrid at the edge of the loading plate does not change significantly, indicating that when L ≥ 3B, the change of L has a low effect on the transverse strain of the geogrid.

### 3.5. Analysis of Failure Modes

When the embankment is broken, the stress direction of the soil around the void is not only affected by the relative displacement of the soil due to the void, but also by the upper load transmitted by the embankment surface, which leads to more complex deflection of the soil around the void. The main stress cross has obvious advantages in monitoring the stress deflection of the soil [30–32], and 4 × 4 measurement circles are distributed in the area of 80 mm × 80 mm on the left and right side of the void, and 10 × 4 measurement circles are distributed in the area of 220 mm × 60 mm in the upper part of the void. The radius of the measuring circle is 15 mm. Figure 16 shows the distribution of the measurement circles around the void.

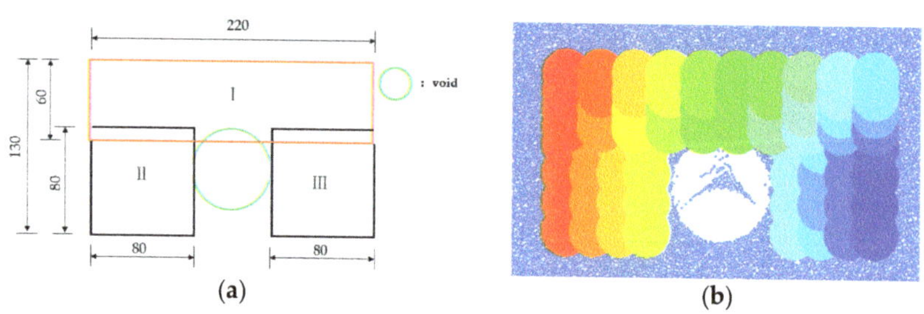


**Figure 16.** Schematic diagram of the measuring circle arrangement: (**a**) layout size chart; (**b**) measurement circle simulation diagram.

Taking L/B = 0 as an example, the failure modes of embankment at different H/B are analyzed. Figure 17 shows the displacement vector field of the embankment during damage at different H/B and the stress deflection distribution around its void (red continuous solid line represents reinforced body; yellow border line represents the main stress measurement area). The length and direction of the line segments on the top right indicate the magnitude and direction of the principal stresses, with the long and short axes indicating the major and minor principal stresses, respectively.

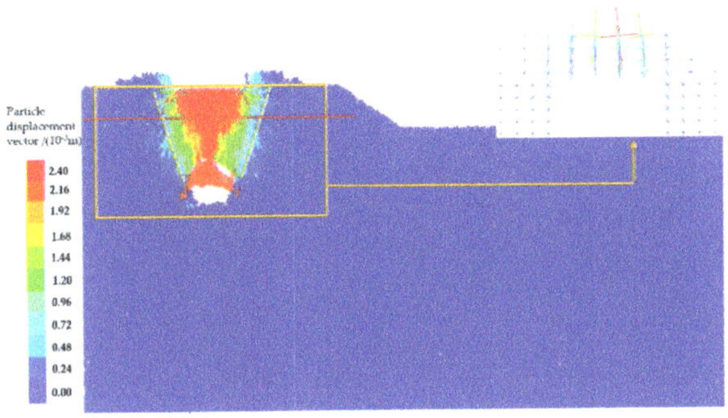


(a)

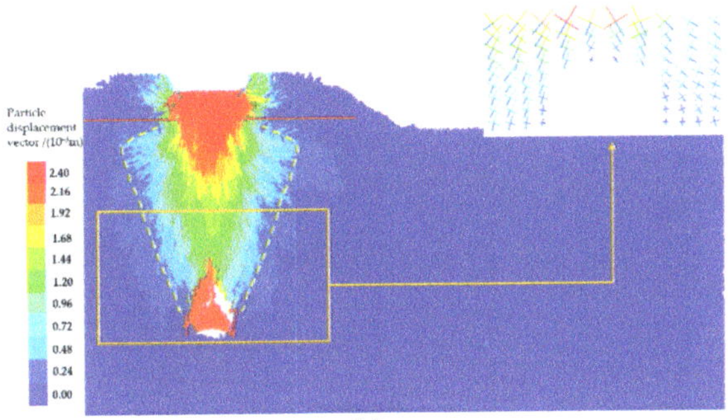


(b)

**Figure 17.** Displacement vector field and deflection of principal stresses around the void at different H/B: (**a**) H = B, L = 0; (**b**) H = 3B, L = 0.

From the displacement vector diagram in Figure 17, we can see that when H = B, perforation failure occurs on the embankment, and the damage surface extends from the embankment surface to the side walls above the left and right of the void. The surface of the damage presents a basin type. The width at the top of the basin damage surface is greater than the width of the loading plate; this is because the loading plate's soil downward pressure under the plate has a horizontal extrusion effect. At this time, because the depth of the void is too small, the main stress direction above the void is mainly affected by the vertical load transmitted by the loading plate, and the main stress deflection is small. When H = 3B, a collapse perforation failure occurs on the embankment; the surface of damage still extends from the embankment surface to the side walls above the left and right of the void. It should be noted that at this time, the surface of damage in the lower side of the reinforced body as a whole is bottle-shaped, and the width of the damage surface in the reinforced body area has a significant narrowing. The main stress direction above the void produces a large deflection due to the relative displacement of the soil body, where there is a significant soil arching effect.

Taking H/B = 2 as an example, the failure modes of embankment at different L/B are analyzed. Figure 18 shows the displacement vector field of the embankment during damage at different L/B and the stress deflection distribution around its void. From the figure, the soil above the geogrid forms a local damage surface extending to the top of the embankment under the action of the embankment surface load. The presence of the geogrid prevents the surface of damage at the upper and lower interfaces of the reinforced soil from being connected, and the width of the damage surface is narrowed in the area of the geogrid.

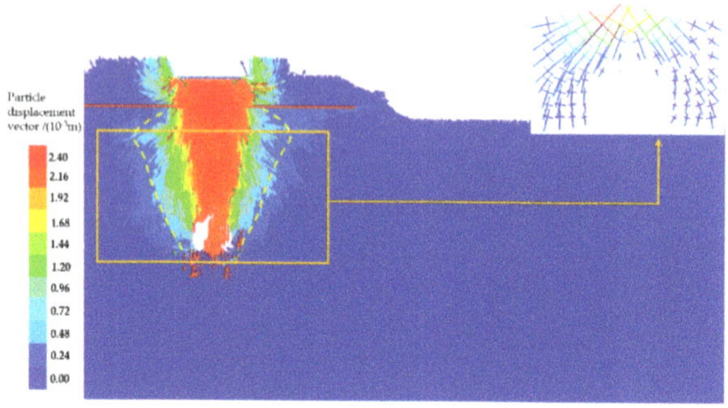


(**a**)

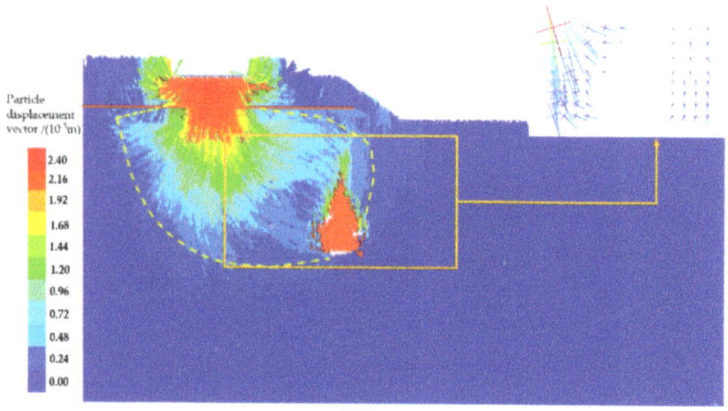


(**b**)

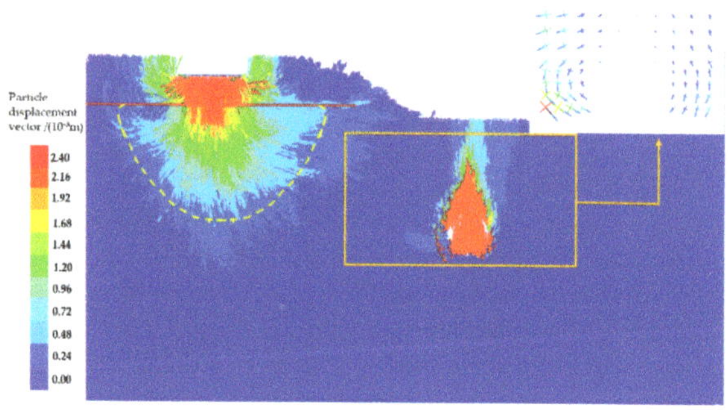


(**c**)

**Figure 18.** Displacement vector field and deflection of principal stresses around the void at different L/B: (**a**) H = 2B, L = 0; (**b**) H = 2B, L = 2B; (**c**) H = 2B, L = 4B.

As shown in Figure 18a, when L = 0, at this time, the embankment is consistent with the failure mode in Figure 17b, i.e., when the collapse perforation failure occurs. The main stress of the soil around the void is deflected around the void, and the maximum stress area is concentrated in the upper part of the void. As shown in Figure 18b, when L = 2B, the damage surface extends from the embankment surface to the collapse area above the void and the lower left area of the void, forming a void side failure. At the same time, the void collapse causes less stress in the soil at the top of the void, and the transfer of embankment load makes the stress maximum area concentrated on the left side of the void, and the main stress direction points to the void location. From Figure 18c, it is found that when L= 4B, the damage surface of the lower part of the geogrid presents a semi-circular arc, but does not extend to the void. At this time, the stress deflection around the void is mainly affected by the void, the main stress around the void deflects around the void and the soil arch effect is relatively obvious, forming a no impact failure.

## 4. Conclusions

(1) The ultimate bearing capacity of the embankment increases as H increases. When $L < 3B$, the ultimate bearing capacity of the embankment increases as L increases. When $L \geq 3B$, the change of L has less influence on the ultimate bearing capacity.

(2) The existence of the void affects the embankment soil displacement. When $0 < L \leq 3B$ and $H = 2B$, the soil on the upper left side of the void is displaced horizontally to the right, while the soil on the upper part of the void and the upper right side of the void is moved downward.

(3) The void hinders the transfer of the load in the soil. When $L = 0$, the load is transferred to both sides of the void, and the contact force chain of the soil below the void is smaller. When $L > 0$, the load is mainly concentrated on the left side of the void, and the contact force chain on the left side of the void is obviously larger than that on the right side.

(4) The maximum vertical deformation of the geogrid increases and then decreases with the increase in the vertical distance H. When $L < 3B$, the maximum vertical deformation of the geogrid decreases with the increase in the horizontal distance L, and when $L \geq 3B$, the change of L has less influence on the maximum vertical deformation of the geogrid. In summary, when the void is located directly below the loading plate, the deformation of the reinforced body is the largest. In the design of the reinforced body, it is safer to consider that the void is located directly below the loading plate.

(5) When $L = 0$, as H increases, the failure mode of the embankment is transformed from perforation failure to collapse perforation failure, and when $L > 0$, as L increases, the failure mode of the embankment is transformed from void side failure to no impact failure.

**Author Contributions:** Writing—original draft preparation, Q.Z.; writing—review and editing, Y.L. All authors have read and agreed to the published version of the manuscript.

**Funding:** This research received no external funding.

**Institutional Review Board Statement:** Not applicable.

**Informed Consent Statement:** Not applicable.

**Data Availability Statement:** Data are contained within the article.

**Conflicts of Interest:** The authors declare no conflict of interest.

## References

1. Villard, P.; Briançon, L. Design of geosynthetic reinforcements for platforms subjected to local. *Can. Geotech. J.* **2008**, *45*, 196–209. [CrossRef]
2. Giroud, J.; Bonaparte, R.; Beech, J.; Gross, B. Design of soil layer-geosynthetic systems overlying voids. *Geotext. Geomembr.* **1990**, *9*, 11–50. [CrossRef]
3. Blight, G.; Barrett, A. Field test of catenary net to protect traffic from mining subsidence. *J. Transp. Eng.* **1990**, *116*, 135–144. [CrossRef]
4. Briancon, L.; Villard, P. Design of geosynthetic-reinforced platforms spanning localized sinkholes. *Geotext. Geomembr.* **2008**, *26*, 416–428. [CrossRef]
5. Sireesh, S.; Sitharam, T.; Dash, S. Bearing capacity of circular footing on geocell-sand mattress overlying clay bed with void. *Geotext. Geomembr.* **2009**, *27*, 89–98. [CrossRef]
6. Yee, T.; Lawson, C. Serviceability limits for basal reinforced embankments spanning voids. *Geosci. Front.* **2011**, *2011*, 3276–3285.
7. Bridle, R.; Jenner, C. Polymer geogrids for bridging mining voids. *Geosynth. Int.* **1997**, *4*, 33–50. [CrossRef]

8. Flum, D.; Roduner, A.; Kalejta, J. Full-scale field tests for bridging sinkholes using flexible steel components as reinforcement. *Carbonates Evaporites* **2012**, *27*, 161–165. [CrossRef]
9. Kinney, T.; Connor, B. Geosynthetics supporting embankments over voids. *J. Cold Reg. Eng.* **1987**, *1*, 158–170. [CrossRef]
10. Huang, J.; Le, V.; Bin-Shafique, S.; Papagiannakis, A. Experimental and numerical study of geosynthetic reinforced soil over a channel. *Geotext. Geomembr.* **2015**, *43*, 382–392. [CrossRef]
11. Lu, W.; Zhang, Y.; Liu, W.; Liu, C.; Wang, H. Evaluation of geomembrane effect based on mobilized shear stress due to localized sinking. *Adv. Civ. Eng.* **2019**, *2019*, 4942578. [CrossRef]
12. Huckert, A.; Briancon, L.; Villard, P.; Garcin, P. Load transfer mechanisms in geotextile-reinforced embankments overlying voids: Experimental and analytical approaches. *Geotext. Geomembr.* **2016**, *44*, 442–456. [CrossRef]
13. Chevalier, B.; Villard, P.; Combe, G. Investigation of load-transfer mechanisms in geotechnical earth structures with thin fill platforms reinforced by rigid inclusions. *Int. J. Geomech.* **2011**, *11*, 239–250. [CrossRef]
14. Benmebarek, S.; Berrabah, F.; Benmebarek, N. Effect of geosynthetic reinforced embankment on locally weak zones by numerical approach. *Comput. Geotech.* **2015**, *65*, 115–125. [CrossRef]
15. Tran, V.; Meguid, M.; Chouinard, L. Three-dimensional analysis of geogrid-reinforced soil using a finite-discrete element framework. *Int. J. Geomech.* **2015**, *15*, 04014066. [CrossRef]
16. Le Hello, B.; Villard, P. Embankments reinforced by piles and geosynthetics-numerical and experimental studies dealing with the transfer of load on the soil embankment. *Eng. Geol.* **2009**, *106*, 78–91. [CrossRef]
17. Li, P.; Su, F. Unidirectional geosynthetic reinforcement design for bridging localized sinkholes in transport embankments. *Math. Probl. Eng.* **2022**, *2022*, 9577348. [CrossRef]
18. Feng, S.; Ai, S.; Chen, H. Membrane effect of geosynthetic reinforcement subjected to localized sinkholes. *Can. Geotech. J.* **2018**, *55*, 1334–1348. [CrossRef]
19. Chen, F.; Lin, Y.; Chen, S. Analytical solutions for geosynthetic-reinforced cohesive subgrade spanning trench voids. *Geotext. Geomembr.* **2020**, *48*, 854–866. [CrossRef]
20. Das, B.; Khing, K. Foundation on layered soil with geogrid reinforcement—Effect of a void. *Geotext. Geomembr.* **1994**, *13*, 545–553. [CrossRef]
21. Wu, G.; Zhao, M.; Zhao, H.; Xiao, Y. Effect of eccentric load on the undrained bearing capacity of strip footings above voids. *Int. J. Geomech.* **2020**, *20*, 04020078. [CrossRef]
22. Zhou, H.; Zheng, G.; He, X.; Xu, X.; Zhang, T.; Yang, X. Bearing capacity of strip footings on c–φ soil with square voids. *Acta. Geotech.* **2018**, *13*, 747–755. [CrossRef]
23. Lai, F.; Chen, F.; Li, D. Bearing capacity characteristics and failure modes of low geosynthetic-reinforced embankments overlying voids. *Int. J. Geomech.* **2018**, *18*, 04018085. [CrossRef]
24. Tahmasebipoor, A.; Noorzad, R.; Shooshpasha, E.; Barari, A. A parametric study of stability of geotextile-reinforced soil above an underground void. *Arab. J. Geosci.* **2012**, *5*, 449–456. [CrossRef]
25. Tafreshi, S.; Khalaj, O.; Halvaee, M. Experimental study of a shallow strip footing on geogrid-reinforced sand bed above a void. *Geosynth. Int.* **2011**, *18*, 178–195. [CrossRef]
26. Fu, Z.; Chen, S.; Liu, S. Discrete element simulations of shallow plate-load tests. *Int. J. Geomech.* **2016**, *16*, 04015077. [CrossRef]
27. Zhang, J.; Liu, X.; Xiong, J.; Liang, L.; Zhang, W. Research on mechanical properties and failure mode of conglomerate based on discrete element method. *Appl. Sci.* **2023**, *13*, 223. [CrossRef]
28. Zhang, G. Researches on Meso-Scale Mechanism of Piping Failure by Means of Model Test and PFC Numerical Simulation. Ph.D. Thesis, Tongji University, Shanghai, China, 2007.
29. Editorial Board of Handbook of Engineering Geology. *Handbook of Engineering Geology*; Architecture and Architecture Press: Beijing, China, 2018.
30. Villard, P.; Huckert, A.; Briançon, L. Load transfer mechanisms in geotextile-reinforced embankments overlying voids: Numerical approach and design. *Geotext. Geomembr.* **2016**, *44*, 381–395. [CrossRef]
31. Rui, R.; Ye, Y.; Han, J.; Wan, Y.; Chen, C.; Zhang, L. Two-dimensional soil arching evolution in geosynthetic-reinforced pile-supported embankments over voids. *Geotext. Geomembr.* **2022**, *50*, 82–98. [CrossRef]
32. Anh Tran, Q.; Villard, P.; Dias, D. Discrete and Continuum Numerical Modeling of Soil Arching between Piles. *Int. J. Geomech.* **2019**, *19*, 04018195. [CrossRef]

applied sciences

MDPI

Article

# Development of High-Performance Fly-Ash-Based Controlled Low-Strength Materials for Backfilling in Metropolitan Cities

Jingyu Han [1], Youngseok Jo [2], Yunhee Kim [3] and Bumjoo Kim [3,*]

1 Department of R&D, Chemius Korea Co., Ltd., Suncheon 57942, Republic of Korea
2 Department of Civil and Environmental Engineering, University of California, Berkeley, CA 94720, USA
3 Department of Civil and Environmental Engineering, Dongguk University, Seoul 04620, Republic of Korea
* Correspondence: bkim1@dongguk.edu

**Abstract:** Controlled low-strength materials (CLSMs) have been developed using various byproducts for backfilling or void-filling around pipelines or culvert boxes. However, these CLSMs have encountered issues related to their inadequate placement around underground facilities, despite satisfying the performance requirements, especially flowability, recommended by the American Concrete Institute (ACI) 229 committee. In this study, a new CLSM is developed to ensure a significantly higher flowability, lower segregation, and faster installation compared with previously developed CLSMs. This is achieved through a series of laboratory tests. To enhance the flowability and prevent segregation, a calcium-sulfoaluminate-based binder and fly ash are used in combination with two types of additives. The measured flowability of the new CLSM is 700 mm, while its compressive strength and bleeding satisfy the general criteria specified by the ACI 229R-13. In addition, the performance of the developed CLSM is compared with that of predeveloped CLSMs. The new CLSM was not only shown to exhibit the highest flowability, but also to satisfy the specified requirements for compressive strength and bleeding. Overall, it is anticipated that the developed CLSM can significantly reduce the costs related to the disposal of old pavements, the installation of new pavements, and other construction expenses compared to the costs related to the conventional method, even though the expenses for the backfill materials could increase due to the higher production costs of CLSMs than soil. In addition, there is a need to investigate its field applicability in order to evaluate the precise costs, maintenance, and long-term stabilities after installation.

**Keywords:** controlled low-strength material; high flowability; backfilling; fly ash; calcium-sulfoaluminate-based binder

**Citation:** Han, J.; Jo, Y.; Kim, Y.; Kim, B. Development of High-Performance Fly-Ash-Based Controlled Low-Strength Materials for Backfilling in Metropolitan Cities. *Appl. Sci.* **2023**, *13*, 9377. https://doi.org/10.3390/app13169377

Academic Editor: José Manuel Moreno-Maroto

Received: 16 July 2023
Revised: 10 August 2023
Accepted: 13 August 2023
Published: 18 August 2023



## 1. Introduction

In metropolitan cities, ground sinking generally occurs due to the deterioration, cracking, and breakage of sewage pipes (Figure S1), as well as the inadequate compaction of backfill materials (Figure S2). These factors can lead to the formation of void spaces around sewage pipes and culvert boxes. Subsequently, ground sinking occurs as void spaces become larger. Jo et al. [1] reported that the vibration transmitted by vehicles in metropolitan cities could cause the collapse of enlarged underground void spaces. The conventional method involves compacting well-graded soils around sewage pipes or culvert boxes using mechanical equipment (e.g., a vibrator) to prevent ground sinking. However, due to the limitation of mechanical equipment, the area adjacent to sewage pipes and culvert boxes is typically left uncompacted, in some cases leading to ground sinking. The foregoing is recognized as one of the causes of ground sinking in metropolitan cities. Accordingly, the utilization of controlled low-strength materials (CLSMs) has emerged as a highly effective solution for backfilling or void-filling around pipelines or culvert boxes and underneath roads.

The American Concrete Institute (ACI) 229 committee defined CLSMs as self-compacting cementitious backfill materials. Other terms used to describe CLSMs include flowable fill, unshrinkable fill, controlled density fill, flowable mortar, plastic soil cement, soil–cement slurry, and K-Krete. CLSMs have been widely used for numerous applications (e.g., backfilling, road bases, pipeline and culvert filling, void filling, tank fills, insulation, and isolation filling) in lieu of conventional compact fillers [2–5]. Table 1 summarizes the guidelines set out by the ACI 229R-13 that need to be satisfied by each CLSM application. In accordance with these guidelines, CLSMs should exhibit a high flowability, low bleeding, early cementation, and sufficient strength [3,6,7]. The ACI 229R-13 document [8] is not only the main guideline used in the USA, but has also been widely referenced in many countries. Ling et al. [9] also noted that apart from the USA, there have been no specific standards or CLSM specifications published by other countries. Therefore, many countries have developed their own CLSM guidelines based on the ACI 229R-13 document.

**Table 1.** General criteria and requirements for CLSM application and target performance in this study.

| References | CLSM Application | Criteria and Requirements to Be Fulfilled |
|---|---|---|
| ACI 229R-13 | General backfilling (void filling, filling abandoned underground structures, etc.) | High degree of flowability (>200 mm spread)<br>Setting time and early strength are not critically essential<br>Twenty-eight-day compressive strength should be less than 0.5 MPa |
| | Excavatable backfilling (underground water, sewer and storm drainage pipelines, roadway trenches, etc.) | High degree of flowability<br>Less subsidence and quick setting time<br>Easy to re-excavate—manually or mechanically<br>Twenty-eight-day compressive strength should be less than 2.1 MPa |
| This study (Target performance) | General and excavatable backfilling (void filling) | Considerably high degree of flowability (>400 mm spread)<br>Less subsidence (<5% bleeding)<br>Quick setting time (>0.1 MPa at 4 h)<br>Easy to re-excavate—manually<br>Twenty-eight-day compressive strength $\leq$ 1.0 MPa |

Since the 2000s, previous researchers have developed CLSMs using various waste materials/industrial byproducts (e.g., fly and bottom ashes, ground granulated blast furnace slag, waste foundry sand, cement kiln dust, steel slag, waterworks and paper sludges, waste rubber tires, and red mud) [7,10–29]. However, the application of these previously developed CLSMs to underground facility backfilling has been challenging due to three primary reasons: (a) previous CLSMs have relatively long durations for hardening; (b) poorer flowability than grout mortar; (c) large equipment is needed for the placement of CLSMs. In addition, these CLSMs only satisfied the acceptable level of flowability recommended by the ACI 229R-13 (i.e., more than 200 mm spread, as listed in Table 1) [5,6,11,23,30–33].

Even though the performance of predeveloped CLSMs satisfied the requirements set out by the ACI 229R-13, especially flowability, those CLSMs could lead to problems of insufficient placement around underground facilities (i.e., pipes and culverts) and cause inconvenience to citizens in metropolitan cities due to the relatively long hardening time [5,6,23,34]. Therefore, there is a need to develop a new CLSM with much higher performance (i.e., high flowability, low bleeding/segregation, and quick setting time) than predeveloped CLSMs.

In this study, a new CLSM with high flowability, low segregation, and quick setting time is developed. The target performance criteria for the developed CLSM are summarized in Table 1. In particular, the water volume increased and the materials (i.e., binder and byproduct) volume was minimized in order to facilitate the placement of the CLSM. To achieve the foregoing, a series of laboratory tests were conducted to measure the flowability, density, bleeding, and compressive strength of the new CLSM in accordance with test standards [35–38]. A sensitivity analysis was performed to compare the flowability, bleeding, and compressive strength of the new CLSM according to the water–material ratio

(W/M). Finally, the performance of the newly developed CLSM was compared with that of predeveloped CLSMs based on soils and bottom ash collected in South Korea.

## 2. Materials and Methods

### *2.1. Preparation of Binder*

A CLSM is typically composed of a binder, cement, byproduct materials (e.g., bottom/fly ashes and other recycled materials), an aggregate, and water. Schmitz et al. [39] reported that the characteristics of CLSMs can vary depending on the type of binder. In this study, calcium sulfoaluminate (CSA) was selected as the primary binder due to its quick hardening and early strength characteristics. Table 2 summarizes the amounts of CSA, anhydrous gypsum (AG), and water used to determine the optimal mixture ratio for the binder. To enhance the early strength and quick hardening, CSA and AG, with relatively high specific surface areas (more than 500 $m^2/kg$), were used. The powders were premixed and then mixed with water for 2 min. The paste was cured in a 50 mm cubic square mold at ambient conditions. The early compressive strength of the binder specimen was tested at 4 h from mixing in the water. The amount of water was fixed, and AG was used to supply sulfate ions ($SO_4^{2-}$), which could produce ettringite by reacting with CSA. Ordinary Portland cement (OPC), classified as a type 1 cement, was added into the CSA mixtures (i.e., CSA-6~10) in order to accelerate the generation of ettringite. Figure 1a shows the binder specimens manufactured based on the mixture proportions in Table 2.

**Table 2.** Mix proportions of CSA-based binder for each specimen.

| Specimen | | CSA (g) | AG (g) | OPC (g) | Water (g) |
|---|---|---|---|---|---|
| Without OPC | CSA-1 | 900 | 100 | - | 350 |
| | CSA-2 | 800 | 200 | - | |
| | CSA-3 | 700 | 300 | - | |
| | CSA-4 | 600 | 400 | - | |
| | CSA-5 | 500 | 500 | - | |
| With OPC | CSA-6 | 720 | 180 | 100 | |
| | CSA-7 | 640 | 160 | 200 | |
| | CSA-8 | 560 | 140 | 300 | |
| | CSA-9 | 490 | 210 | 300 | |
| | CSA-10 | 480 | 120 | 400 | |
| | CSA-11 | 400 | 100 | 500 | |

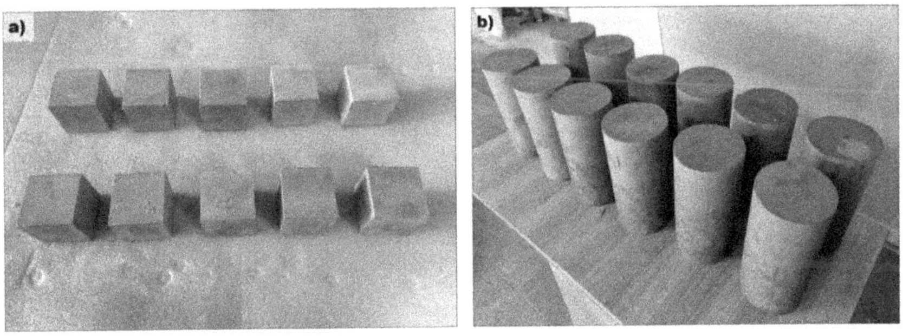


**Figure 1.** Specimens for compressive strength tests: (**a**) CSA-based binders; (**b**) CLSMs.

### *2.2. Properties of the CLSM*

#### 2.2.1. Flowability

High flowability is a crucial characteristic, enabling the CLSM to exhibit self-flowing properties during backfilling. It eliminates the necessity for compaction while effectively filling voids around civil utilities (e.g., pipelines and culvert boxes). The high flowability of the CLSM offers several advantages, including the ability to preserve the alignment of pipes and reduce the likelihood of ground sinking [40–42]. In addition, since the CLSM is injected to fill voids around pipelines, culverts, and underneath roads, it is essential to ensure sufficient flowability. This enables the CLSM to reach specific spots quickly, facilitating the efficient and precise placement of the material.

The flowability of the CLSM is influenced by the water content, with an increase in water content generally resulting in a higher flowability. However, it is crucial to maintain a balance, because an excessive water content can cause segregation and bleeding. To resolve this problem, the inclusion of high volumes of fine particles is recommended. These fine particles aid in reducing the segregation and enhancement of the overall stability of highly flowable CLSM mixtures. Ling et al. [9] reported that adequate materials and mixture proportions were important to achieve the required flowability (>200 mm) without segregation and bleeding. In this study, the flowability of the CLSM was evaluated based on ASTM D 6103. Figure 2 shows the flowability test, and the description and standard limit are listed in Table 3.

**Figure 2.** Flowability test of CLSM developed in this study.

**Table 3.** Test methods and acceptance criteria for determination of CLSM mixtures.

| Property | Method | Description | Standard Limit |
|---|---|---|---|
| Flowability | ASTM D 6103 [34] | 75 × 150 mm openended cylinder modified flow test | Low flowability: <150 mm<br>Normal flowability: 150–200 mm<br>High (good) flowability: >200 mm |
| Compressive strength | ASTM C 109 [35] (For binder) | Determination of compressive strength of hydraulic cement mortars using 50 × 50 × 50 mm cube specimens | Do not consider manifestly faulty specimens<br>Maximum permissible range between specimens from the same mortar batch, at the same test age is 8.7% of the average compressive strength |
| | ASTM D 4832 [36] (For CLSM) | Procedures for the preparation, curing, transporting, and testing of the cylindrical specimen (150 × 300 mm) of CLSM for the determination of compressive strength<br>Special care may be needed because the specimens are often very-low-strength and fragile | Maintaining strengths at a low level and allowing for excavation is an important consideration for CLSMs.<br>Strengths between 0.3 and 2.1 MPa are allowed for future excavation. Even less than 0.3 MPa is also acceptable for future excavation |
| Density | ASTM D 6023 [37] | Test method for unit weight, yield, and air content (gravimetric) of CLSM | Density of normal CLSM in place is in the range of 18.0–22.8 kN/m$^3$, which is greater than most compacted materials. However, a CLSM mixture with only fly ash, cement, and water should have a density of 14.1–15.7 kN/m$^3$ |

2.2.2. Compressive Strength

Compressive strength is one of the essential properties for CLSMs. In general, the ACI 229 committee defined that the 28-day compressive strength of CLSMs should be 8.3 MPa or less to be sufficient. Moreover, they stated that a long-term compressive strength of less than 2.1 MPa is also allowed to enable future excavation. In this study, a compressive strength of 1.0 MPa or less was set as the target value (Table 1). Compressive tests were conducted based on the ASTM C 109 [36] for the binders and ASTM D 4832 [37] for the CLSM. Figure 1 shows the binder and CLSM specimens for the compressive strength tests. The description and standard limit are summarized in Table 3.

2.2.3. Density and Bleeding

The ACI 229 committee specified that the density of a normal CLSM in place should be in the range of 18.0–22.8 $kN/m^3$, which is greater than most compacted materials. However, a CLSM mixture with only fly ash, cement, and water should have a density of 14.1–15.7 $kN/m^3$. The density of the CLSM developed in this study was also measured based on ASTM D 6023 [38], and the measured value was described in Section 4.3. The description and standard limit are shown in Table 3. The bleeding of the CLSM was measured based on ASTM C 940 [43] at 2 h from mixing.

*2.3. Filler*

The specific characteristics of each CLSM can depend on the fillers and binders used. In general, byproducts and coarse or fine aggregates are used as the fillers. In particular, fly ash in a CLSM can improve its flowability and strength. Fly ash can also reduce bleeding, shrinkage, permeability, and volume changes because its particle shape is spherical and it has a low specific gravity and pozzolanic reactivity. Moreover, it functions as a binder or fine aggregate in CLSMs [11,25,44–46].

Therefore, in this study, fly ash was used as the filler material. Table 4 shows the chemical composition, specific surface area, and density of the fly ash and other materials used in this study. The chemical composition was investigated with X-ray fluorescence (XRF) and the specific surface area was determined with the Blaine test.

**Table 4.** Chemical composition and physical characteristics of materials used in this study.

| Materials | Chemical Composition (%) | | | | | | | | | Physical Characteristics | |
|---|---|---|---|---|---|---|---|---|---|---|---|
| | $SiO_2$ | $Al_2O_3$ | $Fe_2O_3$ | CaO | MgO | $SO_3$ | $Na_2O$ | $K_2O$ | LOI | S.S.A. * ($m^2/kg$) | Specific Gravity |
| Fly ash | 58.42 | 18.42 | 0.89 | 17.4 | 0.83 | 1.8 | 0.05 | 0.28 | 2.41 | 342.6 | 2.36 |
| CSA | 8.67 | 33.5 | 1.75 | 42.51 | 1.43 | 8.45 | 0.14 | 0.31 | 0.7 | 535.0 | 2.88 |
| AG | 1.41 | 0.31 | 0.36 | 40.0 | 0.75 | 53.3 | 0.36 | 0.55 | 1.75 | 526.2 | 2.94 |
| OPC | 21.1 | 5.24 | 3.85 | 62.1 | 2.31 | 2.1 | 0.10 | 0.44 | 2.71 | 321.0 | 3.14 |

* Abbreviation of specific surface area.

## 3. Determination of Binder

Binder strength can vary depending on the proportions of ingredients, and can influence the strength of the CLSM. The optimum proportion of binder necessary for the rapid consolidation of the CLSM was determined in this section. In laboratory tests, the targeted initial compressive strength of the binder was 10 MPa in order to minimize the amount of binder in the CLSM composition.

Figure 3 summarizes the 4 h compressive strength of CSA-based binders without OPC (i.e., CSA-1~5 in Figure 3) and with OPC (i.e., CSA-6~11 in Figure 3). The 4 h compressive strength after curing the binder specimens was measured. This was because attaining sufficient compressive strength in the early stage was necessary for quick hardening and a short setting time in order to prevent the segregation of the highly flowable CLSM.

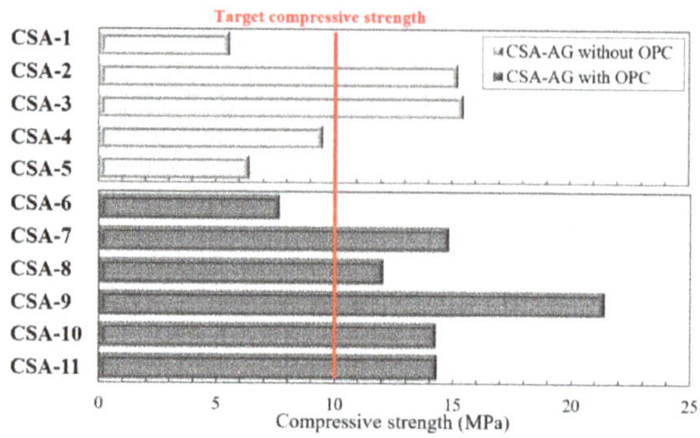


**Figure 3.** Compressive strength of specimens according to the ratio of base materials.

The mixed proportions are summarized in Table 2. The compressive strength of CSA-1, 4, and 5 was below the target strength (i.e., 10 MPa), whereas that of CSA-2 and 3 exceeded 10 MPa. The highest compressive strength was measured (i.e., CSA-3 in Figure 3) as the amount of AG reached 300 g. The reasons related to the foregoing were analyzed as follows: (a) the CSA produced ettringite when sufficient calcium hydroxide and sulfate existed during hydration [47–49]; (b) the compressive strength of CSA-2 and 3 was accelerated due to the supply of sufficient sulfate, facilitating the production of ettringite; (c) specimens CSA-4 and 5 contained a smaller amount of CSA compared with specimens CSA-2 and 3; consequently, the strength increase in CSA-4 and 5 was lower than that of CSA-2 and 3; (d) AG accelerated the hydration of CSA.

The measured compressive strength of the CSA-based binder with OPC (i.e., CSA-6~11 in Figure 3) was larger than the target compressive strength of 10 MPa, except for CSA-6. Overall, the compressive strength of the CSA-based binders with OPC was larger than that of CSA-based binders without OPC. This was because OPC can accelerate the hydration of CSA-based binders through calcium hydroxide ($Ca(OH)_2$), which is induced by OPC. However, due to the short curing time of 4 h, the observed compressive strength of the binders could potentially have been influenced by the free lime content of OPC rather than the hydration products. The highest compressive strength was measured (i.e., CSA-9 in Figure 3) when the amount of CSA, AG, and OPC was 490 g, 210 g, and 300 g, respectively. Subsequently, the compressive strength reduced as the amount of CSA and OPC decreased and increased, respectively. Pelletier et al. [48] stated that CSA was the main component to increase the compressive strength of the binders and an increased amount of OPC could lead to a reduction in compressive strength due to the increase in the porosity of the binders. In particular, their analysis showed that if the OPC content was high and CSA content was low, the production of ettringite and microcrystalline aluminum hydroxide, which are considerably related to the compressive strength of binders, could be reduced. In this study, the mix proportion of CSA-9, shown in Figure 3, was the most adequate for the CLSM development, because the performance of CSA-9 achieved the goal of a quick hardening binder with the highest and early-stage compressive strength.

## 4. Development of CLSM

In general, the mix design of CLSMs has previously been established by engineers based on experience or the trial-and-error method. Mostly, they were designed to achieve a sufficiently low compressive strength (i.e., less than 2.1 MPa) and high flowability (i.e., more than 200 mm spread) to allow re-excavation in the future, as well as to provide sufficient backfilling and void-filling [9,45,50–52].

### 4.1. Compressive Strength, Flowability, and Bleeding

In this section, the optimal mixture ratio between the binder and fly ash was investigated in order to develop the CLSM, satisfying the target fluidity, strength, and bleeding.

Ling et al. [9] reported that CLSMs typically consist of cement (Type 1 Portland), fly ash (Class C or F), a fine aggregate, and water. In this study, fly ash was selected as the filler, and two types of additives were supplemented to reduce the settlement of the mixtures. These materials were Na–montmorillonite as the settlement inhibitor and potassium carbonate ($K_2CO_3$) as the accelerator. As mentioned in Section 3, CSA-9 in Figure 3 was selected as the binder to develop the new CLSM. The water–material ratio was set to be 1:1 (i.e., W/M is equal to 100%).

Figure 4 shows the compressive strength at 4 h, 8 h, 1 d, 3 d, 7 d, and 28 d, the flowability after mixing, and bleeding at 2 h of each CLSM specimen. Detailed information related to the composition of the CLSM is shown in Table 5. The compressive strength ($q_c$) typically increased as the proportion of the binder increased, and the $q_c$ of the entire specimen satisfied the target strength (i.e., 28-day compressive strength had to be less than 1.0 MPa in Table 1). The $q_c$ of CLSM-5 and CLSM-1 showed the highest and lowest compressive strengths (i.e., 0.784 MPa and 0.099 MPa) because the proportion of binder was the highest and lowest, respectively (Figure 4a). As shown in Figure 4b, the flowability of the specimens satisfied the target value (i.e., more than 200 mm spread in Table 1). The flowability increased as the amount of binder reduced. Therefore, the flowability of CLSM-1 and CLSM-5 was the highest and lowest, respectively, contrary to the results of the compressive strength.

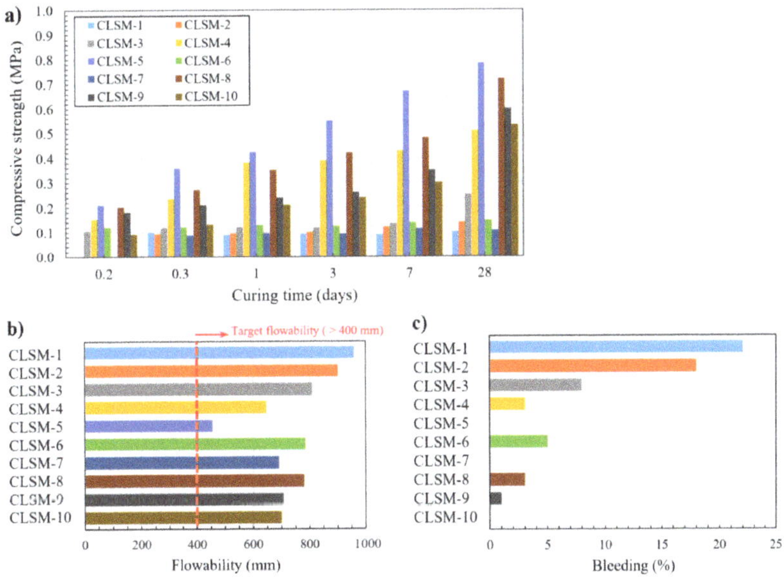


**Figure 4.** (**a**) compressive strength, (**b**) flowability and (**c**) bleeding for each developed CLSM specimen.

Bleeding was generated in seven specimens (i.e., CLSM-1, 2, 3, 4, 6, 8, and 9). The bleeding of CLSM-1 and CLSM-2 was approximately 20% (Figure 4c). This suggested the following: (1) bleeding can increase as the amount of binder decreases; (2) the binder content of more than 10% in CLSMs can prevent segregation caused by a high water content.

Overall, CLSM-10 exhibited the best performance based on laboratory tests. This specimen satisfied the target performance established in this study (Table 1). It showed a high degree of flowability (700 mm), no bleeding (0%), and an adequate 28-day compressive strength (0.533 MPa). The compressive strength of CLSM-5 satisfied the target strength, but the flowability of CLSM-5 was less than that of CLSM-10, even though the bleeding of CLSM-5 was 0%. The flowability and bleeding of CLSM-7 satisfied the target performance, but the compressive strength of CLSM-7 was only 0.2 times that of CLSM-10. The results indicated that fillers composed of small particles (e.g., fly ash) and the quick-hardening binder could prevent bleeding in spite of a high water content (i.e., water content was more than 100%). Furthermore, the incorporation of finer particles had the potential to reduce the required binder content and/or increase the water content within the CLSM mixture. This could increase the flowability of the CLSM and decrease the bleeding, as well as secure a sufficient compressive strength.

**Table 5.** The composition of CLSM specimens.

| Specimen | Composition (g) | | | | |
|---|---|---|---|---|---|
| | Binder | Fly Ash | Additive I | Additive II | Water |
| CLSM-1 | 50 | 950 | - | - | 1000 |
| CLSM-2 | 75 | 925 | - | - | |
| CLSM-3 | 100 | 900 | - | - | |
| CLSM-4 | 125 | 875 | - | - | |
| CLSM-5 | 150 | 850 | - | - | |
| CLSM-6 | 98 | 900 | 2 | - | |
| CLSM-7 | 96 | 900 | 4 | - | |
| CLSM-8 | 98 | 900 | - | 2 | |
| CLSM-9 | 96 | 900 | 2 | 2 | |
| CLSM-10 | 96 | 900 | 3 | 1 | |

### 4.2. Sensitivity Analysis

The compressive strength, flowability, and bleeding of the CLSM were measured depending on the water–material ratio (W/M). In Figure 5, specimen B is CLSM-10. The W/M was changed by varying the amount of water, whereas the amount of the CLSM was fixed. The compressive strength decreased and the flowability and bleeding increased as the proportion of water increased. In addition, the compressive strength and flowability of specimens A and B were similar, and bleeding was equal to 0%. This indicated that the CLSM could be produced by changing the W/M from 90 to 100% depending on the objectives of construction and the geological properties at the sites of interest.

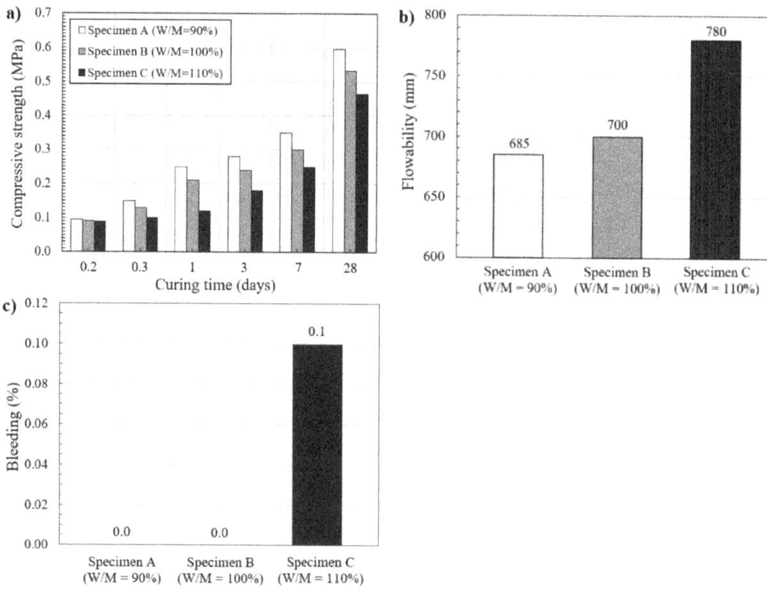


**Figure 5.** The results of sensitivity analysis for the developed CLSM specimens depending on the ratio of water–material; (**a**) compressive strength for each specimen; (**b**) flow for each specimen; (**c**) bleeding for each specimen.

### 4.3. Geotechnical Strength Parameters of the Developed CLSM

Table 6 shows the geotechnical strength parameters (i.e., friction angle, cohesion, and total unit weight) of the CLSM developed in this study (i.e., CLSM-10 in Figure 4). The friction angle and cohesion were determined through a triaxial compression test, and the total unit weight was measured based on ASTM D 6023 [38]. These parameters are planned to be used for future studies (i.e., numerical analysis).

**Table 6.** The results of triaxial compression test for CLSM developed in this study.

| $\phi$ (°) | Cu (kPa) | $\gamma_t$ (kN/m$^3$) |
|---|---|---|
| 12.4 | 69.3 | 13.8 |

## 5. Comparison between New and Predeveloped CLSMs

In this section, the performance of the CLSM developed in this study (i.e., CLSM-10 in Figure 4) was compared with that of predeveloped CLSMs by conducting additional laboratory tests in order to measure their flowability and compressive strength. The predeveloped CLSMs were based on sandy and clayey soils and bottom ash. Based on the unified soil classification system (USCS), each soil was classified into silty sand (SM) and inorganic silt with low-to-medium compressibility (ML). They were collected at a construction site in Boryeng-si, South Korea. The detailed geotechnical characteristics and particle size distribution of this soil are shown in Figure S3 and Table S1, respectively. The bottom ash was collected at a power plant in Pohang-si, South Korea. The chemical properties of the bottom ash are listed in Table S2.

In order to investigate and compare the flowability and compressive strength of the predeveloped CLSMs with those of the new CLSM, the specimens were manufactured based on previous studies by Cho [17] and Saman [52] on soils and Lee [53] on bottom ash (Figure 6).

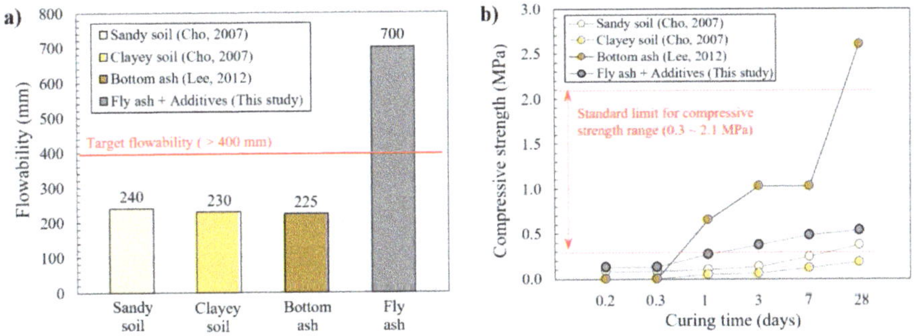


**Figure 6.** Comparison of (**a**) flowability and (**b**) unconfined compressive strength between CLSM developed in this study and predeveloped CLSMs [17,53].

The CLSM developed in this study showed the highest flowability (Figure 6a). This indicated that injecting the developed CLSM into desired locations near underground pipes or culverts could considerably reduce costs. As shown in Figure 6b, the 28-day compressive strength of the CLSM developed in this study was within the standard limit (i.e., from 0.3 to 2.1 MPa). Overall, the developed CLSM possessed a high flowability, which could facilitate placement and re-excavation, as well as provide sufficient compressive strength to support embedded structures, such as pipes, culverts, paved roads, etc.

Even though the new CLSM developed in this study showed good performance, future studies are needed to analyze several uncertainties as follows: (a) due to its exceptional flowability (i.e., the flowability was 700 mm), the precise endpoint and total quantity and installation timeframe of the CLSM injection in fields should be investigated; (b) in terms of the soundness (i.e., volume stability) of the CLSM specimens, it is less likely that the soundness of the binder would be a major concern, because the microstructure of the CLSM was much looser compared to that of concrete and the proportion of binder was relatively small. However, the soundness of the CLSM should be analyzed based on the acceleration test method proposed by Mehta [54] and Kabir [55] for field applicability.

## 6. Conclusions

In this study, a series of laboratory tests were conducted to measure the flowability, density, bleeding, and compressive strength of controlled low-strength materials (CLSMs) to develop a new CLSM that could secure high flowability and low segregation, as well as a quick setting time in order to reduce the inconvenience of citizens in metropolitan cities. In addition, the performance of the developed CLSM was compared with that of three different predeveloped CLSMs. The following conclusions were drawn from the analysis results.

The optimal binder was developed by mixing calcium sulfoaluminate (CSA), anhydrous gypsum (AG), ordinary Portland cement (OPC), and water. As the amounts of CSA and OPC increased, the compressive strength of the binder increased. Eventually, the optimal proportion of binder was determined to be 0.25:0.11:0.15:0.5 in the order CSA, AG, OPC, and water, respectively.

The compressive strength, flowability, and bleeding were measured based on the developed binder, fly ash, and two types of additives. The compressive strength typically increased with the binder's proportion. In contrast, the flowability and bleeding increased as the binder's proportion decreased. Overall, most specimens satisfied the target values. A new CLSM developed in this study showed exceptional flowability (i.e., 700 mm) as well as no bleeding and a suitable 28-day compressive strength. In addition, the CLSM could be manufactured by selecting a W/M in the range from 90 to 100%, depending on the objectives of construction and the geological properties at the sites of interest.

It is anticipated that the developed CLSM could significantly reduce the costs related to the disposal of old pavements, the installation of new pavements, and other construction expenses compared to the costs related to the conventional method, even though the expenses for backfill materials could increase due to the higher production costs for CLSMs than soil. In addition, there is a need to investigate its field applicability in order to evaluate the precise costs, maintenance, and long-term stabilities after installation.

**Supplementary Materials:** The following supporting information can be downloaded at: https://www.mdpi.com/article/10.3390/app13169377/s1.

**Author Contributions:** Conceptualization, J.H. and Y.J.; Methodology, J.H. and B.K.; Validation, B.K.; Investigation: performed the experiments, J.H.; Investigation: data/evidence collection, Y.J. and Y.K.; Writing—original draft preparation, J.H. and Y.J.; Writing—review and editing, Y.K. and B.K.; Supervision, B.K.; Funding acquisition, J.H. All authors have read and agreed to the published version of the manuscript.

**Funding:** This work was supported by the Korea Agency for Infrastructure Technology Advancement (KAIA) grant funded by the Ministry of Land, Infrastructure and Transport (grant RS-2021-KA162543).

**Institutional Review Board Statement:** Not applicable.

**Informed Consent Statement:** Not applicable.

**Data Availability Statement:** Not applicable.

**Conflicts of Interest:** The authors declare no conflict of interest.

## References

1. Jo, Y.S.; Cho, S.H.; Jang, Y.S. Field investigation and analysis of ground sinking development in a metropolitan city, Seoul, Korea. *Environ. Earth Sci.* **2016**, *75*, 1353. [CrossRef]
2. Lee, N.K.; Kim, H.K.; Park, I.S.; Lee, H.K. Alkali-activated, cementless, controlled low-strength materials (CLSM) utilizing industrial by-products. *Construct. Build. Mater.* **2013**, *49*, 738–746. [CrossRef]

3. Kaliyavardhan, S.K.; Ling, T.C.; Guo, M.Z.; Mo, K.H. Waste resources recycling in controlled low-strength material (CLSM) : A critical review on plastic properties. *J. Environ. Manag.* **2019**, *241*, 383–396. [CrossRef]
4. Fauzi, M.A.; Arshad, M.F.; Nor, N.M. Statistical models to develop optimised controlled low-strength materials with wastepaper sludge ash. *Constr. Build. Mater.* **2021**, *286*, 122816. [CrossRef]
5. Dalal, P.H.; Patil, M.; Dave, T.N.; Iyer, K.K.R. An experimental study on controlled low-strength material (CLSM) for utilization as sustainable backfill. *Mater. Today Proc.* **2022**, *65*, 1178–1185. [CrossRef]
6. Nataraja, M.; Nalanada, Y. Performance of industrial byproducts in controlled low-strength materials (CLSM). *Waste Manag.* **2008**, *28*, 1168–1181. [CrossRef] [PubMed]
7. Dev, K.; Robinson, R. Pond ash-based controlled low-strength flowable fills for geotechnical engineering applications. *Int. J. Geosynth. Ground Eng.* **2015**, *1*, 1–13. [CrossRef]
8. *ACI 229R-13*; Report on Controlled Low-Strength Materials, ACI Committee 229. American Concrete Institute (ACI): Farmington Hills, MI, USA, 2013.
9. Ling, T.C.; Kaliyavaradhan, S.K.; Poon, C.S. Global perspective on application of controlled low-strength material (CLSM) for trench backfilling: An overview. *Constr. Build. Mater.* **2018**, *158*, 535–548. [CrossRef]
10. Tikalsky, P.; Gaffney, M.; Regan, R. Properties of controlled low-strength material containing foundry sand. *ACI Mater. J.* **2000**, *97*, 698–702.
11. Du, L.; Folliard, K.J.; Trejo, D. Effects of constituent materials and quantities on water demand and compressive strength of controlled low-strength material. *J. Mater. Civ. Eng.* **2002**, *14*, 485–495. [CrossRef]
12. Hitch, J.L.; Howard, A.K.; Baas, W.P. *Innovations in Controlled Low-Strength Material (Flowable Fill)*; ASTM STP 1459; ASTM International: West Conshohocken, PA, USA, 2002.
13. Lee, Y.S. Recycling of Waste Coal Ash for Controlled low-Strength Materials. Master's Thesis, Hanyang University, Seoul, Korea, 2002. (In Korean)
14. Pierce, C.E.; Tripathi, H.; Brown, T.W. Cement kiln dust in controlled low-strength materials. *ACI Mater. J.* **2003**, *100*, 455–462.
15. Choi, N.H. Analysis of Behavior for Underground Pipe Using Controlled Low Strength Material with Field Soil. Master's Thesis, Hanyang University, Seoul, Korea, 2004. (In Korean).
16. Naik, T.R.; Kraus, R.N.; Ramme, B.W.; Chun, Y.M.; Kumar, R. High-carbon fly ash in manufacturing conductive CLSM and concrete. *J. Mater. Civ. Eng.* **2006**, *18*, 743–746. [CrossRef]
17. Cho, D.H. Reuse of Surplus Soil by Rapid-Setting Properties of Liquefied Stabilized Soil. Ph.D. Dissertation, Joongang University, Seoul, Korea, 2007. (In Korean)
18. Taha, R.A.; Alnuaimi, A.S.; Al-Jabri, K.S.; Al-Harthy, A.S. Evaluation of controlled low-strength materials containing industrial by-products. *Build. Environ.* **2007**, *42*, 3366–3372. [CrossRef]
19. Lachemi, M.; Sßahmaran, M.; Hossain, K.M.A.; Lotfy, A.; Shehata, M. Properties of controlled low-strength materials incorporating cement kiln dust and slag. *Cem. Concr. Compos.* **2010**, *32*, 623–629. [CrossRef]
20. Horiguchi, T.; Fujita, R.; Shimura, K. Applicability of controlled low-strength materials with incinerated sewage sludge ash and crushed-stone powder. *J. Mater. Civ. Eng.* **2011**, *23*, 767–771. [CrossRef]
21. Naganathan, S.; Razak, H.A.; Hamid, S.N.A. Properties of controlled low-strength material made using industrial waste incineration bottom ash and quarry dust. *Mater. Des.* **2012**, *33*, 56–63. [CrossRef]
22. Wang, H.; Chen, B.; Wu, Y. A study of the fresh properties of controlled low-strength rubber lightweight aggregate concrete (CLSRLC). *Construct. Build. Mater.* **2013**, *41*, 526–531. [CrossRef]
23. Do, T.M.; Kim, Y. Engineering properties of controlled low-strength material (CLSM) incorporating red mud. *Int. J. Geo-Eng.* **2016**, *7*, 7. [CrossRef]
24. Wu, H.; Huang, B.; Shu, X.; Yin, J. Utilization of solid wastes/byproducts from paper mills in controlled low-strength material (CLSM). *Construct. Build. Mater.* **2016**, *118*, 155–163. [CrossRef]
25. Chompoorat, T.; Likitlersuang, S.; Jongvivatsakul, P. The performance of controlled low-strength material base supporting a high-volume asphalt pavement. *KSCE J. Civ. Eng.* **2018**, *22*, 2055–2063. [CrossRef]
26. Lin, W.T.; Weng, T.L.; Cheng, A.; Chao, S.J.; Hsu, H. Properties of controlled low-strength material with circulating fluidized bed combustion ash and recycled aggregates. *Materials* **2018**, *5*, 715. [CrossRef]
27. Fang, X.; Lei, W.; Poon, C.S.; Baek, K.; Tsang, D.C.W.; Kwok, S.K. Transforming waterworks sludge into controlled low-strength material: Bench-scale optimization and field test validation. *J. Environ. Manag.* **2019**, *232*, 254–263. [CrossRef]
28. Okuyucu, O.; Jayawickrama, P.; Senadheera, S. Mechanical properties of steel fiber–reinforced self-consolidating controlled low-strength material for pavement base layers. *J. Mater. Civ. Eng.* **2019**, *31*, 4019177. [CrossRef]
29. Kim, Y.; Dinh, B.H.; Do, T.M.; Kang, G. Development of thermally enhanced controlled low-strength material incorporating different types of steel-making slag for ground-source heat pump system. *Renew. Energy* **2020**, *150*, 116–127. [CrossRef]
30. Razak, H.A.; Naganathan, S.; Hamid, S.N.A. Performance appraisal of industrial waste incineration bottom ash as controlled low-strength material. *J. Hazard. Mater.* **2009**, *172*, 862–867. [CrossRef] [PubMed]
31. Kuo, W.T.; Wang, H.Y.; Shu, C.Y.; Su, D.S. Engineering properties of controlled low-strength materials containing waste oyster shells. *Construct. Build. Mater.* **2019**, *46*, 128–133. [CrossRef]
32. Mahamaya, M.; Jain, S.; Das, S.K.; Paul, R. Engineering properties of cementless alkali activated CLSM using ferrochrome slag. *J. Mater. Civ. Eng.* **2023**, *35*, 04022441. [CrossRef]

33. Li, Y.F.; Hsu, Y.W.; Syu, J.Y.; Chen, B.Y.; Song, B. Study on the utilization of waste thermoset glass fiber-reinforced polymer in normal strength concrete and controlled low strength material. *Materials* **2023**, *16*, 3552. [CrossRef] [PubMed]
34. *ASTM STP 1331*; The Design and Application of Controlled Low-Strength Materials (Flowable Fill). ASTM International: West Conshohocken, PA, USA, 1998.
35. *ASTM D 6103*; Standard Test Method for Flow Consistency of Controlled Low-Strength Material (CLSM). ASTM International: West Conshohocken, PA, USA, 2013.
36. *ASTM C 109*; Standard Test Method for Compressive Strength of Hydraulic Cement Mortars (Using 2-in or 50-mm Cube Specimens). ASTM International: West Conshohocken, PA, USA, 2020.
37. *ASTM D 4832*; Standard Test Method for Preparation and Testing of Controlled Low-Strength Material (CLSM) Test Cylinders. ASTM International: West Conshohocken, PA, USA, 2018.
38. *ASTM D 6023*; Standard Test Method for Density (Unit Weight), Yield, Cement Content, and Air Content (Gravimetric) of Controlled Low-Strength Material (CLSM). ASTM International: West Conshohocken, PA, USA, 2016.
39. Schmitz, M.E.; Parsons, R.L.; Ramirez, G.; Zhao, Y. *Use of Controlled Low-Strength Material as Abutment Backfill*; University of Kansas: Lawrence, Kansas, 2004.
40. Abelleira, A.; Barke, N.S.; Pickering, D.G. Corrosion activity of steel in cementitious controlled low-strength materials vs. that in soil. In *The Design and Application of Controlled Low-Strength Materials (Flowable Fill)*; STM STP 1331; Howard, A.K., Hitch, J.L., Eds.; American Society for Testing and Materials: West Conshohocken, PA, USA, 1998.
41. Dockter, B.A. Comparison of dry scrubber and class c fly ash in controlled low-strength materials (CLSM) applications. In *The Design and Application of Controlled Low-Strength Materials (Flowable Fill)*; ASTM STP 1331; Howard, A.K., Hitch, J.L., Eds.; American Society for Testing and Materials: West Conshohocken, PA, USA, 1998; pp. 13–26.
42. Kaneshiro, J.; Navin, S.; Wendel, L.; Snowden, H. Controlled low-strength material for pipeline backfill—Specifications, case histories and lessons learned. In Proceedings of the Pipelines 2001: Advances in Pipelines Engineering and Construction, Pipeline Division Specialty Conference, ASCE, San Diego, CA, USA, 15–18 July 2001; pp. 1–13.
43. *ASTM C 940*; Standard Test Method for Expansion and Bleeding of Freshly Mixed Grouts for Preplaced-Aggregate Concrete in the Laboratory. ASTM International: West Conshohocken, PA, USA, 2022.
44. Hwang, C.L.; Shen, D.H. The effects of blast-furnace slag and fly ash on the hydration of Portland cement. *Cem. Concr. Res.* **1991**, *21*, 410–425. [CrossRef]
45. Trejo, D.; Folliard, K.J.; Du, L. Sustainable development using controlled low-strength material. In Proceedings of the International Workshop on Sustainable Development and Concrete Technology, Beijing, China, 20–21 May 2004; pp. 231–250.
46. Siddique, R.; Noumowe, A. Utilization of spent foundry sand in controlled low-strength materials and concrete. *Resour. Conserv. Recy.* **2008**, *53*, 27–35. [CrossRef]
47. Trauchessec, R.; Mechling, J.M.; Lecornte, A.; Roux, A.; Le Rolland, B. Hydration of ordinary Portland cement and calcium sulfoaluminate cement blends. *Cem. Concr. Compos.* **2015**, *56*, 106–114. [CrossRef]
48. Pelletier, L.; Winnefeld, F.; Lothenbach, B. The ternary system Portland cement-calcium sulphoaluminate clinker-anhydrite: Hydration mechanism and mortar properties. *Cem. Concr. Compos.* **2010**, *32*, 497–507. [CrossRef]
49. Hanic, F.; Kapralik, I.; Gabrisova, A. Mechanism of hydration reactions in the system $C_4A_3S$-CS-CaO-$H_2O$ referred to hydration of sulphoaluminate cement. *Cem. Concr. Res.* **1989**, *19*, 671–682. [CrossRef]
50. Naik, T.R.; Singh, S.S. Flowable slurry containing foundry sands. *J. Mater. Civ. Eng.* **1997**, *9*, 93–102. [CrossRef]
51. NRMCA. *Guide Specification for Controlled Low-Strength Materials (CLSM)*; Report of National Ready Mixed Concrete Association; Specification Guide: Alexandria, VA, USA, 2006.
52. Saman. *Practical Use of Self-Compacting Liquifies Stabilized Method with Surplus Soil*; Construction and Transportation R&D Final Report, No. C15; Saman Corporation: Gwacheon, Korea, 2008. (In Korean)
53. Lee, H.J. Compressive Strength Characteristics of Lightweight Foamed Controlled Low-Strength Material (CLSM) Using Coal Ash. Master's Thesis, Hanyang University, Seoul, Korea, 2012. (In Korean)
54. Mehta, P.K. History and status of performance tests for evaluation of the soundness of cement. In *Cement Standards Evolution and Trends*; ASTM International: West Conshohocken, PA, USA, 1978.
55. Kabir, H.; Hooton, R.D.; Popoff, N.J. Evaluation of cement soundness using the ASTM C151 autoclave expansion test. *Cem. Concr. Res.* **2020**, *136*, 106159. [CrossRef]

*applied sciences*

*Review*

# A Comprehensive Review of Soil Remolding Toughness Determination and Its Use in the Classification of Fine-Grained Soils

**Brendan C. O'Kelly [1,*], Jacinto Alonso-Azcárate [2] and José Manuel Moreno-Maroto [3]**

1 Department of Civil, Structural and Environmental Engineering, School of Engineering, Trinity College Dublin, D02 PN40 Dublin, Ireland
2 Department of Physical Chemistry, Faculty of Environmental Sciences and Biochemistry, University of Castilla-La Mancha, Avenida Carlos III, s/n, 45071 Toledo, Spain
3 Department of Geology and Geochemistry, Faculty of Sciences, Autonomous University of Madrid, Cantoblanco, 28049 Madrid, Spain
* Correspondence: bokelly@tcd.ie

**Abstract:** The remolding toughness property of fine-grained soil has not been investigated that much, mainly because it has not lent easily to direct measurement, with soil toughness usually qualitatively described. In practical terms, as the plastic limit $w_P$ is approached, tougher soils require greater rolling effort during the $w_P$ test, such that plasticity and toughness properties can be used to distinguish those plastic soils having greater deformation resistance for various field applications. This state-of-the-art review paper presents a critical appraisal of soil remolding toughness determination and its limited use, to date, in the classification of fine-grained soils. The recent developments reviewed and critically assessed include mechanical thread rolling for nominal toughness measurement during the $w_P$ rolling-out procedure, various extrusion approaches, and proposed correlations between toughness and the plasticity index to liquid limit ratio. From statistical analysis of previously reported toughness–consistency limits data, some new correlations are introduced in the present paper. Soil classification using the traditional Casagrande plasticity chart is not entirely accurate for certain soil types in that one can observe soils that present high toughness (something typical of *clay*) being incorrectly classified as *silt* soil. From this perspective, a new toughness chart is introduced to augment (or for use instead of) the Casagrande plasticity chart in obtaining more reliable soil classification. This paper concludes with recommendations on future research efforts for routinely obtaining soil toughness measurements.

**Keywords:** extrusion; plastic limit; soil classification; soil plasticity; toughness; workability

**Citation:** O'Kelly, B.C.; Alonso-Azcárate, J.; Moreno-Maroto, J.M. A Comprehensive Review of Soil Remolding Toughness Determination and Its Use in the Classification of Fine-Grained Soils. *Appl. Sci.* **2023**, *13*, 5711. https://doi.org/10.3390/app13095711

Academic Editor: Tiago Miranda

Received: 3 March 2023
Revised: 23 April 2023
Accepted: 28 April 2023
Published: 5 May 2023



## 1. Introduction

Plasticity can be defined as the material property that enables significant deformation to occur without fracture when sufficiently high stresses are applied (i.e., exceeding the material's yield stress) and the ability to retain the deformed shape on removal of those stresses. For soils, and specifically fine-grained soils, plasticity is directly related to the platy habit exhibited by most clay mineral particles that favors their sliding with respect to each other when enough moisture is present (i.e., for water content ($w$) values within the plastic range). The plasticity of these materials is commonly assessed using the manual "rolling of threads" method, with the plastic limit ($w_P$), originally described by Atterberg [1,2], defined as the water content at which the soil threads just begin to crumble (cracking longitudinally and transversely) during the standardized thread rolling (by hand) test [3,4]. In other words, the $w_P$ defines the ductile–brittle states transition. Attempts to improve on the repeatability and reproducibility of the standard hand rolling $w_P$ test (by minimizing the uncertainties associated with the rolling out (by hand) procedure; i.e., rate of rolling, the hand pressure, and/or the initial and final thread diameter criteria) include the thread rolling device technique proposed by Bobrowski and Griekspoor [5]. Their method, which follows the same basic principles as the standard thread rolling (by hand) test, has been standardized and presented as an alternative method for $w_P$ determination in ASTM D4318 [3] and AASHTO T90 [6]. A comprehensive statistical analysis performed on a database of 60 diverse fine-grained soils demonstrated that these hand and device rolling methods produce essentially similar $w_P$ values for a given fine-grained soil tested [7].

For fine-grained soil (and mainly clays), remolding toughness ($T$) represents the mechanical resistance offered when its $w_P$ is about to be reached [8] for reducing water content in the plastic range. For instance, as the $w_P$ is approached, tougher soils require noticeably greater rolling effort for reducing the soil thread diameter from a starting 6 mm dia. down to approx. 3 mm dia. at the crumbling condition [3,9]. Together, the plasticity and toughness properties are used to identify soils that provide greater resistance to deformation while retaining plastic behavior for use in various field applications. For instance, soil remolding toughness is important for the choice and performance of field compaction machinery in earthwork construction and it influences the integrity of earthwork components, e.g., the core of earthen dams and clay liners in water-retaining structures and landfills [8].

Greater plasticity and toughness arise from increases in the quantity of expanding lattice-type clay minerals with greater specific surface area (SSA) and cation exchange, and for decreases in the ion concentration of the pore fluid and the valency of exchangeable cations present [10]. Fundamentally, within the plastic range (of water contents), saturated clayey materials develop suction, such that active/high plasticity clays (characterized by the highest air-entry values) have the highest toughness. For instance, at (just above) the $w_P$, silty, sandy, and peaty soils present slight toughness, whereas low-to-medium plasticity soils (e.g., silty clays) display medium toughness, with active clays exhibiting high toughness [11–13]. As stated on page 805 of the paper by Casagrande [14], "*the higher the position of a soil above the A-line on the plasticity chart (CL, CH), the stiffer are the soil threads near the plastic limit*". Hence, the concepts of plasticity and toughness could be considered synonymous from a practical standpoint. Note that another term often used to denote toughness is workability [8,9,13], perhaps of more importance for the ceramics and brick-making industries and in an agricultural context [8].

Atterberg [1,2] identified the water content range over which soil behaves in a plastic manner using the plasticity index $I_P$ (= $w_L - w_P$), where $w_L$ is the liquid limit water content. These index parameters are ubiquitously used for fine-grained soil classification, typically employing the Casagrande plasticity chart that plots $I_P$ against $w_L$, with the test soil classified as *clay* or *silt* material depending on its position relative to the A-line demarcation boundary in the $I_P$ versus $w_L$ plot [14–16]. However, the size of the water content range over which the soil behaves plastically (i.e., given by its $I_P$) or the upper limit of that range (i.e., $w_L$) are not, by themselves, satisfactory measures of soil toughness. In other words, soil toughness is a more fundamental plasticity property, typically quantified as the amount of work undertaken per unit soil volume necessary to change shape, remold, or, as in the case of the standard $w_P$ test, roll out uniform soil cylinders (threads) applying sufficient finger pressure during the repeated traverses [8,17,18].

The literature presents only a handful of cases employing empirical classifications of soil remolding toughness (these are based on qualitative or subjective assessments/descriptions of the observed soil behavior). A few cases reported in the literature purport to provide semi-quantitative assessments of soil toughness, and apparently, only one reported apparatus and test method [8,17,18] presently exists to provide direct toughness measurements. Neither has the property of soil remolding toughness been used that much in soil classification systems, with all of these points reflected by the sparsity of published work on the subject matter compared to, for instance, the voluminous reported work on the consistency limits (i.e., $w_L$ and $w_P$) determination. This includes improved knowledge and understanding of the fall cone and Casagrande $w_L$ [19–21] and Atterberg $w_P$ [8,17,18,22] state transitions and their experimental determination, methods of relating Casagrande cup and fall cone $w_L$ for a given fine-grained soil investigated [19,23,24], and various alternative proposals for soil plasticity ($w_P$) determination, mostly based on fall cone approaches. The latter includes using a modified fall cone setup with an enhanced cone mass and incorporating a drop height [25,26] (such that the falling cone hits the surface of the soil test specimen with an impact velocity) for the determination of the "plastic strength limit" $PL_{100}$ parameter [19,22,27,28] and various investigations of the fall cone flow curve for soil plasticity determination [29,30]. Significant recent developments in fine-grained soil classification systems based on plasticity include the proposals of Moreno-Maroto and Alonso-Azcárate [11,31] and Vardanega et al. [32]. Moreno-Maroto and Alonso-Azcárate [11,31] developed their proposal according to objective criteria based on the quantitative measurement of properties, such as soil toughness, the bending capacity of formed soil threads, and the observation of adhesive consistency, whereas the Vardanega et al. [32] system requires only fall cone $w_L$ test results to achieve the classification of fine-grained soils to an acceptable degree of accuracy, i.e., the approach does not require $w_P$ measurement, removing the dependence on the thread rolling $w_P$ test that can have high operator variability [25,33]. There is also greater awareness of the limitations of standard/conventional consistency limits test methods and fine-grained soil classification systems for investigations on unconventional soils, including (fibrous) peats [34] and diatomaceous soil [35]. In other words, the measured consistency limits of these soils do not provide reliable information on the soils' likely geotechnical properties as they are conventionally expected to do.

On the subject topic of soil toughness, Casagrande [36] introduced an index of toughness at the $w_P$ (denoted herein as $I_{T(C)}$), defined as the logarithm of the undrained shear strength ($s_u$) ratio determined for the $w_P$-to-$w_L$ water contents. Casagrande [36] also proposed that the value of $I_{T(C)}$ could be determined as the ratio of $I_P$ to the flow index ($I_F$), with $I_F$ determined as the gradient of the $w$ versus $\text{Log}_{10}\, N$ plot obtained from percussion cup test data for $w_L$ determination. However, this approach for $I_{T(C)}$ determination is not reliable as it assumes a linear $\text{Log}_{10}\, s_u$ versus $w$ correlation applies over the full plastic range of fine-grained soil, which is generally not the case in practice, instead often exhibiting highly nonlinear behavior [19,37–39]. Since the early work of Casagrande [36], the property of soil toughness has not been investigated that much in geotechnical engineering, mostly because it does not lend itself easily to direct measurement. In a follow-up paper, Casagrande [14] gave a qualitative classification of soil toughness considering the "cohesiveness" near the $w_P$, using the descriptive terms of "very weak", "weak", "firm", "medium tough", "tough", and "very tough". The ASTM visual–tactile field tests [40] include subjective assessments of the finger pressure required in rolling out uniform soil threads and the stiffness of a soil lump formed from amassing those threads, assigning descriptive terms of "non-plastic", and low, medium, and high toughness accordingly. The US Department of Agriculture Soil Survey Manual [41] considers the relative force required to form 3 mm dia. soil threads at a water content near or at the $w_P$. In other words, soils are classed as having low, medium, and high toughness for the exertion of <8 N, 8–20 N, or >20 N of force applied by the operator's finger pressure during the repeated traverses [41] because when testing tougher soils, such as those with a high montmorillonite content, greater force is required to roll out and cause a reduction in diameter of the soil threads prepared at the $w_P$ water content [9,13]. Various attempts have also been presented for soil toughness classification employing the Casagrande plasticity chart, with different zones or classifications of toughness at the $w_P$ identified in the chart. For instance, the Naval Facilities Engineering Command [42] uses the qualitative terms "slight", "slight to medium", "medium", and "high" when identifying toughness zones for soils plotting above and below the A-line boundary marked in the Casagrande plasticity chart.

Some quasi-toughness tests have been developed specifically for the ceramics industry, with various investigations performed to assess soil remolding toughness. These include a novel proposal by Astbury et al. [43], who defined soil toughness considering the amount of energy absorbed by the test specimen during one cycle of a cyclic torsion test and which was quantified as the area within the experimental stress–strain hysteresis loop. The ASTM

C181-11 standard [44] presents a test to determine a "workability index" for fireclays and high alumina refractory plastics by measuring the plastic deformation of a molded test specimen when subjected to impact. This approach is comparatively similar in function and outcome to the moisture condition apparatus and test [45] that are used to assess the suitability of mainly cohesive soils for incorporation into earthworks. With tougher soil lumps being more resistant to remolding, thereby requiring more compaction energy and giving higher moisture condition values (MCVs), Barnes [8] suggested that the moisture condition test could potentially be used to assess soil toughness.

Fitzjohn and Worral [46] performed extrusion tests on brick clays, measuring their extrusion rates as a function of the applied extrusion pressure ($p_e$) for different clay contents. Subsequently, CERAM Research [47,48] developed a ceramic rheology tester, termed the Martin flow instrument (MFI), which is a modified version of the melt flow indexer used for testing the flow properties of plastics and polymers. For the MFI, a standard length of cored soil sample is inserted into a cylindrical barrel that contains a die at the base. A piston placed inside the barrel applies pressure via a deadweight load to cause the extrusion of the soil to occur through the die orifice. The time period required for a known soil volume to flow through the die orifice is measured, with the flow time (extrusion rate) dependent on the soil toughness, with tougher materials taking longer [47]. The soil sample, from soft earthenware clays to some advanced ceramic materials, is tested within a preferred shear rate range (relevant to the subsequent processing of the material) by employing an appropriate deadweight in the range of 2–30 kg applied to the piston [47].

In soil mechanics, the extrusion approach has also been investigated for the testing of fine-grained soils to possibly assess/measure their toughness [12,13], including proposed power-based extrusion approaches for consistency limits determination [13,49]. These aspects will be discussed in detail later in the paper. Closely replicating Atterberg's hand rolling of threads method, Barnes [8,17,18] developed a mechanical thread rolling apparatus and test that allow nominal applied stress and diametrical strain measurements for a uniform soil cylinder (thread) during the rolling out procedure. The toughness-related soil properties and the value of $w_P$ can be determined from the obtained experimental $T$–$w$ plots (associated methods are elaborated later in Section 2.1). Based on this research, Barnes [17] proposed a tentative toughness classification based on the remolding toughness mobilized at the $w_P$ water content (i.e., the maximum remolding toughness, $T_{max}$). Finally, the work of Moreno-Maroto and Alonso-Azcárate [11] is briefly introduced at this point. They obtained a quantitative relationship between the $I_P/w_L$ ratio and $T_{max}$, thereby allowing toughness estimation for fine-grained soil based solely on consistency limits measurements, allowing them to develop a soil plasticity classification system in which the determining factor is toughness.

The aim of the present paper is to present a state-of-the-art review of soil remolding toughness, with emphasis on its experimental determination and the quantification of soil toughness using various related index properties and coefficients. This review focuses particularly on various extrusion proposals [12,13,47–49] and Barnes' mechanical thread rolling apparatus and methodology [8,17,18], which are presently the only means available for obtaining quantitative soil remolding toughness measurements. Challenges arise in the quantification of, and in assigning units of measurement to, the soil remolding toughness. These aspects are discussed in the context of Barnes' thread rolling approach. Attention then turns to the quantitative assessment of soil toughness via correlations, firstly reporting on the correlation of the $T_{max}$ to $I_P/w_L$ ratio after Moreno-Maroto and Alonso-Azcárate [11]. Then, employing a large dataset assembly from the available literature, new correlations between $T_{max}$ and the consistency limits parameters are deduced in the present paper, facilitating the indirect determination of soil remolding toughness from ubiquitously measured Atterberg (consistency) limits parameters. The toughness-based classification of fine-grained soil is discussed, intending to give a more mechanical perspective to fine-grained soil classification. A novel toughness chart is then introduced to augment (or be used instead of) the traditional Casagrande plasticity chart for obtaining more reliable soil classification when investigating some hitherto challenging fine-grained soil types. For instance, based on the toughness criterion, a previous study by the authors has shown that the Casagrande chart tends to classify soils with intermediate characteristics between clays and silts (e.g., silty clays, clayey silts, or clayey sands) as *clay* if $w_L$ < 50% and as *silt* if $w_L$ > 50%, while a wide range of clays could be classified as *silt* for $w_L$ > 65% when using the aforementioned classification [50]. Next, the small number of fundamentally correct approaches for soil remolding toughness determination are compared based on their advantages and disadvantages. This paper concludes with a discussion on recommended future experimental research efforts for soil remolding toughness determination. Note that this review paper concerns itself with the remolding toughness (of ductile soil); for fracture toughness, which applies to brittle soil (i.e., $w < w_P$), the reader is referred to the papers by Hanson et al. [51] and Wang et al. [52].

## 2. Soil Remolding Toughness Measurement

### *2.1. Barnes' Thread Rolling Apparatus and Method*

Barnes [8,17,18] developed an apparatus (see Figure 1) and test method for the rolling of a uniform soil cylinder (thread) between two configured plates. For each traverse, the applied downward force is controlled and recorded, and the actual tread diameter is recorded, such that the graphs of nominal acting stress against the diametrical strain can be computed for a range of investigated water contents (Figure 2a) that reduce in value to the $w_P$. Based on the areas beneath the produced experimental stress–strain plots, nominal toughness measurements can then be obtained, and are plotted against water content (Figure 2c).

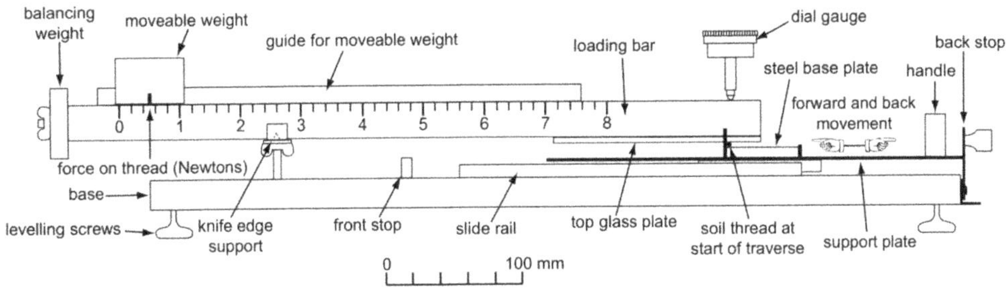


**Figure 1.** Elevation drawing of the Barnes' mechanical thread rolling apparatus [17].

In quantifying remolding toughness, the measurement units of toughness (i.e., work/unit volume) must take into account the number of traverses undergone by the plastic soil thread during the mechanical rolling-out procedure. Barnes [8,17,18] decided that the value of toughness $T$ assigned to the test soil would be based on the cumulative work undertaken per unit soil volume per 100 reversals (i.e., giving units of $kJ/m^3/100$ r) in reducing the soil thread diameter from initially 6 mm down to 4 mm because, over this range of diameters, the soil threads were found to undergo plastic straining in a steady fashion. As described by Barnes [8,17,18], from investigation of the plastic test soil prepared at a range of different water contents, various toughness-related properties can be derived from the obtained experimental $T$–$w$ plot. For reducing water content, these include the toughness limit $w_T$ (i.e., the water content at zero toughness), the stiffness transition $w_S$ (i.e., the water content below which the remolding toughness increases at a greater rate for many soils), and the maximum remolding toughness $T_{max}$, which is mobilized at the $w_P$ and represents the main toughness indicator. Figure 2 shows the experimental sequence carried out by Barnes [8,17,18] in determining the value of $T_{max}$ (i.e., the value of $T$ corresponding to the measured $w_P$ water content).

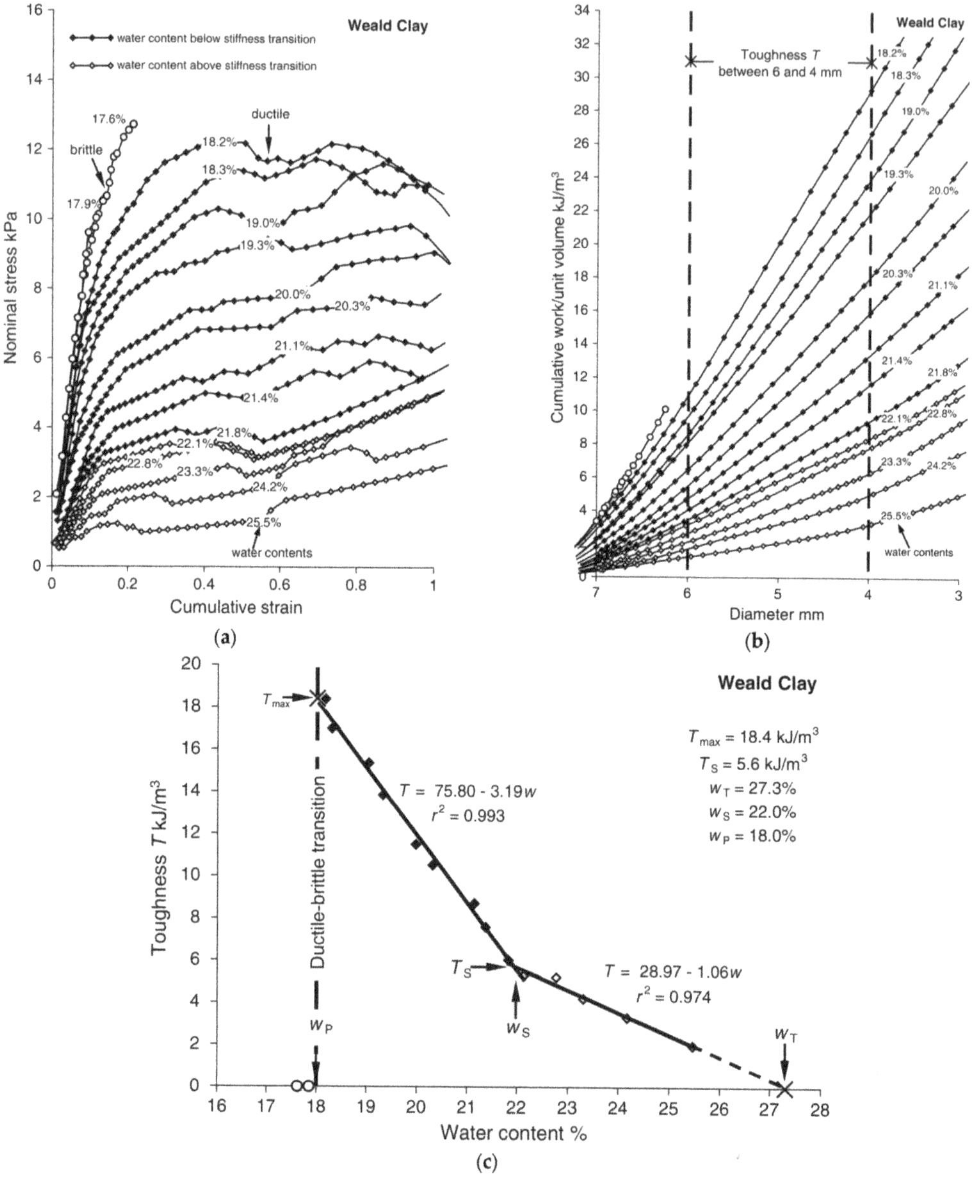


**Figure 2.** Determination of $T_{max}$ according to Barnes' method [8,17,18]: (**a**) graph of nominal stress against cumulative diametrical strain obtained for clay cylinders (threads) rolled out using Barnes' apparatus; (**b**) cumulative work per unit soil volume plotted against thread diameter, with the former obtained as the product of nominal stress and diametrical strain for 100 reversals; (**c**) produced relationship between remolding toughness ($T$) and water content, where $T$ is calculated as the cumulative work per unit volume required to reduce the thread diameter from initially 6 mm down to 4 mm. The presented plots have been extracted and edited from Barnes [18]. Note: $w$, water content; $w_P$, plastic limit; $w_S$, stiffness transition water content; $w_T$, toughness limit; $T_s$, toughness mobilized at $w_S$.

Based on the deduced bi-linear $T$–$w$ relationship (Figure 2c), the plastic range of a given test soil can be subdivided into three distinct regions, namely a workable stiff plastic region for $w_P < w < w_S$, a workable soft plastic region for $w_S < w < w_T$, and an adhesive plastic region for $w_T < w < w_L$ [8,18]. In other words, at the $w_T$, fine-grained soil behavior transforms from workable soft plastic to almost non-workable adhesive plastic. Moreover, for $w > w_T$, the soil would not only be difficult to form (roll) into a thread to display plasticity, but it would also be very sticky. From the authors' independent analysis of the experimental data presented for 55 inorganic fine-grained soils reported by Barnes [8], their $w_S$ water contents corresponded to a liquidity index ($I_L$) of ~0.12 (standard deviation of $\sigma = 0.05$), with measurable toughness evident for $I_L < 0.43$ ($\sigma = 0.14$). In other words, $w_T$ could be approximated as follows:

$$w_T = w_P + 0.43\ I_P \tag{1}$$

Vinod and Pillai [53] showed that $w_T$ shows a good correlation with the optimum water content for compaction. At all energy levels studied, they found that the maximum dry unit weight displayed a good correlation with the dry unit weight at the $w_T$ water content. Additionally, Shimobe et al. [54] investigated $w_T$ and a soil state index parameter defined as SSI = $w_T \times e_0$, where $e_0$ is the initial void ratio, as predictors of the compression index ($C_c$) parameter. The inclusion of $e_0$ means that the SSI takes into account the effects of several index properties, including the initial density and consistency of soils. They concluded that $w_T$ and, more so, the SSI are reasonable predictors of the $C_c$ parameter.

Furthermore, the toughness coefficients, as given by the gradients of the bi-linear $T$–$w$ relationship, the workability index $I_W$ (Equation (2)), and the toughness index $I_{T(B)}$ (= $w_T - w_P$), can be derived [8,17,18]. Here, the toughness index $I_{T(B)}$ simply gives the water content range over which the soil would be plastic and workable:

$$I_W = \frac{w - w_P}{w_T - w_P} \tag{2}$$

where $w$ = soil water content, $w_P$ = plastic limit, and $w_T$ = toughness limit (i.e., water content at zero toughness).

Barnes' workability index $I_W$ [17] relates the soil's actual water content to its "workable" range of $w_P < w < w_T$ (i.e., for $w = w_T$, $I_W = 1 \Rightarrow T = 0$ and for $w = w_P$, $I_W = 0 \Rightarrow T = T_{max}$). In other words, this definition of a workability index does not really relate to the degree of workability (toughness) of the test soil, and, therefore, it does not allow one to discern between *silt* and *clay* materials, whose toughness are very different. Note, Barnes' definition for the toughness index (i.e., $I_{T(B)}$) is different from the $I_{T(C)}$ parameter, after Casagrande [36], previously mentioned in the Introduction. Additionally, as presented in Table 1, Barnes [17] proposed a tentative classification of soil toughness (from "very low" to "extremely high" toughness) based on assigned ranges of $T_{max}$ values mobilized at the $w_P$ water content.

**Table 1.** Proposed classification of soil maximum remolding toughness $T_{max}$ (mobilized at $w_P$), as determined using Barnes' thread rolling apparatus [17].

| Toughness Classification | $T_{max}$ (kJ/m$^3$/100 r) |
| --- | --- |
| Very low | <5 |
| Low | 5–10 |
| Moderate | 10–20 |
| High | 20–30 |
| Very high | 30–50 |
| Extremely high | >50 |

Note that Barnes' definition of and measurement units for soil toughness represent one of many possible ways of quantifying toughness and, therefore, could be viewed as providing relative measures (based on the 100 reversals (i.e., 100 r) definition) of the soil's toughness properties/characteristics. Barnes could have chosen to base his work calculations on any set number of reversals (e.g., 20 r, 50 r, 120 r, etc.), thereby giving a different definition and hence producing an altered value of $T_{max}$ for investigating the same fine-grained soil. For instance, compared to the 100 r definition, a greater value of $T_{max}$ would arise with the toughness calculation defined for a larger number of reversals. From the authors' viewpoint, if one knows the distance traveled for each reversal during the experimental testing, a possible solution could be to translate those reversals into the distance traveled (in meters), simplifying the measurement units to something like $kJ/m^4$, which is the same as $kN/m^3$. Using Barnes' rolling apparatus, the travel distance in each reversal could be fixed (by the apparatus set-up), or alternatively, it could also be recorded during the testing.

### *2.2. Proposed Extrusion Approaches*

The extrusion method has been recently investigated for the determination of the consistency limits and also possibly of the remolding toughness of fine-grained soils [12,13,28,49,55]. Extrusion involves a reduction in the cross-sectional area of the soil billet (test specimen) by forcing it to flow through a die orifice under the action of a force $F_e$ (extrusion pressure $p_e$), employing either direct extrusion (DE) or reverse extrusion (RE) approaches [12] (refer to Figure 3). Note, RE is conventionally assumed to require a steady-state $p_e$ (Figure 3d), whereas this is not the case for DE (Figure 3c), which is affected by friction effects at the soil billet–chamber-sidewall interface [12]. Typically, a constant ram/die displacement rate (and hence extrusion velocity of the soil worm) is employed for consistency limits determination and $s_u$ testing. The resulting plot of extrusion force (pressure) against ram/die displacement (Figure 3c,d) can be regarded as the work diagram of extrusion [12].

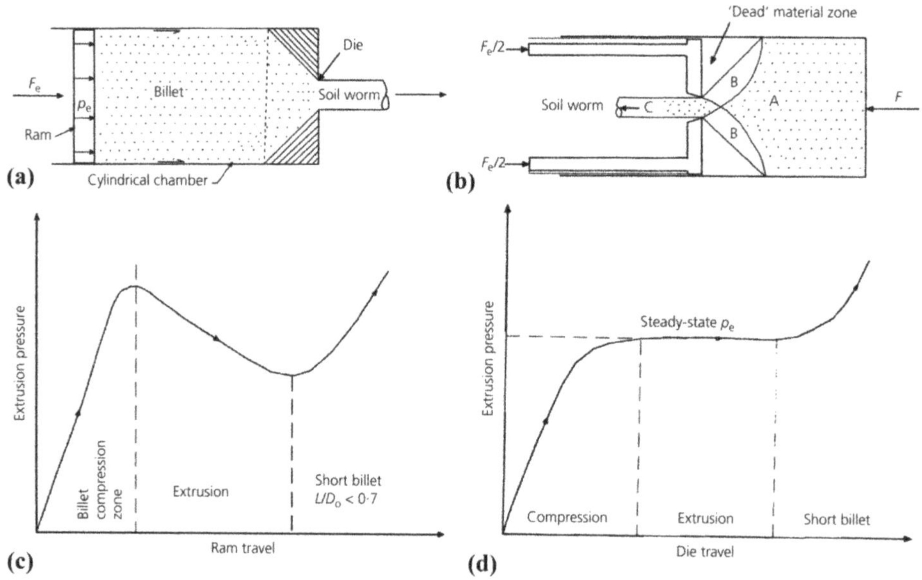


**Figure 3.** Schematic diagram of extrusion approaches ((**a**) direct extrusion, (**b**) reverse extrusion) and the characteristic extrusion pressure against ram/die displacement trace ((**c**) direct extrusion, (**d**) reverse extrusion) (after O'Kelly [12]). Note: $D_o$, billet diameter; $F_e$, ram/die force; $L$, initial billet length; $p_e$, extrusion pressure.

#### 2.2.1. Extrusion and Remolding Toughness

Based on a reassessment of an extensive RE dataset compiled for many hundreds of different fine-grained soils reported in the existing literature, O'Kelly [12] found that for a given extrusion apparatus with a set die orifice configuration, the $p_e$ value corresponding to the $w_P$ (i.e., $p_{e(PL)}$) appeared to show a marginally increasing trend for increasing plasticity ($I_P$) investigated over a very wide $I_P$ range. Considering that extrusion involves the plastic flow of the test soil, O'Kelly [12] postulated that, whereas $s_u$ is the soil property governing the extrusion resistance capacity for the adhesive plastic region (with zero toughness), when considering the workable soft and stiff plastic regions (i.e., $w_P < w < w_T$), it is the remolding toughness. Compared with low-plasticity soil, it would follow that for water contents in the workable soft plastic and especially the stiff plastic regions, the superior toughness of fine-grained soils with greater plasticity would require higher ram/die force (i.e., energy) application to produce extrusion. On this basis, the steady-state $p_e$ value mobilized using the RE approach for the workable soft- and stiff-plastic water content ranges would provide a relative measure of the soil toughness ($T$) [12]. That is, at the same $I_L$, high values of $p_e$ are needed for tougher soils. Having determined the $w_P$ value for the test soil using the standard thread rolling method [3,4], the associated value of $p_{e(PL)}$ can then be deduced from the regression analysis of the obtained experimental steady-state $p_e$—$w$ correlation [12]. To apply this approach for assessing relative values of $T_{max}$ mobilized between fine-grained soils of different plasticity (toughness) levels, the $p_e$ data employed in the $T_{max}$ determinations should be for water contents limited to the workable stiff plastic region ($w_P < w < w_S$), given the characteristic bi-linear nature of the $T$ ($p_e$)—$w$ relationship [12].

#### 2.2.2. Power-Based Approaches for Consistency Limits Determination

Employing purpose-built DE apparatus, O'Kelly [13] described an investigation of the power concept (i.e., the work undertaken for the extrusion of a known soil volume in a given time period) to potentially determine the consistency limits. Referring to Figure 4, the self-weight of the moving ram assembly applies a constant vertical force (pressure $p_e$) to the remolded soil sample that initially fills the cylindrical chamber (internally 35 mm dia. × 50 mm deep), causing its downward extrusion to occur via die orifices in the chamber base. This experimental method is analogous to the MFI tester [47,48] approach described in the Introduction, with them both measuring the time period required for the known volume of soil to flow through the die orifice(s) under constant deadweight load action. In other words, for these particular tests, the extrusion process is load/pressure controlled (i.e., not displacement rate controlled). Note that a larger and more complex DE device that essentially follows the same testing methodology as described in [13] is presented by Manafi et al. [49], wherein the power-based extrusion testing approach is termed as a "workability" measurement. O'Kelly [13] hypothesized that the power-based extrusion methodology could potentially serve as an alternative means for the classification of fine-grained soils based on the values of power required to cause extrusion at their $w_P$ water contents.

However, with a given extrusion apparatus and die–orifice combination, the power-based extrusion criterion [13,49] that employs a defined (set) value of power in the determination of the $w_P$ for all inorganic fine-grained soils is fundamentally not correct [9,13,56]. This results from different fine-grained soils invariably having dissimilar values of $T_{max}$ at their $w_p$ water contents [8,11,12,17,18]. In other words, with the extrusion rate varying according to the soil toughness (e.g., [47]), tougher soils would take longer to flow through the die orifice(s) and, therefore, they would require relatively lower power inputs to cause soil extrusion, and vice versa. The power-based extrusion criterion could be validly applied for the determination of the $w_L$, as it is defined based on a small (measurable) $s_u$ value (e.g., [19–21]), with the experimental $p_e$ for the adhesive plastic region related to the soils' $s_u$ [12]. However, the $w_L$ can already be reliably determined, e.g., using the simpler fall-cone liquid limit method [4,19,56].

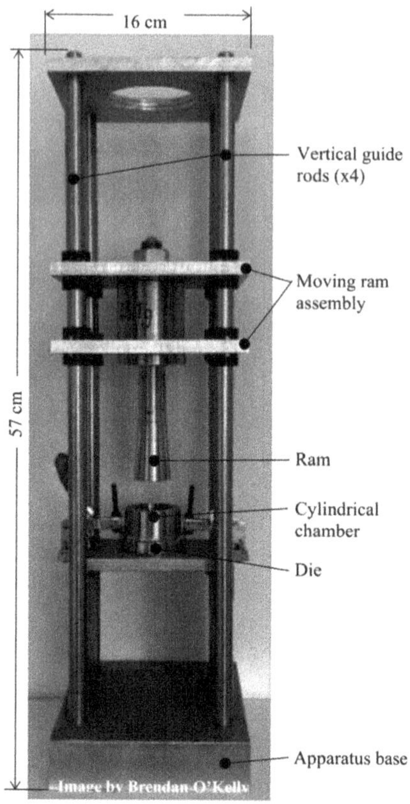


**Figure 4.** Trinity College Dublin (TCD) direct-extrusion apparatus for testing fine-grained soils (O'Kelly [13]).

## 3. Quantitative Assessment of $T_{max}$ (after Moreno-Maroto and Alonso-Azcárate [11])

As the consistency limits are strongly dependent on the soil mineralogical properties (of the clay fraction) [19], it would be expected for $T_{max}$ to correlate with the values of $w_L$, $w_P$, and hence $I_P$. Direct experimental measurements of soil remolding toughness were reported by Barnes [8] for 59 inorganic fine-grained soils, including natural clays, brick and ceramic clays, and for different mixtures of clay, silt, and sand materials with pure kaolinite and montmorillonite samples, with these test soils encompassing a very wide range of plasticity/toughness characteristics. Extracting Barnes' [8] reported data of $T_{max}$, the fall cone $w_L$ (i.e., $w_{L\text{-}FC}$, as obtained using the 80 g/30° fall cone device [4]) and $I_P$ (= $w_{L\text{-}FC} - w_P$), Moreno-Maroto and Alonso-Azcárate [11] plotted the $I_P/w_{L\text{-}FC}$ ratio against $T_{max}$ for the 59 investigated soils (Figure 5). In this analysis, they choose the $I_P/w_L$ ($I_P/w_{L\text{-}FC}$) ratio as it relates to the $I_P$ against $w_L$ plot form of the traditional Casagrande plasticity chart. A good correlation was found between $I_P/w_{L\text{-}FC}$ and $T_{max}$ (Equation (3) and Figure 5). In particular, based solely on consistency limits measurements, Equation (3) provides a simple means of identifying those fine-grained soils with measurable remolding toughness and then estimating their values of $T_{max}$ (as defined for Barnes' thread rolling apparatus in units of kJ/m$^3$/100 r):

$$\frac{I_P}{w_{L-FC}} = 0.0077\ T_{max} + 0.3397\ (n = 59,\ R^2 = 0.76) \tag{3}$$

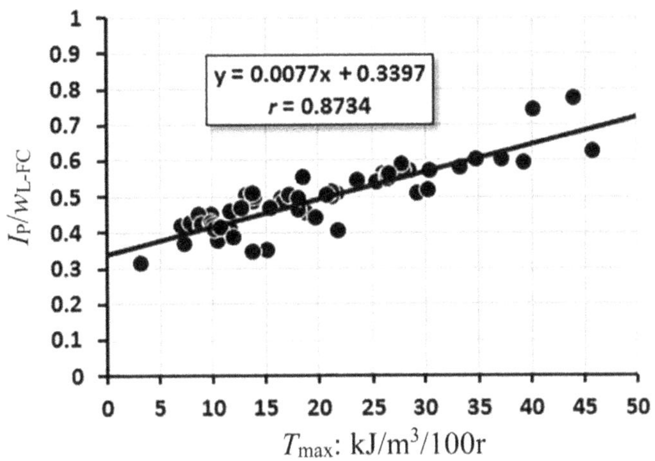


**Figure 5.** Relationship between maximum remolding toughness $T_{max}$ and the plasticity index to liquid limit ($I_P/w_{L\text{-}FC}$) ratio (after Moreno-Maroto and Alonso-Azcárate [11], prepared using data extracted from Barnes [8]).

## 4. Correlating $T_{max}$ with the Consistency Limits

Using tabulated data presented for inorganic fine-grained soils reported on pages 187–191 of Barnes [8], this section investigates new correlations between the consistency limits and $T_{max}$ (Figure 6). As expected, there are clear trends of $T_{max}$ increasing overall for increasing $w_{L\text{-}FC}$, $w_P$, and $I_P$ (Figure 6a–c), with the *clay* soils mobilizing significantly greater toughness than the *silt* soils. Compared to the data plot against $I_P$ (Figure 6c), there appears to be a strong demarcation between *clay* and *silt* evident for plotting the data in the $T_{max}$ versus $w_P$ chart of Figure 6b. The authors have included a tentative demarcation line (i.e., $T_{max} = 1.1 \times w_P - 20$) in Figure 6b for distinguishing between *clay* and *silt* soils, whose original classification, as reported in Barnes [8], was based on the Casagrande plasticity chart. As indicated in Figure 6c, the decision on the positioning of the *clay–silt* demarcation line in the $T_{max}$ versus $w_P$ chart was aided by the fact that, when the data were plotted in the Casagrande plasticity chart, three of the investigated soils were plotted on the A-line boundary. Figure 6d presents a plot of $T_{max}$ against the $w_P/w_{L\text{-}FC}$ ratio, termed the "plasticity ratio" by Shimobe and Spagnoli [57]. They showed that the $w_P/w_L$ ratio inversely correlates with SSA, which is consistent with greater remolding toughness for increasing SSA [10]. From an inspection of Figure 6d, with an increasing plasticity class level from CL to CE, the value of $T_{max}$ reduces due to the increasing $w_P/w_{L\text{-}FC}$ ratio, according to Equation (4):

$$T_{max} = 70 - 98 \times w_P/w_{L\text{-}FC} \ (n = 55, R^2 = 0.76) \quad (4)$$

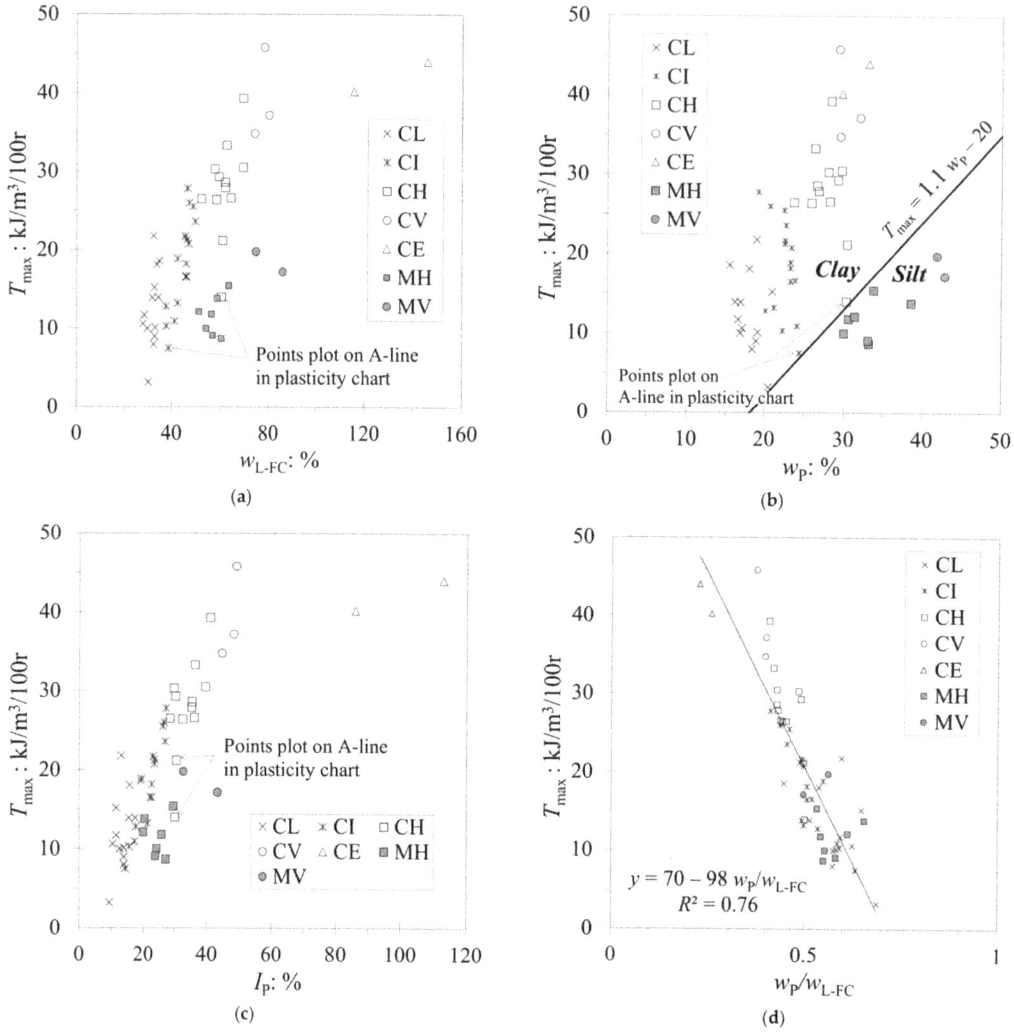


**Figure 6.** Maximum remolding toughness ($T_{max}$) of inorganic fine-grained soils plotted against (**a**) fall cone liquid limit, $w_{L\text{-}FC}$; (**b**) thread rolling plastic limit, $w_P$; (**c**) plasticity index, $I_P$; and (**d**) the $w_P/w_{L\text{-}FC}$ ratio (plots prepared using data extracted from Barnes [8]). Soil classes (CL, CI, CH, CV, CE, MH, MV) are referring to the British Standard soil plasticity classification obtained using the Casagrande chart [16].

## 5. Distinguishing between *Clay* and *Silt* Soils Based on the Value of $T_{max}$

Based on Equations (3) and (4), there is a real influence of the clay mineral content on soil toughness for $w_P/w_{L\text{-}FC} < 0.71$ and $I_P/w_{L\text{-}FC} > 0.3397$ (i.e., ~0.33), with $T_{max} > 0$ kJ/m³/100 r mobilized. Conversely, considering that plasticity (toughness) is a property exclusive to clayey materials [14], this means fine-grained soils cannot be considered as *clay* for $w_P/w_{L\text{-}FC} > 0.71$ or $I_P/w_{L\text{-}FC} \leq 0.33$ (i.e., Equations (3) and (4) give a computed $T_{max} = 0$); that is, those inorganic soils would have no workable soft- and stiff-plastic water content ranges. Owing to their lack of toughness, silts and sandy soils produce low plasticity results, only requiring slight finger pressure to be applied during the rolling out of the soil threads for $w_P$ determination, and they also do not present the typical sticky consistency of clays, while non-plastic soils can be considered as materials with an $I_P/w_L$ ($I_P/w_{L\text{-}FC}$)

ratio < 0.1896 (rounded to 0.2) [11]. Relying on Barnes' toughness classification (see Table 1), "high", "very high", and "extremely high" plasticity *clay* should mobilize a value of $T_{max}$ $\geq$ 20 kJ/m$^3$/100 r, which, according to Equation (3), corresponds to $I_P/w_{L\text{-}FC} \geq 0.4937$ (i.e., ~0.5). Hence, Moreno-Maroto and Alonso-Azcárate [11] concluded that those fine-grained soils with clay minerals present but which do not exert a significant influence on plasticity would exhibit $I_P/w_{L\text{-}FC}$ ratios of between 0.33 and 0.5.

## 6. Toughness-Based Classification of Fine-Grained Soils

Soil classification using the Casagrande plasticity chart is not entirely accurate, being inefficient in discerning toughness, e.g., some soils with high toughness (typical of *clay*) can be incorrectly classified as *silt*, and vice versa [50]. This observation supports a toughness-based soil classification to augment (or be used instead of) the Casagrande plasticity chart in obtaining more reliable soil classification for some hitherto challenging fine-grained soil types.

Based on fundamental experimental observations described in the previous section and directly connecting plasticity with toughness, Moreno-Maroto and Alonso-Azcárate [11] proposed an updated soil plasticity classification method in which the determining factor is toughness. For measured $w_{L\text{-}FC}$ and $I_P$, the soil is non-plastic for $I_P/w_{L\text{-}FC} < 0.2$ and low plastic for $0.2 \leq I_P/w_{L\text{-}FC} \leq 0.33$ (i.e., silt, organic soil, sandy soil, etc., with computed $T_{max}$ of ~ 0 kJ/m$^3$/100 r). Soil with a low or moderate influence of clay minerals (i.e., clayey silt, sandy clay, etc., whose $T_{max}$ < 20 kJ/m$^3$/100 r) has $0.33 < I_P/w_{L\text{-}FC} < 0.5$, while *clay* (i.e., plastic soil with $T_{max} \geq$ 20 kJ/m$^3$/100 r) has $I_P/w_{L\text{-}FC} \geq 0.5$ [11]. The developed soil classification chart of Moreno-Maroto and Alonso-Azcárate [11] is presented in Figure 7a. Moreno-Maroto et al. [50] investigated the correctness of soil classifications assigned to 31 different inorganic fine-grained soils by applying six of the main plasticity-based soil classification proposals (charts), including the Moreno-Maroto and Alonso-Azcárate [11] and Casagrande plasticity charts, with their research results ranking the Moreno-Maroto and Alonso-Azcárate [11] chart (Figure 7a) as having the strongest predictive capacity among the examined proposals.

An original contribution of the present paper is the chart presented in Figure 7b, which provides a simple and expedient way of estimating the value of $T_{max}$ from the measured consistency limits values. Taking into account Equation (3), Figure 7b was produced by calculating the $T_{max}$ isolines (of 0 to 80 kJ/m$^3$/100 r) and then plotted them in the typical plasticity chart (of $I_P$ vs. $w_{L\text{-}FC}$).

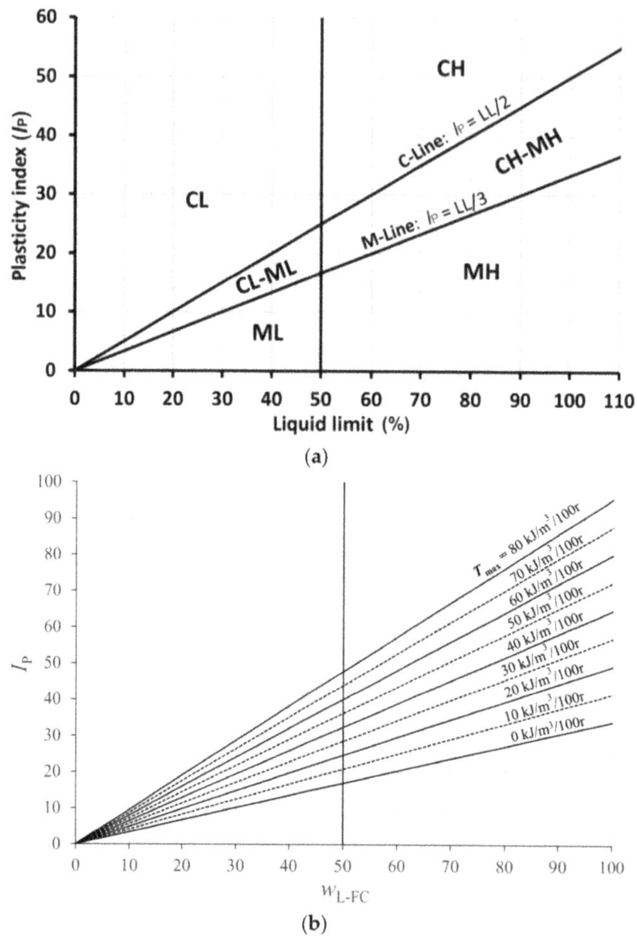


**Figure 7.** Toughness-based classification of fine-grained soil: (**a**) classification chart of Moreno-Maroto and Alonso-Azcárate [11]; (**b**) chart for estimating maximum remolding toughness, $T_{max}$, from consistency limits measurements (present study). Note: CH, clay of high plasticity; CL, clay of low plasticity; MH, silt of high plasticity; ML, silt of low plasticity; $I_P$, plasticity index; $w_L$, liquid limit water content; $w_{L\text{-}FC}$, fall-cone liquid limit water content.

## 7. Advantages/Limitations of the Existing Methods for Toughness Determination

In the preceding sections, four different approaches were highlighted as the fundamental ones for soil toughness assessment/determination: i.e., manual rolling of soil threads [3,4,40,41], Barnes' thread rolling apparatus [8,17,18], extrusion approaches [12,13,46–49], and via various correlations deduced in the present paper and reported in [11] between the consistency limits and values of Barnes' $T_{max}$. Their advantages and disadvantages are summarized in Table 2. Manual and mechanical thread rolling are classed as direct methods as they conform to the definition of toughness and the original protocol for its determination.

**Table 2.** Advantages and weaknesses of the main methods employed for soil toughness determination. Note: $w_L$, liquid limit; $w_P$, plastic limit; $T_{max}$, maximum remolding toughness (mobilized at $w_P$).

| Method | Type | Advantage | Weakness |
|---|---|---|---|
| Manual thread rolling [3,4,41] | Direct | Meets original definition of toughness. | Qualitative (does not allow obtaining numerical values). Subjectivity; high operator dependence. |
| Barnes' thread rolling apparatus [8,17,18] | Direct | Meets original definition of toughness. Objectivity (mechanically determined). | According to [50]: relatively complex measurement approach; slowness (many experimental points required); great skill on operator's part to reach value of $T_{max}$. |
| Extrusion [12,13,46–49] | Indirect | Objectivity (mechanically determined). Good correlation between extrusion resistance and toughness [12,13,46,47]. | Non-standardized (features of extruder device and extrusion conditions may give different results). Equipment cost. |
| Estimation from consistency limits ([11], this paper) | Indirect | $w_L$ and $w_P$ obtained using standardized and widely known methods. Good correlation with Barnes' values of $T_{max}$. Speed and low cost. | Of the parameters associated with consistency limits, only the $w_P$ test meets toughness definition. Based on estimates according to statistical criteria, outliers may not fit estimated values. |

Considering their various advantages and weaknesses, as listed in Table 2, it is not a simple task to determine which of these four approaches may prove to be the more effective for toughness determination. When the aim is to get accurate results, the use of relatively complex measurement devices, such as Barnes' thread rolling and various extrusion apparatuses, seems to be a good alternative, particularly the Barnes' approach, as it is a direct method. However, along with associated equipment costs and complexity, these mechanized methods can present certain operational disadvantages during the testing, as well as a long test duration with associated time/cost implications. On the other hand, manual thread rolling would be the simplest and fastest of all four approaches and the one that most closely approximates the original definition of toughness. However, being qualitative, subjective, and hence highly dependent on the operator's performance [25,33], the manual thread rolling approach is deemed not suitable when trying to get quantitative and precise results. Therefore, the statistical approximations obtained via consistency limits results (e.g., [11]) do seem to find a balance with what has been said previously, with the obtained toughness predictions conforming satisfactorily to the actual $T_{max}$ values measured for dozens of different fine-grained soils investigated by Barnes [8,17,18]. Given that the consistency limits are basic index tests performed on fine-grained soils, the use of their parameters for obtaining $T_{max}$ estimations does not entail any additional testing cost or time (i.e., from the authors' perspective, the purpose of soil toughness determination is not to replace the ubiquitous consistency limits testing used to obtain $w_L$ and $w_P$). As it is based on statistical criteria, probably one weakness of this approach is that it could potentially give erroneous $T_{max}$ predictions for soils with atypical behavior. In this regard, the authors consider that the observations of an experienced operator in judging the relative force and effort required to form and roll out uniform soil cylinders for the thread rolling test method would serve as quality control of the mathematical $T_{max}$ estimation. Therefore, continued use of standard consistency limits testing along with the estimation of $T_{max}$ via correlations with $I_P/w_{L\text{-}FC}$ or $w_P/w_{L\text{-}FC}$ (i.e., using Equations (3) and (4)) is judged as the most satisfactory option.

## 8. Recommendations on Future Research Efforts for Soil Toughness Determination

The authors' recommendations for future research work on the topic include investigations of soil remolding toughness employing (i) the MFI apparatus and testing approach [47,48] and (ii) the moisture condition test and obtained MCV results [45]. As described earlier in the paper, the MFI tester measures the time period required for a known soil volume to flow through a die orifice under a constant deadweight load action. The mobilized toughness for a particular fine-grained soil (and water content) is related to the flow time (soil extrusion rate), with tougher materials taking longer periods [47]. For the moisture condition test, tougher soils (and for lower water content) give higher MCVs, as more compaction effort is needed to remold the lumps of clay in performing the testing, with Barnes [8] envisaging that the MCV should correlate satisfactorily with the values of toughness. As such, increased research efforts should be directed toward investigating the MFI and/or moisture condition tester for soil toughness determination. Other ways of soil classification, based on remolding toughness parameters or incorporating the concept to complement existing plasticity-based classification systems, also merit investigation for obtaining more reliable soil classification when investigating some hitherto challenging fine-grained soil types.

## 9. Summary and Conclusions

This review paper covered the basic index (i.e., Atterberg/consistency limits) testing of fine-grained soil, the fundamental understanding of soil plasticity and remolding-toughness properties, and their determination/measurement and use in the classification of fine-grained soils. Plasticity and toughness properties can be used to distinguish plastic soils that provide greater deformation resistance for various field applications. The maximum remolding toughness $T_{max}$ correlates well to the consistency limits parameters, particularly with the $I_P/w_L$ and $w_P/w_L$ ratios, and the presented Equations (3) and (4) allow indirect $T_{max}$ estimations via these ratios from consistency limits measurements for fine-grained soils. Being inefficient in discerning toughness, soil classification using the traditional Casagrande plasticity chart system is not entirely accurate (e.g., some soils with high toughness can be incorrectly classified as *silt*, something typical of *clay*, and vice versa). As part of the original research work in the present paper, a strong demarcation between *clay* and *silt* soils was found in the plot of $T_{max}$ against $w_P$ for 59 dissimilar fine-grained soils (Figure 6b), giving a more mechanical perspective to soil classification. A new toughness chart, in which the presented $T_{max}$ isolines were deduced via correlations with the consistency limits parameters (i.e., using Equations (3) and (4)), was introduced as a graphical and simple way to estimate the tenacity of fine-grained soils. This chart, with the inclusion of the demarcation zones of $I_P/w_{L\text{-}FC} \leq 0.33$ for *silt* (ML, MH), $I_P/w_{L\text{-}FC} \geq 0.5$ for *clay* (CL, CH), and $0.33 < I_P/w_{L\text{-}FC} < 0.5$ for intermediate soils ($0.33 < I_P/w_{L\text{-}FC} < 0.5$), augments (or can be used instead of) the Casagrande plasticity chart for obtaining more reliable soil classification. Various useful toughness coefficients and indices can be measured using Barnes' thread rolling apparatus and method, but this approach is relatively complex and time-consuming. Accordingly, the less complicated and quicker MFI and moisture condition test approaches are recommended for future research investigations in regard to obtaining routine soil toughness measurements.

**Author Contributions:** Conceptualization, B.C.O.; methodology, B.C.O.; validation, B.C.O. and J.A.-A.; J.M.M.-M.; formal analysis, B.C.O.; investigation, B.C.O.; Data curation, B.C.O.; writing—original draft preparation, B.C.O.; writing—review and editing, B.C.O., J.A.-A. and J.M.M.-M.; visualization, B.C.O. and J.M.M.-M. All authors have read and agreed to the published version of the manuscript.

**Funding:** This research received no external funding.

**Data Availability Statement:** All data generated or analyzed during this study are included in this published article.

**Conflicts of Interest:** The authors declare no conflict of interest.

## Abbreviations

| | |
|---|---|
| CE | clay of extremely high plasticity |
| CH | clay of high plasticity |
| CI | clay of intermediate plasticity |
| CL | clay of low plasticity |
| CV | clay of very high plasticity |
| DE | direct extrusion |
| MCV | moisture condition value |
| MFI | Martin flow instrument |
| MH | silt of high plasticity |
| ML | silt of low plasticity |
| MV | silt of very high plasticity |
| RE | reverse extrusion |
| SSA | specific surface area |
| SSI | soil state index |

**Notations**

| | |
|---|---|
| $e_0$ | initial void ratio |
| $D_o$ | billet diameter |
| $F_e$ | ram/die force |
| $I_F$ | flow index |
| $I_L$ | liquidity index |
| $I_P$ | plasticity index |
| $I_{T(B)}$ | toughness index (after Barnes) |
| $I_{T(C)}$ | toughness index (after Casagrande) |
| $I_W$ | workability index (after Barnes) |
| $L$ | initial length of billet |
| $n$ | number of observations or data points |
| $N$ | number of blows |
| $p_e$ | extrusion pressure |
| $p_{e(PL)}$ | extrusion pressure at plastic limit water content |
| $PL_{100}$ | plastic strength limit |
| $R^2$ | coefficient of determination |
| $s_u$ | undrained shear strength |
| $T$ | remolding toughness |
| $T_{max}$ | maximum remolding toughness (at $w_P$ water content) |
| $T_s$ | remolding toughness at stiffness transition water content |
| $w$ | water content |
| $w_L$ | liquid limit water content |
| $w_{L\text{-}FC}$ | fall-cone liquid limit water content |
| $w_P$ | plastic limit water content |
| $w_S$ | stiffness transition |
| $w_T$ | toughness limit |
| $\sigma$ | standard deviation |

## References


1. Atterberg, A. Die plastizität der tone. *Internationale Mitteilungen Bodenkunde* **1911**, *1*, 4–37.
2. Atterberg, A. Lerornas forhållande till vatten, deras plasticitetsgränser och plasticitetsgrader. *Kungliga Lantbruksakademiens Handlingar och Tidskrift* **1911**, *50*, 132–158.
3. *ASTM D4318-17e1*; Standard Test Methods for Liquid Limit, Plastic Limit, and Plasticity Index of Soils. ASTM International: West Conshohocken, PA, USA, 2017.
4. *BS EN ISO 17892-12:2018+A2:2022*; Geotechnical Investigation and Testing. Laboratory Testing of Soil. Part 12: Determination of Liquid and Plastic Limits. British Standards Institution: London, UK, 2022.
5. Bobrowski, L.J.; Griekspoor, D.M. Determination of the plastic limit of a soil by means of a rolling device. *Geotech. Test. J.* **1992**, *15*, 284–287. [CrossRef]
6. *AASHTO T90*; Standard Method of Test for Determining the Plastic Limit and Plasticity Index of Soils. American Association of State Highway and Transportation Officials (AASHTO): Washington, DC, USA, 2020.
7. Soltani, A.; O'Kelly, B.C. Reappraisal of the ASTM/AASHTO standard rolling device method for plastic limit determination of fine-grained soils. *Geosciences* **2021**, *11*, 247. [CrossRef]

8. Barnes, G.E. The Plastic Limit and Workability of Soils. Ph.D. Thesis, University of Manchester, Manchester, UK, 2013.
9. O'Kelly, B.C.; Moreno-Maroto, J.M.; Alonso-Azcárate, J. Discussion of "Determining Soil Plasticity Utilizing Manafi Method and Apparatus" by Masoud S. G. Manafi, An Deng, Abbas Taheri, Mark B. Jaksa, and Nagaraj HB, published in 45, no. 4 (2022): 797–818. *Geotech. Test. J.* **2022**, *45*, 1144–1150. [CrossRef]
10. Prakash, K.; Sridharan, A. Critical appraisal of the cone penetration method of determining soil plasticity. *Can. Geotech. J.* **2006**, *43*, 884–888. [CrossRef]
11. Moreno-Maroto, J.M.; Alonso-Azcárate, J. What is clay? A new definition of "clay" based on plasticity and its impact on the most widespread soil classification systems. *Appl. Clay Sci.* **2018**, *161*, 57–63. [CrossRef]
12. O'Kelly, B.C. Reappraisal of soil extrusion for geomechanical characterisation. *Geotech. Res.* **2019**, *6*, 265–287. [CrossRef]
13. O'Kelly, B.C. Appraisal of novel power-based extrusion methodology for consistency limits determinations of fine-grained soils. In Proceedings of the Civil Engineering Research in Ireland 2022 Conference, Dublin, Ireland, 25–26 August 2022; Holmes, N., de Paor, C., West, R.P., Eds.; Civil Engineering Research Association of Ireland: Dublin, Ireland, 2022; Volume 1, pp. 317–322.
14. Casagrande, A. Classification and identification of soils. *Proc. Am. Soc. Civ. Eng.* **1947**, *73*, 783–810. [CrossRef]
15. *ASTM D2487-17*; Standard Practice for Classification of Soils for Engineering Purposes (Unified Soil Classification System). ASTM International: West Conshohocken, PA, USA, 2017.
16. *BS EN ISO 14688-2:2017*; Geotechnical Investigation and Testing—Identification and Classification of Soil. Part 2: Principles for a Classification. British Standards Institution: London, UK, 2017.
17. Barnes, G.E. An apparatus for the plastic limit and workability of soils. *Proc. Inst. Civ. Eng.—Geotech. Eng.* **2009**, *162*, 175–185. [CrossRef]
18. Barnes, G.E. An apparatus for the determination of the workability and plastic limit of clays. *Appl. Clay Sci.* **2013**, *80–81*, 281–290. [CrossRef]
19. O'Kelly, B.C.; Vardanega, P.J.; Haigh, S.K. Use of fall cones to determine Atterberg limits: A review. *Géotechnique* **2018**, *68*, 843–856. [CrossRef]
20. Haigh, S. Consistency of the Casagrande liquid limit test. *Geotech. Test. J.* **2016**, *39*, 13–19. [CrossRef]
21. O'Kelly, B.C. Fallacy of wide undrained strength range at the Casagrande liquid limit. *Geotech. Res.* **2019**, *6*, 205–217. [CrossRef]
22. Haigh, S.K.; Vardanega, P.J.; Bolton, M.D. The plastic limit of clays. *Géotechnique* **2013**, *63*, 435–440. [CrossRef]
23. O'Kelly, B.C.; Vardanega, P.J.; Haigh, S.K. Discussion of "Mohajerani method: Tool for determining the liquid limit of soils using fall cone test results with strong correlation with the Casagrande test" by E. Hrubesova, B. Lunackova and M. Mohyla [Engineering Geology 278 (2020) 105852]. *Eng. Geol.* **2022**, *302*, 106623. [CrossRef]
24. O'Kelly, B.C.; Soltani, A. Machine learning techniques for relating liquid limit obtained by Casagrande cup and fall cone test in low-medium plasticity fine grained soils [Eng. Geol. (2021) 294, 106381]. *Eng. Geol.* **2022**, *306*, 106746. [CrossRef]
25. Sivakumar, V.; O'Kelly, B.C.; Henderson, L.; Moorhead, C.; Chow, S.H. Measuring the plastic limit of fine soils: An experimental study. *Proc. Inst. Civ. Eng.—Geotech. Eng.* **2015**, *168*, 53–64. [CrossRef]
26. O'Kelly, B.C. Discussion of "Advancement in estimation of undrained shear strength through fall cone tests" by Abhishek Ghosh Dastider, Santiram Chatterjee, and Prasenjit Basu. *J. Geotech. Geoenviron. Eng.* **2022**, *148*, 07022005. [CrossRef]
27. Sivakumar, V.; O'Kelly, B.C.; Henderson, L.; Moorhead, C.; Chow, S.H.; Barnes, G.E. Discussion: Measuring the plastic limit of fine soils: An experimental study. *Proc. Inst. Civ. Eng.—Geotech. Eng.* **2016**, *169*, 83–85. [CrossRef]
28. O'Kelly, B.C. Review of recent developments and understanding of Atterberg limits determinations. *Geotechnics* **2021**, *1*, 59–75. [CrossRef]
29. Soltani, A.; O'Kelly, B.C. Reappraisal of fall-cone flow curve for soil plasticity determinations. *Geotech. Test. J.* **2022**, *45*, 225–243. [CrossRef]
30. O'Kelly, B.C.; Soltani, A. Discussion: Determining the plasticity properties of high plastic clays: A new empirical approach [Arab J Geosci (2020) 13(11), 394]. *Arab. J. Geosci.* **2021**, *14*, 715. [CrossRef]
31. Moreno-Maroto, J.M.; Alonso-Azcárate, J. Plastic limit and other consistency parameters by a bending method and interpretation of plasticity classification in soils. *Geotech. Test. J.* **2017**, *40*, 467–482. [CrossRef]
32. Vardanega, P.J.; Haigh, S.K.; O'Kelly, B.C. Use of fall-cone flow index for soil classification: A new plasticity chart. *Géotechnique* **2022**, *72*, 610–617. [CrossRef]
33. Sherwood, P.T. *The Reproducibility of the Results of Soil Classification and Compaction Tests*; Transport and Road Research Laboratories Report LR 339; Department of Transport: London, UK, 1970.
34. O'Kelly, B.C. Atterberg limits are not appropriate for peat soils. *Geotech. Res.* **2015**, *2*, 123–134. [CrossRef]
35. Vardanega, P.J.; Haigh, S.K.; O'Kelly, B.C.; Zhang, X.; Liu, X.; Chen, C.; Wang, G. Discussion: Use of fall-cone flow index for soil classification: A new plasticity chart. *Géotechnique* **2023**. *ahead of print*. [CrossRef]
36. Casagrande, A. Research on the Atterberg limits of soils. *Public Roads* **1932**, *13*, 121–136.
37. Wasti, Y.; Bezirci, M.H. Determination of the consistency limits of soils by the fall-cone test. *Can. Geotech. J.* **1986**, *23*, 241–246. [CrossRef]
38. O'Kelly, B.C. Atterberg limits and remolded shear strength–water content relationships. *Geotech. Test. J.* **2013**, *36*, 939–947. [CrossRef]
39. Vardanega, P.J.; Haigh, S.K. The undrained strength–liquidity index relationship. *Can. Geotech. J.* **2014**, *51*, 1073–1086. [CrossRef]

40. *ASTM D2488-17e1*; Standard Practice for Description and Identification of Soils (Visual-Manual Procedures). ASTM International: West Conshohocken, PA, USA, 2017.
41. Soil Science Division Staff. Chapter 3—Examination and description of soil profiles. In *Soil Survey Manual—United States Department of Agriculture Handbook 18*; Ditzler, C., Scheffe, K., Monger, H.C., Eds.; Government Printing Office: Washington, DC, USA, 2017. Available online: https://www.nrcs.usda.gov/wps/portal/nrcs/detail/soils/ref/?cid=nrcs142p2_054261 (accessed on 22 July 2022).
42. Naval Facilities Engineering Command (NAVFAC). *Soil Mechanics Design Manual 7.01*; Naval Facilities Engineering Command: Alexandria, VA, USA, 1986.
43. Astbury, N.F.; Moore, F.; Lockett, J.A. A cyclic torsion test for study of plasticity. *Trans. J. Br. Ceram. Soc.* **1966**, *65*, 435–461.
44. *ASTM C181-11*; Standard Test Method for Workability Index of Fireclay and High-Alumina Refractory Plastics. ASTM International: West Conshohocken, PA, USA, 2018.
45. Parsons, A.W. Moisture condition test for assessing the engineering behaviour of earthwork material. In Proceedings of the Conference on Clay Fills, London, UK, 14–15 November 1978; Institution of Civil Engineers: London, UK, 1978; pp. 169–175. [CrossRef]
46. Fitzjohn, W.H.; Worrall, W.E. Physical properties of raw brick clays. *Trans. J. Br. Ceram. Soc.* **1980**, *79*, 74–81.
47. CERAM Research. Rheology Testing for Ceramics. Version Jun 12, 2001. Available online: https://www.azom.com/article.aspx?ArticleID=547 (accessed on 28 February 2023).
48. Kessel, S. Ceramic rheology tester gets the thumbs up. *Mater. World* **1998**, *16*, 474–475.
49. Manafi, M.S.G.; Deng, A.; Taheri, A.; Jaksa, M.B.; HB, N. Determining soil plasticity utilizing Manafi method and apparatus. *Geotech. Test. J.* **2022**, *45*, 797–818. [CrossRef]
50. Moreno-Maroto, J.M.; Alonso-Azcárate, J.; O'Kelly, B.C. Review and critical examination of fine-grained soil classification systems based on plasticity. *Appl. Clay Sci.* **2021**, *200*, 105955. [CrossRef]
51. Hanson, J.A.; Hardin, B.O.; Mahboub, K. Fracture toughness of compacted cohesive soils using ring test. *J. Geotech. Eng.* **1994**, *120*, 872–891. [CrossRef]
52. Wang, J.J.; Zhu, J.G.; Chiu, C.F.; Zhang, H. Experimental study on fracture toughness and tensile strength of a clay. *Eng. Geol.* **2007**, *94*, 65–75. [CrossRef]
53. Vinod, P.; Sreelekshmy Pillai, G. Toughness limit: A useful index property for prediction of compaction parameters of fine grained soils at any rational compactive effort. *Indian Geotech. J.* **2017**, *47*, 107–114. [CrossRef]
54. Shimobe, S.; Karakan, E.; Sezer, A. Evaluation of dependency of compression index on toughness limit for fine-grained soils. *Neural Comput. Appl.* **2023**, *35*, 11183–11205. [CrossRef]
55. O'Kelly, B.C. Discussion of "Strength and consolidation characteristics for cement stabilized cohesive soil considering consistency index" by Ahmed, F. Zidan, published in Geotechnical and Geological Engineering, https://doi.org/10.1007/s10706-020-01367-6. *Geotech. Geol. Eng.* **2021**, *39*, 4659–4662. [CrossRef]
56. Haigh, S.K.; O'Kelly, B.C.; Vardanega, P.V. Discussion of "Determining soil plasticity utilizing Manafi method and apparatus" by Masoud, S. G. Manafi, An Deng, Abbas Taheri, Mark B. Jaksa, and Nagaraj HB, published in Geotechnical Testing Journal 45, no. 4 (2022): 797–818. *Geotech. Test. J.* **2023**, *46*. [CrossRef]
57. Shimobe, S.; Spagnoli, G. Some generic trends on the basic engineering properties of fine-grained soils. *Environ. Earth Sci.* **2019**, *78*, 281. [CrossRef]

MDPI
St. Alban-Anlage 66
4052 Basel
Switzerland
www.mdpi.com

*Applied Sciences* Editorial Office
E-mail: applsci@mdpi.com
www.mdpi.com/journal/applsci

www.ingramcontent.com/pod-product-compliance
Lightning Source LLC
Chambersburg PA
CBHW061515210926
51850CB00008B/1921
* 9 7 8 3 7 2 5 8 0 8 6 1 8 *